Motorcycle Accident Reconstruction

Second Edition

Motorcycle Accident Reconstruction

Second Edition

NATHAN A. ROSE
Principal Accident Reconstructionist
Explico Engineering
nathan@explico.com

Warrendale, Pennsylvania, USA

400 Commonwealth Drive
Warrendale, PA 15096-0001 USA
E-mail: CustomerService@sae.org
Phone: 877-606-7323 (inside USA and Canada)
724-776-4970 (outside USA)
FAX: 724-776-0790

Copyright © 2022 SAE International. All rights reserved.

No part of this publication may be reproduced, stored in a retrieval system, or transmitted, in any form or by any means, electronic, mechanical, photocopying, recording, or otherwise, without the prior written permission of SAE International. For permission and licensing requests, contact SAE Permissions, 400 Commonwealth Drive, Warrendale, PA 15096-0001 USA; e-mail: copyright@sae.org; phone: 724-772-4028.

Library of Congress Catalog Number 2021948804
http://dx.doi.org/10.4271/ 9781468603989

Information contained in this work has been obtained by SAE International from sources believed to be reliable. However, neither SAE International nor its authors guarantee the accuracy or completeness of any information published herein and neither SAE In ternational nor its authors shall be responsible for any errors, omissions, or damages arising out of use of this information. This work is published with the understanding that SAE International and its authors are supplying information but are not attempting to render engineering or other professional services. If such services are required, the assistance of an appropriate professional should be sought.

ISBN-Print 978-1-4686-0397-2
ISBN-PDF 978-1-4686-0398-9
ISBN-ePub 978-1-4686-0399-6

To purchase bulk quantities, please contact: SAE Customer Service

E-mail: CustomerService@sae.org
Phone: 877-606-7323 (inside USA and Canada)
724-776-4970 (outside USA)
Fax: 724-776-0790

Visit the SAE International Bookstore at books.sae.org

Chief Growth Officer
Frank Menchaca

Publisher
Sherry Dickinson Nigam

Director of Content Management
Kelli Zilko

Production and Manufacturing Associate
Erin Mendicino

Table of Contents

About the Author — xi
Preface — xiii
Acknowledgments — xv

CHAPTER 1
Introduction to Accident Reconstruction — 1

The Approach Used in Accident Reconstruction — 2
The Context of Reconstruction — 2
Investigation and Analysis — 2
Analysis by Phases — 3
Theoretical and Empirical Modeling — 4
Uncertainty Analysis — 4
Incorporating Witness Statements and Testimony — 6
Causation — 8
Analyzing Avoidance Scenarios — 10

Physical Principles Used in Accident Reconstruction — 11
Conservation of Energy — 11
Newton's Second Law and the Principle of Work and Energy — 12
Principle of Impulse and Momentum (Conservation of Momentum) — 18

References — 21

CHAPTER 2
Motorcycle Characteristics — 25

Motorcycle Types — 26
Motorcycle Dimensions and Inertial Properties — 31
Motorcycle Controls — 37
Motorcycle Tires — 38
Motorcycle Tire Markings — 38

Motorcycle Tire Friction Coefficients — 40
References — 41

CHAPTER 3

Braking and Acceleration — 43

Braking Dynamics — 43
    Braking Dynamics Example — 46
Motorcyclist Braking and Deceleration Capabilities — 48
    Motorcycles with Integrated, Linked, and Antilock Braking Systems — 57
    Speed and Avoidance Calculations — 60
Engine Braking — 61
Motorcycle Acceleration Capabilities — 62
Determining Speed Based on Gear — 63
    Case Study: Determining Speed Based on Gear — 64
Motorcycle Autonomous Emergency Braking (MAEB) — 67
References — 67

CHAPTER 4

Cornering, Swerving, and Changing Lanes — 71

Analysis of a Motorcycle Traversing a Curve — 71
    Incorporating Roadway Superelevation — 72
    Example: Lean Angle Calculation — 74
    Assumptions — 75
    Case Study: Application of the Lean Angle Equations at the Apex of a Curve — 76
    Case Study: Application of the Lean Angle Equations in the Time Domain — 81
    Friction-Limited Speed — 86
    Geometric Limit on Speed — 87
    Willingness to Lean — 90
Lane Change, Swerve, and Turn-Away — 90
    Example: Turn-Away and Swerve Calculation — 96
References — 97

CHAPTER 5

Physical Evidence from Motorcycle Crashes — 99

Scene Evidence — 99
    Skid Marks — 101

Table of Contents   vii

Gouges, Scrapes, Scuffs, and Tire Marks	104
Site Inspection Checklist	106

Evidence on the Motorcycle — 107

Motorcycle Inspection Checklist	110

Damage to the Struck Vehicle — 112

Evidence Documentation Methods — 114

Mapping with LIDAR	114
Photogrammetry	115
Camera Reverse Projection	119
Removing Lens Distortion	121
Case Study: Photogrammetric Analysis of Video of a Motorcycle Accident	123
Case Study: Photogrammetric Analysis of Video of Another Motorcycle Accident	129
Image-Based Scanning	132
Mapping with Small Unmanned Aerial Vehicles	138

References — 139

CHAPTER 6

Sliding and Tumbling of the Motorcycle and Rider — 143

Decelerations for a Sliding or Tumbling Motorcycle — 143

Decelerations for a Sliding or Tumbling Rider — 150

Determining the Initial Speed for a Sliding Motorcycle or Rider — 152

How to Calculate Speed Loss for a Sliding Motorcycle	152

References — 153

CHAPTER 7

Motorcycle Falls — 157

Low-Side Falls — 157

Case Study: A Low-Side Fall	158

High-Side Falls — 161

Case Study: A High-Side Fall	162
Case Study: A High-Side Fall	164

Falls Because of Front-Wheel Over-Braking and Lock-Up — 166

Impact-Induced Capsize — 167

Motorcycle Interactions with Potholes, Roadway Deterioration, and Debris ... 167
- A Motorcycle Encountering a Longitudinal Pavement Edge ... 167
- A Motorcycle Encountering a Pothole ... 171

Analyzing the Motion of Projected Riders ... 173

References ... 174

CHAPTER 8

Motorcycle Collisions with Vehicles and Roadside Barriers ... 177

Analysis Based on Motorcycle Wheelbase Reduction ... 177

Determining Motorcycle Speed from the Struck Vehicle Postimpact Translation and Rotation ... 191
- Planar Impact Mechanics ... 191
- Relating the Rotational Displacement to the Angular Velocity ... 196
- Ogden and Kloberdanz Equation ... 203

WREX 2016 Crash Tests ... 207
- Case Study: WREX 2016, Test #3 ... 208
- Case Study: WREX 2016, Test #5 ... 212
- Case Study: WREX 2016, Test #8 ... 216
- Case Study: WREX 2016, Test #11 ... 220
- Case Study: WREX 2016, Test #22 ... 224
- Case Study: WREX 2016, Test #23 ... 227
- Case Study: WREX 2016, Test #24 ... 232
- Discussion of WREX 2016 Tests ... 235

How Much of the Rider's Weight Should be Included? ... 238

Impacts into Moving Vehicles ... 240

Motorcycle Collisions with Roadside Barriers ... 241
- Motorcycle-to-Roadside Barrier Crash Tests ... 243
  - Steel Guardrail: Upright ... 243
  - Concrete Barrier: Upright ... 245
  - Modified Steel Guardrail: Upright ... 245
  - Steel Guardrail: Capsized ... 246
  - Concrete Barrier: Capsized ... 246
  - Modified Steel Guardrail: Capsized ... 248

References ... 248

CHAPTER 9

Event Data Recorders and Speedometers in Motorcycle Accidents — 253

Event Data from the Struck or Striking Vehicle — 253
- Precrash Data — 253
- Crash Data (ΔV) — 254
  - Potential Error Sources — 255
  - Potential Magnitude of the Error—Longitudinal ΔV — 256
  - Potential Magnitude of the Error—Lateral ΔV — 260

Accounting for Tire Forces When Incorporating EDR Data — 261
- WREX 2016 Test #3 — 262
- WREX 2016 Test #5 — 266

Case Study #1: Utilizing EDR Data from the Passenger Vehicle — 271

Case Study #2: Utilizing EDR Data from the Passenger Vehicle — 282

Event Data from the Motorcycle — 289

Video as a Source of Data from the Motorcycle — 291

A Residual Speedometer Reading as a Source of Data from the Motorcycle — 292
- Case Study: Speedometer Testing on a 2019 Harley-Davidson Heritage Softail Classic — 296

References — 300

CHAPTER 10

Human Factors in Motorcycle Crashes — 305

Physical Factors Affecting the Visibility of Motorcycles — 306

Why do Drivers Sometimes Fail to "See" Motorcyclists? — 309

The Size-Arrival Effect — 316

Effectiveness of Daytime Running Lights on Motorcycles — 318

The Perception-Response Process — 321

Motorcyclists' Responses to a Laterally Incurring Vehicle — 325
- A Numerical Example of the Left Turn Across Path Scenario — 328
- A More Nuanced Example — 332
- Evidence of Ambiguity in Path Intrusion Scenarios — 333

Characteristics of Passenger Car Left Turns and Intersection Crossings — 335

Are Training and Licensing Effective? ... 336
References ... 339

CHAPTER 11

Visualization of Motorcycle Crashes ... 345

View from the Motorcyclist's Perspective ... 345
    Mounting Equipment on the Motorcycle ... 345
    Mounting Equipment on the Operator ... 346
    Mounting Equipment on Another Vehicle ... 347

View of the Motorcycle from Other Vehicles ... 350

Camera Setup and Representing a Driver's View ... 351
    Camera Height ... 351
    Field of View ... 354
    Scaling the Images for Playback ... 355

Computer Visualization of a Motorcycle Crash ... 356
    Case Study: CG Visualization of a Truck Versus Motorcycle Crash ... 359

References ... 369

Index ... 371

About the Author

Nathan Rose is a Principal Accident Reconstructionist at Explico Engineering. Prior to that, from 2019 to 2021, he was a founder and Principal Accident Reconstructionist at Luminous Forensics. From 2005 to 2019, he was a partner, director, and principal accident reconstructionist at a Denver-based accident reconstruction firm that he helped found in 2005. Before that, he held positions as an engineer (1998 to 2003) and a senior engineer (2003 to 2005) at Knott Laboratory. He holds a bachelor's degree in engineering with a civil specialty from the Colorado School of Mines (1998) and a master's degree in mechanical engineering from the University of Colorado at Denver (2003).

Nathan is accredited as a traffic accident reconstructionist by the Accreditation Commission for Traffic Accident Reconstruction (ACTAR), and he has offered expert testimony as a reconstructionist in courts around the United States. During his graduate studies, he specialized in dynamics and impact mechanics, and he has published numerous technical articles, reports, and books related to vehicular accident reconstruction. Nathan holds a motorcycle endorsement in the State of Colorado, and he has completed the Motorcycle Safety Foundation's Basic and Advanced RiderCourses and their Street Strategies course.

In addition to this book, *Motorcycle Accident Reconstruction*, he is also the author of another book published by the SAE International, *Rollover Accident Reconstruction*. You can reach Nathan at (720)839-1995 and read his blog at www.nathanarose.com.

Preface to the Second Edition

In this book, my aim is a treatment of motorcycle accident reconstruction that is thorough, systematic, and scientific. When the first edition of this book was published in late 2018, such a book did not exist, and I attempted to fill that gap, pulling together as much of the relevant accident reconstruction literature and science as possible. But, of course, the field has continued to progress! This new edition contains:

- Additional theoretical models, examples, case studies, and new test data from me and others.
- An updated bibliography that cites the newest studies.
- Expanded coverage of the braking capabilities of motorcyclists.
- Updated, refined, and expanded discussion of the decelerations of motorcycles sliding on the ground.
- A thoroughly rewritten and expanded discussion of motorcycle impacts with passenger vehicles.
- Updated coefficients of restitution for collisions between motorcycles and cars.
- A new and expanded discussion of using passenger car event data recorder (EDR) data in motorcycle accident reconstruction.
- A new section covering recently published research on postcollision "frozen" speedometer readings on motorcycles.
- A new section on motorcycle interactions with potholes, roadway deterioration, and debris and expanded coverage of motorcycle falls.
- And finally, I have corrected several typographical errors that were present in the first edition. I am sure I have created new ones in the second edition!

A word about terminology: There is sometimes controversy surrounding the use of words such as *accident*, *crash*, *collision*, and *impact*. For example, the *British Medical Journal* has banned the use of the word *accident*, stating that "an accident is often understood to be unpredictable—a chance occurrence or an 'act of God'—and therefore unavoidable. However, most injuries and their precipitating events are predictable and preventable."[1] The California Motorcycle Handbook states: "*Accident* implies an unforeseen event that occurs without anyone's fault or negligence. Most often in traffic, this is not the case. In fact, most people involved in a collision can usually claim some responsibility for what takes place." David Hough writes: "Back in the days of the Hurt Report, unfortunate events were called accidents—as if no one could predict what was happening or do anything about it.

---

[1] Davis, R.M. and Pless, B., "BMJ Bans 'Accidents'," *British Medical Journal*, 322. June 2001.

But today unfortunate events are called crashes. We're getting away from the concept of a motorcyclist being an accident victim because in most situations the people involved can observe what's happening and take evasive action."[2] Elsewhere, he has written: "The right attitude begins with thinking of crashes as *crashes*, not *accidents*. The word *accident* implies that no one could have done anything to have avoided what happened. I believe, however, that someone always has some ability to change the outcome."[3]

I generally agree with these sentiments. However, the word *accident* is still widely used, not the least in the title of my profession: *accident* reconstruction. I still use the word *accident*. I also use the words *crash*, *impact*, and *collision*. To me, the words *collision* and *impact* are synonyms, but their scope is more limited than the words *accident* or *crash*. When I use the words *collision* or *impact*, it will refer to the actual contact between two vehicles or a vehicle and another object (although, I probably stray from that narrow definition at times!). I have encountered other accident reconstructionist using the word *collision* the way I use the word crash. At any rate, I agree that *crashes* are usually preventable. So are *accidents*, despite the underlying semantics of that word. Beyond that, I still consider myself an *accident* reconstructionist, a term that is much more common and useful for Google search results than *crash* reconstructionist. When in doubt, look to the substance of my opinions for how I am using these words.

Nathan Rose
Parker, Colorado
September 2021

[2.] Hough, D.L., *Proficient Motorcycling: The Ultimate Guide to Riding Well*, 2nd ed., Fox Chapel Publishers, 2008, ISBN 978-1-62008-119-8. Hough does not always follow his own terminology rules, though, writing that, "The ultimate purpose of a helmet is to prevent brain injuries during an *accident*." (emphasis added)

[3.] Hough, D.L., *More Proficient Motorcycling: Mastering the Ride*, 2nd ed., I-5 Press, 2012.

Acknowledgments

I would like to thank Andrew O'Donnell, Jordan Dickinson, and Janna Webb for their contributions to the illustrations and graphics contained in the first edition of this book. Many of those illustrations carry over to this second edition. Jordan Dickinson has continued assisting me with illustrations and graphics for this second edition. Also, Neal Carter, Connor Smith, Tilo Voitel, Nathan McKelvey, David Pentecost, Alireza Hashemian, Tomas Owens, Sean McDonough, Martin Randolph, and Daniel Koch all made helpful contributions to the testing and analysis reported in the first edition of this book. Much of that material carries over to this second edition. Neal Carter has been a valuable and insightful sounding board throughout the development of both the first and second editions of this book. His efforts were integral to many sections, including the cornering and physical evidence documentation sections. Louis Peck made helpful and much appreciated contributions by providing data related to motorcycle braking and motorcycle event data recorders (EDRs) and by reading and commenting on portions of the first edition. Ken Strohmeyer of the Scottsdale Police Department provided a helpful reading and editing of a draft version of the first edition. He also provided me with the notes from a presentation he gave in 2018 on motorcycle accident reconstruction. Jarrod Carter made much needed corrections to the grammar, and his good humor and friendship has been much appreciated as I have navigated the rough and tumble world of accident reconstruction. Shaun Helman of the Transportation Research Laboratory directed me to research he conducted related to motorcycle conspicuity, which significantly improved the chapter on human factors. Ed Fatzinger's great research has made both this book and the industry better. Any errors or omissions in this second edition, of course, remain mine.

Introduction to Accident Reconstruction

Accident reconstruction utilizes principles of physics and empirical data to analyze the physical, electronic, video, audio, and testimonial evidence from a crash, to determine how and why the crash occurred or to determine whose description of the crash is most accurate. The ways in which a crash could have been avoided may also be analyzed. Crash reconstruction draws together aspects of mathematics, physics, engineering, materials science, human factors, and psychology and combines analytical models with empirical test data. Different types of crashes—vehicle-to-vehicle collisions, single-vehicle rollover crashes, pedestrian and bicycle crashes, motorcycle crashes, or heavy truck crashes, for instance—produce different types of evidence and call for different analysis methods. Still, the basic philosophical approach is the same from crash type to crash type, as are the physical principles that are brought to bear on the analysis.

This introductory chapter covers a basic approach to crash reconstruction along with the underlying physical principles employed. Brief comments are offered related to the application of these principles to motorcycle crash reconstruction, but the chapters that follow give a more detailed account of how each approach and physical principle is applied to motorcycle crashes.

The Approach Used in Accident Reconstruction

The Context of Reconstruction

Accident reconstruction is carried out in several contexts. For instance, reconstruction in a research context can evaluate or improve safety systems for vehicles and drivers. In this context, reconstruction seeks to understand and characterize the real-world conditions under which vehicle operators and occupants are injured. This could include quantifying the severity of an impact, determining speeds and accelerations, analyzing accident-avoidance scenarios, and analyzing the real-world performance of occupant restraint systems, vehicle structures, and driver assistance technologies.

In other instances, crash reconstruction is carried out in a forensic setting and is aimed at answering technical questions relevant to litigation. In this context, crash reconstruction aims to assist lawyers, judges, and juries in their roles of assessing responsibility for a crash and the injuries that result. Rule 702 of the Federal Rules of Evidence states that "a witness who is qualified as an expert by knowledge, skill, experience, training, or education may testify in the form of an opinion or otherwise if: (a) the expert's scientific, technical, or other specialized knowledge will help the trier of fact to understand the evidence or to determine a fact at issue; (b) the testimony is based on sufficient facts or data; (c) the testimony is the product of reliable principles and methods; and (d) the expert has reliably applied the principles and methods to the facts of the case." The notes of the Advisory Committee on this rule state that "an intelligent evaluation of facts is often difficult or impossible without the application of some scientific, technical, or other specialized knowledge. The most common source of this knowledge is the expert witness, although there are other techniques for supplying it." Thus, in a forensic setting, the reconstructionist steps into the role of a teacher, instructing the trier of fact and helping them to understand the technical issues relevant to the facts of the crash that is the subject of litigation.

Investigation and Analysis

Reconstruction is a scientific activity that employs an understanding of physical principles obtained through study, testing, and observation, as well as an intellectual understanding through reason and logic. Reconstruction typically involves both investigation and analysis. The goal of the *investigation* phase is to gather the available evidence relevant to the analysis (physical, electronic, video, audio, and testimonial). This will often include: (1) Reviewing investigative reports and photographs from any police investigation that was conducted. (2) Reviewing statements and testimony from involved drivers and witnesses. (3) Documenting, mapping, and diagramming geometrical conditions of the crash location, like slope, cross-slope, lane widths, shoulder width, roadway surface type, and off-road surface types. This often involves visiting the crash site, but could also rely on aerial imagery [1, 2] and photogrammetric analysis. This may also include documenting changes that have occurred at an accident site to striping, signage, pavement surfaces, or vegetation. (4) Documenting, mapping, and diagramming the physical evidence deposited at the scene. Some of this evidence may still be present during a crash site inspection. Other evidence locations may need to be reconstructed during the analysis phase based on measurements taken by police or using methods of photogrammetry. (5) Documenting, mapping, and diagramming the precrash geometry of the vehicle. At times, this involves obtaining manufacturer specifications for the vehicles involved in the crash or inspecting an exemplar. (6) Documenting and diagramming the physical evidence deposited on the vehicle. This often involves physically inspecting the vehicle but could also rely on photogrammetric analysis and analysis of photographs or video.

The *analysis* phase typically includes: (1) photographic or photogrammetric analysis to locate physical evidence that was no longer present during the site and vehicle or equipment inspections; (2) determining the motion that accounts for the physical evidence identified at the scene or damage identified on components, vehicles, or equipment involved in the crash; (3) applying principles of physics to interpret the physical evidence and add the quantitative elements to the reconstruction—speeds, times, and distances; (4) applying scientific principles to incorporate any electronic, video, audio, or testimonial evidence into the analysis; (5) analyzing the precollision motion and scenarios under which drivers, riders, or pedestrians could have avoided a crash; and (6) evaluating factors that may have contributed to the crash.

Analysis by Phases

Accident reconstruction often involves analyzing a crash in phases. For example, vehicle-to-vehicle collisions can be segmented into the preimpact, impact, and postimpact phases. Single-vehicle rollover crashes can be segmented into the loss-of-control, trip, and roll phases. Single-vehicle motorcycle crashes can be segmented into the loss-of-control, capsizing, and sliding phases. Pedestrian crashes can be segmented into the preimpact, impact, airborne, and sliding or tumbling phases. For each of these crash types, the analysis of each phase will draw on different evidence and analysis techniques. There will also be a human factors phase preceding the other phases of the crash. In this phase, a hazard (or some other stimulus) appears that may initiate a response by a driver, motorcycle rider, or pedestrian. Human factors may also be relevant to the other phases. Crash reconstruction often involves analyzing human factors—evaluating the driver or rider's perception, response, decision-making, and behavior. This analysis will lead the reconstructionist to an understanding of what the driver or rider's response was and if that response was typical, appropriate, or within the normal range of expected human driver responses.

As an example of analyzing crashes in phases, a single-vehicle motorcycle crash could be segmented into the loss-of-control, capsizing, and sliding phases. This is analogous to a single-vehicle rollover crash that could be separated into the loss-of-control, trip, and rolling phases. Conceptually, this can be represented with Equation (1.1), an energy balance equation representing a single-vehicle motorcycle crash. Calculating the initial speed of the motorcycle involves determining the energy dissipated during each phase. The energy loss during each phase of the crash is calculated based on the distance traveled during that phase (d_{loc}, $d_{capsize}$, and d_{slide}) and the corresponding decelerations or drag factors (f_{loc}, $f_{capsize}$, and f_{slide}). In this equation, m_{mc} is the mass of the motorcycle, W_{mc} is the weight of the motorcycle, and v_i is the initial velocity of the motorcycle. This equation could be amended to include energy loss because of an impact and the initial kinetic energy for a vehicle involved in a collision with the motorcycle. The concept of parsing the crash into phases for analysis would not change, though.

$$\frac{1}{2}m_{mc}v_i^2 = f_{loc}W_{mc}d_{loc} + f_{capsize}W_{mc}d_{capsize} + f_{slide}W_{mc}d_{slide} \qquad (1.1)$$

Theoretical and Empirical Modeling

Crash reconstruction utilizes both theoretical and empirical modeling. As an example, the reconstruction of vehicle–pedestrian collisions has historically drawn on both theoretical and empirical models [3]. On the theoretical side, researchers have developed projectile models describing the motion of pedestrians following a collision. Searle's models are, perhaps, the best known of these models [4, 5], but others have been derived by Aronberg [6], Eubanks [7], Han and Brach [8], and others. On the empirical side, several authors have generated equations relating vehicle impact speed to pedestrian throw distance by fitting curves to experimental data [9, 10]. This area of reconstruction benefits from empirical models because, as Toor [10] notes, the theoretical models "are very difficult to apply to real world collisions, because the data necessary to solve the mathematical equations is only partly available from the real-world collisions." This difficulty largely arises from the complex interaction that can occur between the pedestrian and the striking vehicle. Theoretical projection models have continued to be successfully applied to various areas of reconstruction, though, including occupant ejection modeling for rollover crashes [11, 12] and motorcycle rider projection analysis. Beyond that, even for pedestrian crash reconstruction, the theoretical models are helpful for identifying which parameters are significant to a model and to a reconstruction, and thus which parameters should be considered in the empirical modeling or in setting up an experiment. Within motorcycle accident reconstruction, empirical models have been developed that relate the motorcycle impact speed to the postcrash deformation of the motorcycle and the car. The theoretical crush analysis models that are often used for analyzing car-to-car crashes sometimes provide a framework for these empirical models, but for motorcycle crashes, the full theoretical crush analysis models are usually not applied.

Uncertainty Analysis

Accident reconstruction calculations will often involve taking measurements or selecting reasonable inputs for a formula. For example, calculating a vehicle speed from skid marks will require the analyst to measure the length of the skid marks and to select a reasonable value, or range of values, for the coefficient of friction. The coefficient of friction could be selected based on test data in the literature, or testing could be conducted at a specific site. Either way, there will be uncertainty in the skid mark distance and the coefficient of friction. This uncertainty propagates to uncertainty in the calculated speed.

Various authors have discussed methods for quantifying the uncertainty in accident reconstruction calculations. Brach and Dunn [13], for instance, published a treatise covering analytical methods of uncertainty analysis in a forensic science setting. In a 1994 article, Brach [14] covered some of the same techniques of uncertainty analysis specifically in a crash reconstruction context. Several other treatments have appeared in the accident reconstruction literature, including Kost and Werner [15], Wood and O'Riordain [16], Tubergen [17], and Rose [18].

Among these sources, three methods of uncertainty analysis are often mentioned. First, a simple high-low approach can be used, where the analyst combines the high and low ends of the input ranges to produce the highest and lowest results from the formula. This approach yields an overestimate of the uncertainty because it is improbable that the actual values of the inputs would all fall at an extreme of the ranges simultaneously. Second, an analytical approach that utilizes differential calculus can be utilized. This approach involves first taking partial derivatives of the formula with respect to each of the variables. Then, the uncertainty in the dependent variable can be calculated using the following formula [14]:

$$dy = \frac{\partial y}{\partial u} du + \frac{\partial y}{\partial v} dv + \cdots \qquad (1.2)$$

In this equation, dy is the uncertainty in the dependent variable y, and du and dv are the uncertainties in the independent variables u and v. This equation can be extended to accommodate any number of independent variables. When calculating the uncertainty with this equation, the partial derivatives are evaluated at a nominal or reference set of values, typically the values at the middle of the range for each variable. This approach has been applied in an accident reconstruction setting [19], but it can become cumbersome with many of the formulas employed by crash reconstructionists. One advantage of this approach, though, is that it allows for comparison of the relative contributions of uncertainty in each independent variable to the overall uncertainty in the dependent variable.

Third, reconstructionists have often employed a statistical technique called Monte Carlo analysis [18, 20]. Brach [14] referred to this technique as "a brute force randomized simulation on a computer of a mathematical model using appropriate statistical distributions for each of the variables." Kost and Werner [15] noted that, in Monte Carlo simulation, "appropriate probability distributions are assigned to the desired input parameters, and the analyses are repeatedly performed with values of the input parameters selected in accordance with the probability distributions. The results are expressed in the form of probability distributions of each of the desired output parameters, which then allows the analyst to determine the probability of the results falling within selected ranges." Wood and O'Riordain [16] noted that "Monte Carlo simulation methods are well established in many fields and are successfully used where non-linear relationships between variables occur." Several software packages for performing Monte Carlo simulation are commercially available.

Bartlett et al. [21] published a study quantifying the uncertainty in various measurements that are commonly used in crash reconstruction. Much of the data for their study was collected through measurements taken by participants at the World Reconstruction Exposition in 2000 (WREX 2000) in College Station, Texas. As an example, this study included distance measurements utilizing a 25-ft carpenter's tape measure, a flexible measuring tape, and a roller wheel. Using the flexible measuring tape, two "short" measurements of 36.06 and 38.50 ft had standard deviations of 0.017 and 0.025 ft (0.2 and 0.3 in.). Two "long" measurements of 90.60 and 91.60 ft had standard deviations of 0.061 and 0.060 ft (0.73 and 0.72 in.). The distribution of measurements was found to be normal with a mean that coincided closely with the actual measurement. Using a single-wheel roller wheel, a "short" measurement had a standard deviation of 0.081 ft (0.97 in.) and a "long" measurement had a standard deviation of 0.116 ft (1.39 in.). Using a dual-wheel roller wheel, a "short" measurement had a standard deviation of 0.076 ft (0.91 in.) and a "long" measurement had a standard deviation of 0.160 ft (1.92 in.). This study also quantified uncertainties for measurements of the radius of an arced tire mark, for angle measurements, for frictional drag measurements, and for vehicle crush measurements.

In addition to the choice of methods for quantifying uncertainty, how a reconstructionist addresses uncertainty in their analysis will also be influenced by the context in which they are carrying out a reconstruction. In a civil litigation context, the criteria for the reconstruction will typically be what is most probable—perhaps stating conclusions on a more-probable-than-not basis or to a reasonable degree of certainty or probability. In a criminal context, on the other hand, the standard for the reconstruction is an analysis that reaches conclusions beyond a reasonable doubt. In a civil litigation context, ranges on inputs can be considered, and the reconstructionist can state that the speed of the vehicle is within some range that represents reasonable consideration of the input uncertainties. The reconstructionist could state, for example: "The speed of the vehicle was between 52 and 62 mph." In a criminal context, on the other hand, the question may be something like, "Was the driver speeding?" If the range on speeds in this context was 52 to 62 mph and the speed limit was 55 mph, then the reconstructionist would not be able to say, beyond a reasonable doubt, that the driver was speeding because part of the range falls below the speed limit.

Incorporating Witness Statements and Testimony

A reconstruction typically considers the statements of involved parties and witnesses. For a motorcycle crash, these statements can often provide valuable information about surrounding traffic, motorcycle lane positioning, group riding formation, rider experience, rider skill level, and rider clothing and gear. However, Robins [22] and others have noted that witnesses often provide poor estimates of crash variables such as speed, time, and distances. Rose and Carter showed that this inaccuracy extends to distance estimates given in car lengths [23]. As another example, Cummings [24] examined the accuracy of stationary witnesses when estimating the speed of a passing motorcycle. Their study involved 40 participants who each provided estimates of the speed of a motorcycle for 20 individual runs (five runs with each of four different motorcycles). This process resulted in 799 useable speed estimates. The results of this study demonstrated "that individual motorcycle speed estimates were often unreliable, even under these ideal experimental conditions." About 27% of the time, the error in the estimates was greater than 10 mph. The maximum errors in speed estimation were an underreporting of the speed by 26 mph and an overreporting of the speed by 41 mph.

This is consistent with what Fricke states in Chapter 12 of *Traffic Crash Reconstruction* [25]: "One caution that should be passed on from experience is that eyewitness accounts of speed in motorcycle collision cases, just as in other vehicle collision cases, tend to be somewhat questionable …. Eyewitness testimony may suggest that a motorcycle was traveling 'at a high rate of speed.' Such a statement may or may not be true. The eyewitness may be influenced by the lack of experience in observation and knowledge of motorcycles. Eyewitnesses may also be influenced by the existence of a modified motorcycle exhaust system (louder than original equipment). Thus, such information must be judged against the available physical evidence."

However, that witness estimates of speeds, times, and distances are not always reliable is not grounds for dismissing or ignoring what a witness says. Witness testimony is not likely to be all right or all wrong. Witness statements provide information, some of which may be important, particularly related to the events leading up to a crash for which there may be no physical evidence. For those practicing crash reconstruction in a legal context, their role is often to bring physics and physical evidence into the conversation and to inform our clients or a jury what the physics and physical evidence say about how a crash occurred. In this regard, eyewitnesses and involved drivers sometimes make *testable statements*—a statement where the veracity of what they say can be tested against physics, physical evidence, or logic and reasoning. One valuable role a crash reconstructionist can play is to test statements from witnesses and involved drivers and to reach a conclusion about which of the stories is more consistent with scientific principles and physical evidence.

A few additional points are worth considering when a reconstructionist is evaluating witness statements. First, there are instances when a witness's estimate of speed can be accurate. As an example, suppose a witness was following the vehicle that was involved in a crash. Such a witness could accurately report the speed of the involved vehicle because they were following the involved vehicle and the witness has a measuring device (a speedometer) in their own vehicle that reports speed in an accurate way. That is not to say that the witness was necessarily looking at their speedometer just prior to the crash, but often drivers monitor their speed through time and have a reasonable sense for the speed they are traveling. This is also not to say that drivers always tell the truth about their own speed. They do not. But this is to say that there are contexts where speed estimates should not be dismissed simply because the one giving the estimate is human and prone to error.

Second, the reconstructionist should consider what first alerted a witness to something unusual and how much time the witness had to observe the events they are describing. As Robins [22] has noted: "You cannot remember what you never consciously processed in the first place." If a witness's attention was drawn to a crash by the sound of collision itself, this witness may not have much first-hand knowledge about what occurred prior to the collision. When recounting the events, though, this witness may still talk about what happened prior to the collision, for as Robins has also observed, "witnesses are exposed to a variety of postimpact sources of information which are for the most part uncontrolled. Witnesses may be questioned by police and other investigators, they may talk with and share information with other participants and witnesses, they may be exposed to a variety of reports of the original events provided by radio, television, and print media." Second, a witness who is driving will be allocating at least part of their attention to their own driving. This will distract from their focus from the events related to a crash. Third, a witness's position and vantage point will also influence what they can see and when they see it. Accident reconstructionists can often evaluate the vantage point of a witness and determine based on geometric considerations if there is any reason to doubt their description.

Loftus [26] notes that there are three stages that an eyewitness progresses through in giving a description of the event they witnessed—acquisition, retention, and retrieval. The acquisition phase is the witness's original perception of the event during which information is stored in memory. The retention phase is the period between the event and the eventual recollection and recounting of information about that event. The retrieval phase is the witness's actual recalling and recounting of information about the event. The level of accuracy present in a witness statement depends on factors that play out in each of these phases, factors related to the event itself and factors related to the witness. For instance, according to Loftus, the following factors *related to the event* influence the success of the acquisition stage: (1) the duration of the event ("… the less time a witness has to look at something, the less accurate the perception … an eyewitness should be better able to recall an event when the event transpired and was observed over a longer period of time" (p. 23).); (2) the number of times the event occurs; (3) the detail salience ("Some things just catch our attention more readily than others" (p. 25).); and (4) the type of fact being acquired ("Is the witness being asked to remember the height or weight of a criminal, the amount of time an incident lasted, the speed of a car before an accident, the details of a conversation, or the color of the traffic signal? These different types of facts are not equally easy to perceive and recall" (p. 27).). Similarly, the following factors *related to the witness* can influence the success of the acquisition stage: (1) stress, (2) expectations, (3) distraction, and (4) what the witness knows before the event.

An older article by Loftus and Palmer [27] raised an additional concern. They conducted two experiments in which the test subjects watched films of automobile accidents and then answered questions related to these accidents. For example, some subjects were asked "About how fast were the cars going when they *smashed* into each other?" This question resulted in higher estimates of speed from test subjects than the estimates from subjects who were asked the same question using the words *collided*, *bumped*, *contacted*, or *hit*, rather than *smashed*. Even on a retest, a week later, those who were originally asked this question with the word *smashed* were more likely to say they had seen broken glass, even though the film did not depict any glass breaking. Loftus and Palmer observed that "these results are consistent with the view that the questions asked subsequent to an event can cause a reconstruction in one's memory of that event."

Causation

Fricke [25] observed that "reconstruction does not try to explain why a collision happened. That would require describing the entire combination of conditions that would produce another collision." Harris [28], on the other hand, stated that "the ultimate goal of a traffic accident investigation or reconstruction is to determine the events of the accident or what caused the accident." Fricke states elsewhere that "a *cause* is whatever is required to produce a *result*," and so his point seems to be that describing the *entire* set of factors that make up the cause of a crash is beyond the scope of most crash reconstructions. Harris agrees, observing that "determining all the factors that were present in any single accident would be a monumental undertaking … there are simply too many variables, factors, modifiers and circumstances present that may have disappeared long before the investigator becomes involved. Other circumstances may simply never be revealed, or realized, by the parties involved."

Still, though, many reconstructionists and those that hire them would not consider their work complete until they had at least examined the actions of the involved drivers to determine how those actions contributed to the crash. In many instances, the reconstructionist will also evaluate factors related to the involved vehicles and the environment. Along these lines, Harris argues that "cause in law is different from cause in the rest of the world. In the eyes of the courts, cause is an issue of policy and not an instrument of factual analysis. The issue for the court is whether, as a policy decision, a defendant should be held liable for a plaintiff's injuries and the resolution of that issue in an accident case depends on whether the crash was a reasonably foreseeable result of the defendant's negligence … the essential test for proximate cause under the law is the accident must be the natural and probable result of the negligent act or omission and be of such a character as an ordinarily prudent person ought to have foreseen as likely to occur as a result of negligence … This definition makes the work of the reconstructionist easier in that not all the factors present in an accident may be sufficiently relevant to the cause for consideration. Mere presence is insufficient, it must have in some significant way contributed to the result … With a good understanding of the relationships of accident factors, circumstances and modifiers, and sufficient data, an accident can be analyzed, causation determined, and the conclusions effectively presented."

Harris goes on to note that, when accident reconstruction is practiced in a legal setting, "the work of the reconstructionist is preliminary to the analysis performed by the jury. The jury is bound to consider the evidence and will consider the quality and completeness of the investigator's presentation in coming to a conclusion." Thus, the reconstructionist's role in a legal setting is to inform the jury as they make their determination about the degree to which a defendant will be held responsible for a plaintiff's injuries. This may involve reconstructionists testifying about factors they have identified which contributed to or caused a crash.

Within the context of crash reconstruction, epidemiological studies are useful in illuminating factors that have contributed to crashes in the past, and thus may have contributed to a crash under consideration. However, these studies cannot reveal which factors *actually* contributed to any particular crash. Determining which factors contributed to a particular crash will involve evaluating the evidence and facts specific to that crash. For example, epidemiological studies related to motorcycle crashes have demonstrated that, after being involved in a crash with a motorcycle, passenger car drivers sometimes report not having "seen" the motorcycle. Researchers have identified some factors that could contribute to the motorcycle not being seen—the small and narrow profile of some motorcycles, the passenger car driver not expecting to see a motorcyclist, the motorcycle being occluded by other traffic or some geometric feature of the site, a lack of lighting to make the motorcycle detectable, or a lack of contrast between the rider and surrounding environment, for instance. Researchers have also proposed modifications to design features of motorcycles and the operator's clothing to increase the probability of passenger car drivers recognizing the presence of motorcyclists.

For a particular crash, though, it must be determined through reconstruction what specific factors contributed to the crash. The reconstructionist will need to evaluate the evidence, perform testing, or engage in some other scientific process to determine if a common explanation for many crashes happens to also be the correct explanation for a specific crash. It is possible that none of the common factors contributed to a crash and that the physical evidence will show that the driver did see the motorcyclist (there may be preimpact skid marks, for instance, even though the driver reported not having seen the motorcyclist). It is also possible that some common factors were present, but that the actual cause was an inattentive or distracted driver. The principle here is that a reconstructionist's conclusions related to any crash should be driven by the evidence and facts related to that case, not by the findings of epidemiological studies.

A study conducted by the Association of European Motorcycle Manufacturers (ACEM) and referred to as the Motorcycle Accidents in Depth Study (MAIDS) examined the causes of motorcycle accidents in five European countries (France, Germany, the Netherlands, Spain, and Italy) [29]. This study utilized a well-defined methodology for evaluating accident causation for specific motorcycle accidents, which included classifying each potential contributing factor into one of the following categories:

- The factor was present but did not contribute.
- The factor was the precipitating event that initiated the accident sequence.
- The factor was the primary contributing factor in causing the accident.
- The factor was a contributing factor that was present in addition to other contributing factors.
- The factor was not present.

The MAIDS report notes that "the last portion of the investigative process was to determine the contribution of a given factor (e.g., human, vehicle or environmental factor) in the causation of the accident. Typically, this was done at a team meeting, where all the investigative specialists were able to provide input on the accident's causation." The MAIDS researchers defined the following categories of *human* factors:

a) *Perception failure*: One of the drivers failed to detect a dangerous condition.
b) *Comprehension failure*: One of the drivers detected a condition, but failed to comprehend the danger associated with that condition.
c) *Decision failure*: The dangerous condition was detected and comprehended, but a driver or motorcycle operator failed to make the correct decision to avoid the dangerous condition.

These researchers also included a category for "reaction failure," which they defined as a driver or motorcycle operator "failed to react to the dangerous condition, resulting in a continuation or faulty collision avoidance." This seems closely related and potentially indistinguishable from the comprehension and decision failure categories. Any of these four types of human factors could also be an indication of an *attention failure*, which was defined as "any activity of the vehicle operator that distracted him or her from the normal operation of the vehicle … including the normal observation of traffic both in front of and behind the vehicle operator." These researchers further noted that "a proper assessment of the presence of an attention failure depends upon the interview skills of the investigator, since in most cases, the rider or … driver must admit to being distracted …." The MAIDS researchers also mentioned other human failure types, including traffic-scan errors and failing to account for visual obstructions. The MAIDS researchers defined the following categories of *environmental factors*: roadway design or maintenance defects, traffic hazards present from maintenance or construction, and weather. They also noted that *vehicle* component failures could contribute to the occurrence of an accident.

Analyzing Avoidance Scenarios

A reconstruction often culminates in an evaluation of how a crash could have been avoided. When conducting this analysis, the reconstructionist will need to ensure that a proposed alternative course of action was feasible for a typical driver. This will often involve evaluating a reasonable range of perception-response times for a driver, rider, or pedestrian, determining what level of deceleration a driver or rider could produce with braking, or assessing what lateral acceleration a driver or rider could produce with swerving. To ensure that they are imposing reasonable expectations, the reconstructionist will need to consider human variability.

Olson [30] has observed: "There are great differences in raw ability from one individual to the next, and great differences in how a given individual will respond to an identical situation on different occasions. Human variability is a recognized fact that is, sadly, often ignored by crash investigators when offering opinions concerning the performance of a particular individual … it is important to remember that the average is but a single point in a distribution. The average tells us nothing about the scatter in performance, the shape of the distribution, or how that distribution relates to other distributions that may be of interest. In addition, the average is of virtually no help in predicting what can be expected from one randomly-selected individual … Reconstructionists are generally interested in assessing the performance of a given individual, who is drawn from an unknown part of the distribution of interest. In such cases the average is virtually useless as a guide and may be seriously misleading." In evaluating a driver's or motorcyclist's ability to avoid a collision, the reconstructionist should not simply assume a 50th-percentile perception-response time for the driver or motorcyclist. This amounts to expecting the driver or operator to be faster than 50% of the population. It would be more reasonable to assume an 85th-percentile perception-response time, some other percentile that effectively considers the variability of the larger group of people, or to use a range.

Another issue in evaluating a driver's or motorcycle operator's ability to avoid a collision is that the reconstructionist will need to decide where in space and time to invoke a hypothetical change in driver behavior. For example, consider an intersection collision where a driver accelerates away from a stop sign and their vehicle is struck by an oncoming motorcycle traveling through the intersection at a 90° angle to the intruding vehicle. Even if the oncoming motorcycle had the right-of-way through the intersection, there may be a question of how the speed of the motorcycle contributed to the crash. In running a hypothetical scenario where the speed of the oncoming motorcycle is different from what it was in the crash, one option is to invoke that change at the time the other vehicle begins to pull away from the stop sign or stop bar. Another option would be to invoke the change at the time the driver of the turning vehicle begins making their decision to turn [31, 32]. The first option is more focused on the avoidance process the motorcyclist would go through, and the second could be more focused on the decision process of the intruding driver.

Physical Principles Used in Accident Reconstruction

Conservation of Energy

One physical principle that is used frequently in accident reconstruction is *conservation of energy* [33]. Application of this principle to determine the initial speed of a car or a motorcycle in a single-vehicle crash involves identifying the mechanisms by which the initial kinetic energy of the vehicle was dissipated, quantifying how much energy was dissipated by each mechanism, and then adding those dissipated energies up to arrive at an estimate of the initial kinetic energy of the vehicle. The initial speed of the vehicle can then be calculated from this initial kinetic energy. This process can be illustrated with Equation (1.3). In this equation, the initial kinetic energy of the vehicle is determined by adding up the energy dissipated by various mechanisms, including crushing of the structure of the car or motorcycle or frictional-type energy losses that occur during braking, yawing, sliding, tripping, or tumbling and rolling.

$$KE_{initial} = \Delta E_{crushing} + \Delta E_{braking} + \Delta E_{yawing} + \Delta E_{sliding} + \Delta E_{tripping} + \Delta E_{rolling} \quad (1.3)$$

For analysis of the impact phase of a two-vehicle collision, the energies of the two vehicles need to be considered together, as illustrated in Equation (1.4). This equation indicates that the total kinetic energy that the two vehicles bring to the collision is equal to the energy dissipated by crushing of the vehicle structures plus the total kinetic energy of the two vehicles following the collision. These kinetic energies can include both translational and rotational motions. The postcollision kinetic energies for each vehicle would be calculated with an expression like Equation (1.3).

$$KE_{initial,1} + KE_{initial,2} = \Delta E_{crushing,1} + \Delta E_{crushing,1} + KE_{final,1} + KE_{final,2} \quad (1.4)$$

Newton's Second Law and the Principle of Work and Energy

The process of determining how much energy was dissipated by each mechanism typically utilizes the *principle of work and energy* [34], which equates a change in kinetic energy to the work performed by the force that accomplishes or causes the change in kinetic energy. *Work* is defined as the action of a force through a distance. This principle can be derived from *Newton's second law*, another physical principle utilized by accident reconstructionists. Newton's second law is given by Equation (1.5). It states that a body will be accelerated in proportion to the sum of the forces applied to the body. The acceleration will occur along the line of action of the vector sum of the forces and the mass of the body, m, acts as the proportionality constant.

$$\sum \vec{F} = m\vec{a} \tag{1.5}$$

Newton's second law is also one of the underlying physical principles utilized in crash simulation packages, such as PC-Crash, HVE, and Virtual Crash. If one can determine the forces applied to a vehicle at any instant in time, then one can calculate the acceleration the vehicle is experiencing. Thus, reconstructionists have developed models to calculate instantaneous tire forces, suspension forces, and collision forces.

The *principle of work and energy* can be derived from Newton's second law, by first considering Equation (1.5) along one coordinate axis, as follows:

$$F_i = ma_i \tag{1.6}$$

In this equation, i designates the coordinate direction being considered—x, y, or z in a Cartesian coordinate system. Because acceleration is equal to a differential change in velocity (dv_i) over a differential change in time (dt), and velocity is equal to a differential change in position (ds_i) over a differential change in time, Equation (1.5) can be rewritten as follows:

$$F_i = m\frac{dv_i}{dt} = m\frac{dv_i}{ds_i}\frac{ds_i}{dt} = mv_i\frac{dv_i}{ds_i} \tag{1.7}$$

Equation (1.7) can be rewritten as follows:

$$F_i ds_i = mv_i dv_i \tag{1.8}$$

Equation (1.8) can be integrated along the coordinate direction i along a distance from Points 1 to 2, as follows:

$$\int_1^2 F_i ds_i = \int_1^2 mv_i dv_i = \frac{1}{2}m\left(v_{i,2}^2 - v_{i,1}^2\right) \tag{1.9}$$

The left side of Equation (1.9) is the work performed by the force acting through the distance from 1 to 2 and the right side is the change in kinetic energy experienced by the body. This means that the ΔE terms in Equations (1.3) and (1.4) will take the form of forces acting through a distance. For example, for a frictional-type energy loss, the change in energy will be calculated as follows:

$$\Delta E_{frictional} = \mu W d \tag{1.10}$$

In this equation, μ is the coefficient of friction and W is the vehicle weight. The coefficient of friction multiplied by the weight is the frictional force that acts through the distance, d.

Another example is the work done in compressing a spring, an analogy that is used by crash reconstructionists to model the crushing behavior of vehicle structures. The energy expended in compressing a spring from its undeformed position through a distance, Δx, is given by the following equation:

$$\Delta E_{spring} = k \cdot \Delta x^2 \tag{1.11}$$

Figure 1.1 contains a frame of video from a motorcycle-to-vehicle collision conducted at the World Reconstruction Exposition (WREX) in 2016. This frame shows the motorcycle impacting the side of the passenger car and the motorcycle forks and front wheel deforming, along with the side of the passenger car. The deformation to both vehicles in this collision could be modeled with the spring analogy.

FIGURE 1.1 Video frame from motorcycle-to-car crash test conducted at WREX 2016.

In this collision, the motorcycle impacted the car with an initial velocity, V_A, and thus, with an initial kinetic energy. All or most of this kinetic energy is absorbed through the crushing of the motorcycle and vehicle structures. During the first phase of the impact—the approach or compression phase—the deformations of both the car and the motorcycle reach a maximum level. Figure 1.2 depicts an idealized force-crush curve that could be used to model the force that builds up as each vehicle deforms [35]. This idealized force-crush curve begins where there is no crush and no force. The curve ascends in a linear fashion (constant stiffness, K_1) to the maximum dynamic force, which is achieved coincident with the maximum dynamic crush (C_m). Per Newton's third law, the collision force applied to first vehicle will be equal in magnitude to the collision force applied to the other vehicle, although each vehicle will likely have a unique stiffness, and thus, the collision partners will experience different levels of crush. The absorbed energy (E_A) for each vehicle is equal to the area under the portion of the force-crush curve from Points 1 to 2.

After the approach phase is completed, the impact force drops quickly as the structure of each vehicle experiences a partial rebound from the maximum crush (Points 2 to 3). This structural rebound has the effect of imparting a velocity to each vehicle that leads to separation. Some kinetic energy is restored to each vehicle. When the vehicles separate, the collision force drops to zero and the vehicle structures finish their rebound, settling in on the final residual crush (C_R) from Points 3 to 4 (wheelbase reduction and wheel deformation for the motorcycle). The phase during which the vehicles rebound from the collision and experience partial structural restoration is referred to as the *restitution or rebound phase* of the collision. Again, this is assumed to occur with a constant stiffness (K_2) for each vehicle.

FIGURE 1.2 Absorbed, restored, and dissipated energies.

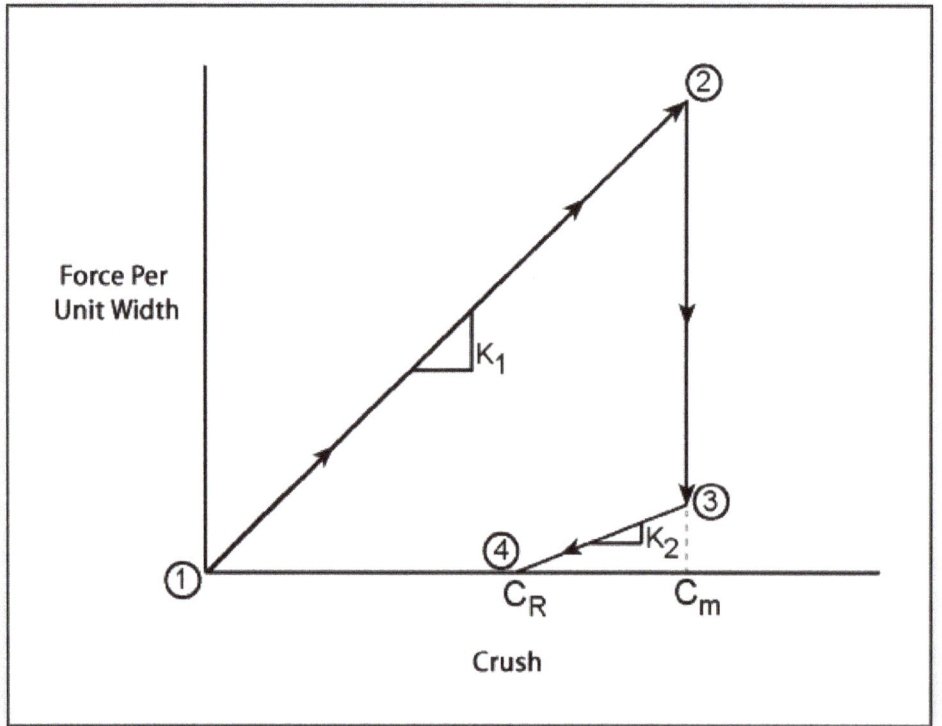

The energy imparted to the vehicle during the restitution phase is referred to as the restored energy, E_R. This energy is the area under the line from Points 3 to 4.

The coefficient of restitution for the impact characterizes this restoration of energy and can be calculated with the following equation:

$$e = \sqrt{\frac{E_R}{E_A}} \qquad (1.12)$$

In addition to the absorbed and restored energies, the idealized force-crush curve of Figure 1.2 can be used to illustrate an additional energy value—the dissipated energy (energy loss). This is the difference between the absorbed and restored energies, or the area between the line for the approach phase and the line for the rebound phase.

Crash reconstructionists have developed methods for calculating the absorbed, dissipated, and restored energies. These methods typically utilize measurement of the post-crash crush to each of the vehicles along with application of stiffness characteristics of those vehicles determined from crash tests. These together enable the reconstructionists to calculate both the collision force and the absorbed energy. Equation (1.13) can then be used to calculate the relative speed of impact between the vehicles, along the line of action of the collision force. In this equation, $E_{A,total}$ is the total absorbed energy for the two vehicles involved in the collision and γ_1 and γ_2 are the effective mass multipliers, which account for collision forces that do not pass through the center of mass of the vehicles (noncentral collisions) [36, 37, 38].

$$V_A = (1+e) \cdot \sqrt{\frac{\gamma_1 m_1 + \gamma_2 m_2}{\gamma_1 m_1 \cdot \gamma_2 m_2} \cdot 2 E_{A,total}} \qquad (1.13)$$

The effective mass multipliers are defined as follows, where k_i is the radius of gyration of the vehicle under consideration and h_i is the moment arm of the collision force about the center of gravity of the vehicle under consideration.

$$\gamma_i = \frac{k_i^2}{k_i^2 + h_i^2} \qquad (1.14)$$

McHenry developed a method for quantifying the absorbed energy based on vehicle crush and implemented it in a software package called CRASH computer program (Computer Reconstruction of Automobile Speeds on the Highway) [39, 40]. The CRASH program was created in conjunction with the SMAC program (Simulation Model of Automobile Collisions), an accident simulation program that required the user to input an estimate of the initial speeds of the vehicles and then to iteratively change those speeds to achieve a match with physical evidence. The CRASH program was originally intended to be a preprocessor for SMAC, providing initial estimates of the vehicle impact speeds for use with the SMAC program. However, the CRASH program has often been used within the field of accident reconstruction as a stand-alone method.

The CRASH method assumed that the crushing vehicle structure could be represented as a series of springs, each with a linear force-crush relationship. McHenry represented the stiffness properties for these linear springs by the two parameters, A and B. The graphic of Figure 1.3 depicts how the A and B values define the linear force-crush relationship in CRASH. The horizontal axis in the figure represents the residual crush depth and the vertical axis represents the force per unit width of damage. The B value, the slope of the linear force-crush relationship, governs the proportionality between residual crush and impact force. The A value, the force intercept on the force-crush plot, represents the force that can be applied to the vehicle structure before the onset of permanent crush.

FIGURE 1.3 *Linear force-crush characteristics used in CRASH.*

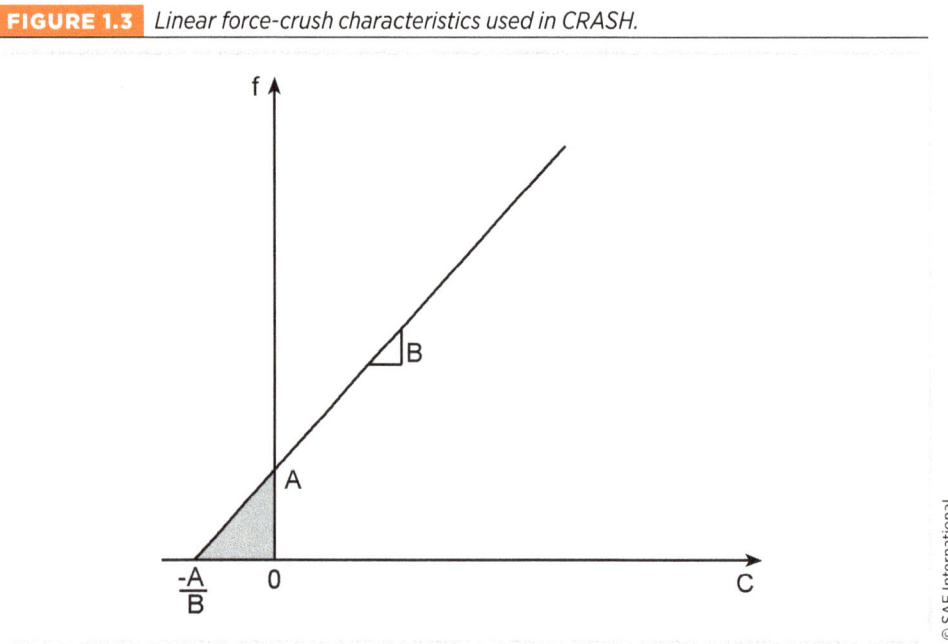

The force-crush line intersects the crush axis at a value of −A/B. This negative intercept on the crush axis has traditionally been interpreted as the elastic recovery of the vehicle structure, which is the difference between the dynamic crush and the residual crush. With this interpretation, the A value arises because of the use of residual crush in the model and shifting the force-residual crush line over a distance of A/B would yield the dynamic force-crush line. This interpretation of the A value is sound if the difference between the dynamic maximum crush and the residual maximum crush is constant and independent of the severity of the impact.

Equation (1.15) describes the force-crush curve of Figure 1.3.

$$f = A + BC_R \tag{1.15}$$

Using Equation (1.15) to describe the force, McHenry integrated over the crush depth and damage width (w_0) to determine the deformation energy, and obtained the following equation:

$$E_A = \left(\frac{B}{2}C_R^2 + AC_R + \frac{A^2}{2B}\right) \cdot w_0 \qquad (1.16)$$

While, in general, a crush profile will not have a constant crush depth, a general crush profile can be represented by an equivalent damage width and average crush depth value. Woolley [41] has discussed methods for calculating a representative damage width and crush depth for a general crush profile.

A method for obtaining A and B stiffness coefficients for CRASH can be developed by recognizing that Equation (1.16) is quadratic in the residual crush depth, C_R. Application of the quadratic formula to Equation (1.16) yields the following equation:

$$\sqrt{\frac{2E_A}{w_0}} = \sqrt{B}C_R + \frac{A}{\sqrt{B}} \qquad (1.17)$$

The left side of Equation (1.17) is a normalized energy term, which has been referred to as the Energy of Approach Factor (EAF). For a barrier impact, the damage energy in Equation (1.17) would be set equal to the initial kinetic energy of the vehicle. For other impact types, say a vehicle-to-vehicle frontal impact, the damage energy in Equation (1.17) would be set equal to the damage energy absorbed by the vehicle of interest during the impact. Each crash test will provide one point for constructing an EAF versus residual crush plot. With multiple points, a line can be fit to the data. The slope and the intercept of this line will yield the A and B coefficients.

When residual crush measurements are taken perpendicular to the original shape of the impacted vehicle side, the damage energy that results from Equation (1.16) will only include the work done by the component of the collision force that acted normal to the vehicle side. In many cases, the collision force will not have acted perpendicular to one of the vehicle sides and, therefore, crush measured perpendicular to the original vehicle surface will not represent the crush along the actual collision force direction. The absorbed damage energy, thus, may need to be adjusted to include the work done by the component of the collision force that acted tangential to the vehicle surface.

One approach for adjusting the damage energy is to adjust the residual crush depth so that it represents the distance the structure was displaced along the direction of the collision force. This approach is used in CRASH3, and it results in the calculated damage energy being multiplied by the following multiplier:

$$1 + \tan^2\alpha \qquad (1.18)$$

In Equation (1.18), α is the angle between the normal to the vehicle side and the actual direction of the collision force. Woolley [41] and Fonda [42] proposed that the width of the damaged region should also be adjusted to reflect the direction of the collision force. When both the crush depth and the crush width are adjusted, the crush energy of Equation (1.16) would be adjusted using the following multiplier:

$$\frac{1}{\cos\alpha} \qquad (1.19)$$

Principle of Impulse and Momentum (Conservation of Momentum)

In addition to the principle of work and energy, crash reconstructionists also utilize the *principle of impulse and momentum*, which can be utilized to relate the initial (immediately prior to the collision) and final (immediately after separation) velocities. Brach's book *Mechanical Impact Dynamics* contains a thorough treatment of the application of the principle of impulse and momentum to collision analysis [43]. The application of this principle will be illustrated here with the two-dimensional (planar) particle impact model of Figure 1.4. This figure depicts two particles colliding, with velocities $\vec{v}_1$ and $\vec{v}_2$. A normal axis and a tangential axis are depicted, and the velocities of the particles are assumed to lie entirely in the plane defined by these axes. The force generated by deformation occurs on the normal axis and frictional-type forces occur along the tangential axis. Often, when a particle impact model is introduced, collision forces along the tangential axis are assumed negligible, but that assumption is not invoked here. The following assumptions will be invoked: (1) the duration of contact is assumed short enough to be approximated as instantaneous (this equates to no change in position or orientation of the particles during the collision); (2) the deforming region is assumed to be small relative to the size of the particle (this equates to no change in the particle inertial properties during the collision); (3) angular velocities are assumed negligible, both before and after the collision (thus, a central collision, where the resultant collision force passes through the center of mass of both particles, is assumed); and (4) the collision force is the only force considered. External forces are assumed negligible (for vehicle collisions, this would be tire forces, forces from scraping and gouging, and aerodynamic forces acting during the collision).

FIGURE 1.4 Two-dimensional particle impact model.

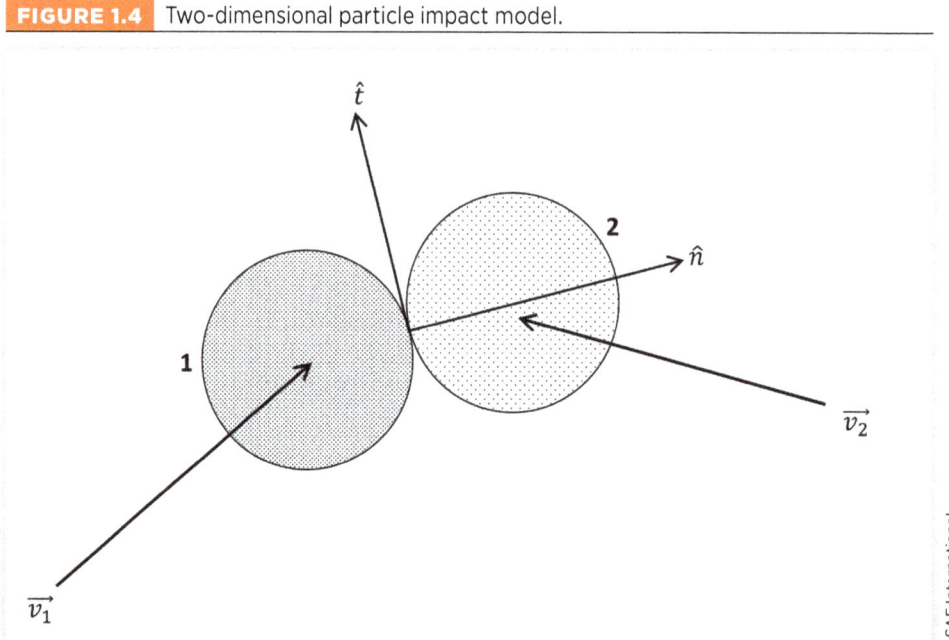

According to the principle of impulse and momentum:

$$P_n = m_1\left(v_{1n,f} - v_{1n,i}\right) = -m_2\left(v_{2n,f} - v_{2n,i}\right) \quad (1.20)$$

$$P_t = m_1\left(v_{1t,f} - v_{1t,i}\right) = -m_2\left(v_{2t,f} - v_{2t,i}\right) \quad (1.21)$$

In Equations (1.20) and (1.21), P_t is the tangential impulse, P_n is the normal impulse, $v_{1n,i}$ is the normal-direction velocity of particle 1 immediately prior to the collision, and $v_{1n,f}$ is the normal-direction velocity of particle 1 immediately following the collisions. The normal velocities for particle 2 and the tangential velocities for particles 1 and 2 are identified with similar subscripts. Equations (1.20) and (1.21) are statements of *conservation of momentum* along each coordinate axis and they provide two equations for relating the initial and final velocities. Two additional equations are needed.

One equation is provided by the coefficient of restitution (e), which relates the initial and final relative velocities of the particles along the normal axis and thus controls the energy loss along the normal axis, as follows:

$$e = -\frac{v_{2n,f} - v_{1n,f}}{v_{2n,i} - v_{1n,i}} \quad (1.22)$$

The coefficient of restitution will typically fall between 0 and 1, though there are instances in which a negative coefficient of restitution makes physical sense. For instance, Brach mentions the example of an "errant baseball passing through a window screen." In the context of vehicular crash reconstruction, a similar situation could occur if a corner-to-corner collision occurs between two vehicles and the structures give way before a common velocity can be reached normal to the contact surface. Coefficients of restitution for a given collision type can be determined either experimentally or analytically. Both approaches have been used by crash reconstructionists. Another additional equation is provided by the impulse ratio (μ), which describes the ratio of the tangential to normal impulses generated by this collision and controls the energy loss along the tangential axis, as follows:

$$\mu = \frac{P_t}{P_n} \quad (1.23)$$

In some instances, the impulse ratio can be interpreted as a friction coefficient between the colliding objects, but it is not limited to this interpretation [43–47]. In addition to the effects of friction, the impulse ratio can also include the effects of forces generated by snagging between the vehicles. The "available friction" can be set at a value that reflects such snagging or furrowing when it occurs. Brach observes that "for an oblique impact when the particles have a nonzero initial relative tangential velocity component, sliding must always occur, at least during initial stages of contact. A question that must be answered is if relative tangential motion continues through separation or ceases prior to separation. Although the laws of Coulomb friction do not apply during all impacts, the concept and mathematical form of a coefficient of friction prove to be very convenient." Brach also notes that "the tangential force and its impulse usually are found to be dissipative, such as from friction or plastic deformation, and cannot add energy to the system. For a particle, the tangential impulse always opposes initial relative tangential motion." This implies that the sign of μ can be either positive or negative and is determined with the following equation:

$$sgn(\mu) = sgn\left(\frac{v_{2t,i} - v_{1t,i}}{v_{2n,i} - v_{1n,i}}\right) \quad (1.24)$$

Brach also derived the following expression for the critical impulse ratio, which is the impulse ratio necessary to cause sliding to cease during the collision. The impulse ratio should not be set at a value that exceeds the critical impulse ratio.

$$\mu_c = \frac{1}{1+e} \cdot \frac{v_{2t,i} - v_{1t,i}}{v_{2n,i} - v_{1n,i}} \quad (1.25)$$

The relative magnitude of the available friction coefficient to the critical impulse ratio will have physical significance for an impact. The available friction coefficient represents the magnitude of friction force that *can* be recruited during the impact. The critical impulse ratio represents the magnitude of friction force that *must* be recruited for relative motion to cease along the contact surface. When the critical impulse ratio is greater than the available friction coefficient, all the available friction will be recruited during the impact, but that friction will be insufficient to cause sliding to cease in the contact region. When the available friction coefficient exceeds the critical impulse ratio, only a portion of the available friction will be recruited, and sliding will cease in the contact region. In such cases, the value of the impulse ratio for the impact model should be set at the value of the critical impulse ratio, not the available friction coefficient. This is because recruitment of the available friction depends on relative velocity being present along the contact surface. Once this relative motion ceases, no additional friction can be recruited.

Brach continues, "in specific applications the appropriate value of μ must be determined experimentally or estimated by means such as analytical modeling of the mechanism of tangential force generation. If Coulomb friction is appropriate, then μ is related to the coefficient of dynamic friction. Otherwise, it can be considered to be an equivalent coefficient of friction." Brach demonstrates that Equations (1.20), (1.21), (1.22), and (1.23) can be solved for the final velocity components, which results in the following set of equations relating the initial velocities to the final velocities:

$$v_{1n,f} = v_{1n,i} + (1+e)\frac{m_2}{m_1+m_2}(v_{2n,i} - v_{1n,i}) \quad (1.26)$$

$$v_{2n,f} = v_{2n,i} - (1+e)\frac{m_1}{m_1+m_2}(v_{2n,i} - v_{1n,i}) \quad (1.27)$$

$$v_{1t,f} = v_{1t,i} + \mu(1+e)\frac{m_2}{m_1+m_2}(v_{2n,i} - v_{1n,i}) \quad (1.28)$$

$$v_{2t,f} = v_{2t,i} - \mu(1+e)\frac{m_1}{m_1+m_2}(v_{2n,i} - v_{1n,i}) \quad (1.29)$$

When applying conservation of momentum to a motorcycle-to-car collision, the reconstructionist will need to be aware of potential sensitivity in the solution because of a significant discrepancy between the weight of the car and the weight of the motorcycle and rider. If the struck car is significantly heavier than the motorcycle and the rider, then small errors or uncertainties in some of the inputs to the analysis can lead to significant errors or uncertainties in the calculated results.

References

[1] Wirth, J., Bonugli, E., and Freund, M., "Assessment of the Accuracy of Google Earth Imagery for use as a Tool in Accident Reconstruction," SAE Technical Paper 2015-01-1435, 2015, doi:10.4271/2015-01-1435.

[2] Harrington, S., Teitelman, J., Rummel, E., Morse, B. et al., "Validating Google Earth Pro as a Scientific Utility for Use in Accident Reconstruction," SAE Technical Paper 2017-01-9750, 2017, doi:10.4271/2017-01-9750.

[3] Toor, A. and Araszewski, M., "Theoretical vs. Empirical Solutions for Vehicle/Pedestrian Collisions," SAE Technical Paper 2003-01-0883, 2003, doi:10.4271/2003-01-0883.

[4] Searle, J. and Searle, A., "The Trajectories of Pedestrians, Motorcycles, Motorcyclists, etc., Following a Road Accident," SAE Technical Paper 831622, 1983, doi:10.4271/831622.

[5] Searle, J., "The Physics of Throw Distance in Accident Reconstruction," SAE Technical Paper 930659, 1993, doi:10.4271/930659.

[6] Aronberg, R., "Airborne Trajectory Analysis Derivation for Use in Accident Reconstruction," SAE Technical Paper 900367, 1990, doi: 10.4271/900367.

[7] Eubanks, J.J. and Hill, P.F., *Pedestrian Accident Reconstruction*, 2nd ed., Lawyers & Judges Publishing Company, Tucson, AZ, 1999, ISBN 0-913875-56-2.

[8] Han, I. and Brach, R., "Throw Model for Frontal Pedestrian Collisions," SAE Technical Paper 2001-01-0898, 2001, doi:10.4271/2001-01-0898.

[9] Happer, A., Araszewski, M., Toor, A., Overgaard, R. et al., "Comprehensive Analysis Method for Vehicle/Pedestrian Collisions," SAE Technical Paper 2000-01-0846, 2000, doi:10.4271/2000-01-0846.

[10] Toor, A., Araszewski, M., Johal, R., Overgaard, R. et al., "Revision and Validation of Vehicle/Pedestrian Collision Analysis Method," SAE Technical Paper 2002-01-0550, 2002, doi:10.4271/2002-01-0550.

[11] Funk, J. and Luepke, P., "Trajectory Model of Occupants Ejected in Rollover Crashes," SAE Technical Paper 2007-01-0742, 2007, doi:10.4271/2007-01-0742.

[12] Funk, J. and Rose, N., "Occupant Ejection Trajectories in Rollover Crashes: Full-Scale Testing and Real-World Cases," *SAE International Journal of Passenger Cars—Mechanical Systems*, 1(1): 43–54, 2009, doi:10.4271/2008-01-0166.

[13] Brach, R.M. and Dunn, P.F., *Uncertainty Analysis for Forensic Science*, Lawyers and Judges Publishing Company, Tucson, AZ, 2004, ISBN 1-930056-20-6.

[14] Brach, R., "Uncertainty in Accident Reconstruction Calculations," SAE Technical Paper 940722, 1994, doi:10.4271/940722.

[15] Kost, G. and Werner, S., "Use of Monte Carlo Simulation Techniques in Accident Reconstruction," SAE Technical Paper 940719, 1994, doi:10.4271/940719.

[16] Wood, D. and O'Riordain, S., "Monte Carlo Simulation Methods Applied to Accident Reconstruction and Avoidance Analysis," SAE Technical Paper 940720, 1994, doi:10.4271/940720.

[17] Tubergen, R., "The Technique of Uncertainty Analysis as Applied to the Momentum Equation for Accident Reconstruction," SAE Technical Paper 950135, 1995, doi:10.4271/950135.

[18] Rose, N. and Hughes, C., "Integrating Monte Carlo Simulation, Momentum-Based Impact Modeling, and Restitution Data to Analyze Crash Severity," SAE Technical Paper 2001-01-3347, 2001, doi:10.4271/2001-01-3347.

[19] Rose, N.A., "Quantifying the Uncertainty in the Coefficient of Restitution Obtained with Accelerometer Data from a Crash Test," SAE Technical Paper 2007-01-0730, 2007, doi:10.4271/2007-01-0730.

[20] Wach, W. and Unarski, J., "Uncertainty Analysis of the Preimpact Phase of a Pedestrian Collision," SAE Technical Paper 2007-01-0715, 2007, doi:10.4271/2007-01-0715.

[21] Bartlett, W., Wright, W., Masory, O., Brach, R. et al., "Evaluating the Uncertainty in Various Measurement Tasks Common to Accident Reconstruction," SAE Technical Paper 2002-01-0546, 2002, doi:10.4271/2002-01-0546.

[22] Robins, P.J., *Eyewitness Reliability in Motor Vehicle Accident Reconstruction and Litigation*, Lawyers and Judges Publishing Company, Tucson, AZ, 2001, ISBN 0-913875-92-9.

[23] Rose, N. and Carter, N., "How Accurate Are Witness Distance Estimates Given in Car Lengths?" *Collision: The International Compendium for Crash Research*, 11(1): 74–86, 2016.

[24] Cummings, J.R., Fletcher, H.J., Biller, B.A., Scanlan, S. et al., "Estimates of Motorcycle Speed Made by Eyewitnesses Under Ideal Experimental Conditions," *Accident Reconstruction Journal*, 26(1): 12–17, January/February 2016, ISSN: 1057-8153.

[25] Fricke, L.B., *Traffic Crash Reconstruction*, 2nd ed., Northwestern University Center for Public Safety, Evanston, IL, 2010, ISBN 0-912642-03-3.

[26] Loftus, E.F., *Eyewitness Testimony*, Harvard University Press, Cambridge, MA, 1996, ISBN 0-674-28777-0.

[27] Loftus, E.F. and Palmer, J.C., "Reconstruction of Automobile Destruction: An Example of the Interaction Between Language and Memory," *Journal of Verbal Learning and Verbal Behaviour*, 13: 585–589, 1974, doi:10.1016/S0022-5371(74)80011-3.

[28] Harris, J.O., "Cause and Contributing Factors," *Accident Reconstruction Journal*, 2(6): 16–17, November/December 1990, ISSN 1057-8153.

[29] Association of European Motorcycle Manufacturers (ACEM), "MAIDS: Motorcycle Accidents In Depth Study," Final Report 2.0, April 2009, http://www.maids-study.eu/pdf/MAIDS2.pdf, accessed on April 12, 2018.

[30] Olson, P.L., *Forensic Aspects of Driver Perception and Response*, 1st ed., Lawyers and Judges Publishing Company, Tucson, AZ, 1996, ISBN 0-913875-22-8.

[31] Robinson, G.H., Erickson, D.J., Thurston, G.L., and Clark, R.L., "Visual Search by Automobile Drivers," *Human Factors*, 14(4): 315–323, 1972.

[32] Rahimi, M., "A Task, Behavior, and Environmental Analysis for Automobile Left-Turn Maneuvers," *Proceedings of the Human Factors Society, 33rd Annual Meeting*, Denver, CO, 1989.

[33] Van Ness, H.C., *Understanding Thermodynamics*, Dover Publications Inc., New York, 1969, ISBN 0-486-63277-6.

[34] Beer, F.P. and Johnston, E.R., *Vector Mechanics for Engineers: Dynamics*, 5th ed., McGraw-Hill, New York, 1988, ISBN 0-07-079926-1.

[35] Emori, R.I., "Analytical Approach to Automobile Collisions," SAE Technical Paper 680016, 1968, doi:10.4271/680016.

[36] Rose, N., Fenton, S., and Ziernicki, R., "An Examination of the CRASH3 Effective Mass Concept," SAE Technical Paper 2004-01-1181, 2004, doi:10.4271/2004-01-1181.

[37] Rose, N., Fenton, S., and Ziernicki, R., "Crush and Conservation of Energy Analysis: Toward a Consistent Methodology," SAE Technical Paper 2005-01-1200, 2005, doi:10.4271/2005-01-1200.

[38] Kerkhoff, J., Husher, S., Varat, M., Busenga, A. et al., "An Investigation into Vehicle Frontal Impact Stiffness, BEV and Repeated Testing for Reconstruction," SAE Technical Paper 930899, 1993, doi:10.4271/930899.

[39] McHenry, R.R., "The CRASH Program – A Simplified Collision Reconstruction Program," ZQ-5731-V-1, Calspan, 1975, http://www.mchenrysoftware.com/McHenry%201975%20The%20CRASH%20Program.pdf, accessed on April 12, 2018.

[40] McHenry, R.R. and McHenry, B.G., "A Revised Damage Analysis Procedure for the CRASH Computer Program," SAE Technical Paper 861894, 1986, doi:10.4271/861894.

[41] Woolley, R., "Non-Linear Damage Analysis in Accident Reconstruction," SAE Technical Paper 2001-01-0504, 2001, doi:10.4271/2001-01-0504.

[42] Fonda, A., "Principles of Crush Energy Determination," SAE Technical Paper 1999-01-0106, 1999, doi:10.4271/1999-01-0106.

[43] Brach, R.M., *Mechanical Impact Dynamics: Rigid Body Collisions*, Revised edition, Brach Engineering, South Bend, IN, 2007, ISBN 978-1-4028-9462-6.

[44] Brach, R.M. and Brach, R.M., "A Review of Impact Models for Vehicle Collision," SAE Technical Paper Number 870048, 1987, doi: 10.4271/870048.

[45] Brach, R.M. and Brach, R.M., "Energy Loss in Vehicle Collision," SAE Technical Paper Number 871993, 1987, doi 10.4271/871993.

[46] Brach, R.M. and Brach, R.M., *Vehicle Accident Analysis and Reconstruction Methods*, 2nd ed., SAE International, Warrendale, PA, 2005, ISBN 978-0-7680-3437-0.

[47] Marine, M.C., "On the Concept of Inter-Vehicle Friction and Its Application in Automobile Accident Reconstruction," SAE Technical Paper Number 2007-01-0744, 2007, doi:10.4271/2007-01-0744.

2
Motorcycle Characteristics

This chapter covers motorcycle characteristics that could be relevant when reconstructing a motorcycle crash. These characteristics could be associated with a particular type of motorcycle (standard, cruiser, touring, sport, etc.), or they could be features of a particular make and model or after-market modifications to a specific motorcycle. As an example, the characteristics of a motorcycle braking system can influence the deceleration a rider can achieve with heavy braking (i.e., the presence or absence of antilock brakes or the independence or integration of the front and rear brakes). Geometric features, such as crash bars, footrests or pegs, or panniers can limit the maximum lean angle achievable by a motorcycle, and thus influence the lateral acceleration a rider can achieve during cornering or swerving. These geometric features can also influence the deceleration of a motorcycle that has capsized and is sliding on the ground. The style of motorcycle can influence the rider's posture, and thus their eye height when sitting on the motorcycle. This can influence what they are able to see. Mass and inertial properties such as the center of gravity (CG) height can be relevant to calculations related to straight-line braking or to the lean angle of the motorcycle while cornering. The mass and inertial properties can also influence the motorcycle motion in a multibody simulation.

Another item worth considering: some motorcycles have multiple modes and settings in which they can be ridden. For example, a 2020 KTM Adventure 790 motorcycle has the following modes which affect the responsiveness of the throttle: street, rain, and off-road. This motorcycle also has traction control that can be disabled and antilock brakes that can be set in on-road or off-road modes or turned off entirely. In the on-road ABS mode, both front and rear ABS are active. In the off-road ABS mode, the rear ABS is disabled, but front ABS remains active. These settings will alter the way the motorcycle behaves under conditions that could be relevant to a reconstruction. The owner's manual for individual motorcycles can help in understanding what settings and modes are available on a particular motorcycle, and online forums can be useful for understanding how the various modes operate. In some instances, testing of the various modes could be useful. It may also be useful during a physical inspection of the motorcycle to determine what settings and modes are active on the motorcycle. Keep in mind, though, that many motorcycles are set up to return to the default settings (safety systems ON) when the power is turned off and then cycled back on. For example, on the KTM Adventure 790, if the ABS is turned off, it will automatically turn back on if the motorcycle is cycled off and then back on.

Motorcycle Types

Standard motorcycles are the most basic and versatile category of on-road motorcycles. Figure 2.1 depicts a typical standard motorcycle. These motorcycles are sometimes referred to as *traditional* or *naked* motorcycles. The term naked refers to their lack of fairings. Standard motorcycles are characterized by their upright handlebars and upright, neutral riding posture. They are also typically equipped with a gas tank that is too small to go long distances without stopping to refuel. Illustratively, standard motorcycles often have wheelbases in the range of 53 to 63 in. and CG heights in the 18- to 25-in. range. Lean angles achievable by these motorcycles often depend on the placement of the footrests or pegs and the geometry of the exhaust. Examples of standard motorcycles include the Honda CB1000R, the Triumph Bonneville and Street Twin, and the Suzuki SV650.[1]

Cruisers are sized between touring and standard motorcycles (Figure 2.2). They are styled for looks and performance, without the long-range cargo carrying capabilities. They are characterized by their pulled-back handlebars and a relaxed, feet-forward riding position. The Highway Loss Data Institute states that "cruiser motorcycles mimic the style of earlier American motorcycles from the 1930s to the early 1960s, such as those made by Harley-Davidson and Indian." Illustratively, cruisers often have wheelbases in the range of 62 to 67 in. and CG heights in the 15- to 22-in. range. Lean angles achievable by these

FIGURE 2.1 Typical standard motorcycle.

© SAE International

[1.] https://www.motorcyclecruiser.com/standard-motorcycles/

FIGURE 2.2 Typical cruiser motorcycle.

© SAE International

motorcycles often depend on the placement of the footrests or pegs, the geometry of the exhaust, and the size and placement of crash bars or panniers. Examples of cruisers include the Indian Scout, the Harley-Davidson Street Bob, and the Honda Fury.

Touring motorcycles are typically the largest of the on-road motorcycles and are designed for comfort and distance. Figure 2.3 depicts a typical touring motorcycle. These motorcycles are large in overall size and engine size and usually accommodate riders with bucket seats and provide cargo carrying areas large enough for long trips. Illustratively, touring motorcycles often have wheelbases in the range of 66 to 68 in. and CG heights in the 15- to 20-in range. Lean angles achievable by these motorcycles often depend on the placement of the footrests or pegs, the geometry of the exhaust, and the size and placement of crash bars or panniers. Examples of touring motorcycles include the Indian Challenger or Roadmaster, the Harley-Davidson Road King, and the Honda Goldwing.[2]

FIGURE 2.3 Typical touring motorcycle.

© SAE International

---

[2] https://motorbikewriter.com/10-best-touring-motorcycles/

FIGURE 2.4 Typical sport motorcycle.

Sport motorcycles (Figure 2.4) are focused on capabilities for cornering, maneuverability, braking, and acceleration. Their engines sizes have a large range but in general are geared and designed for speed and handling. Illustratively, sport motorcycles often have wheelbases in the range of 54 to 59 in. and CG heights in the 20- to 25-in. range. These motorcycles can typically lean more significantly than other motorcycle types, perhaps about 50°, before encountering any geometric interaction with the roadway. Sport motorcycles typically have dropped handlebars, raised foot pegs, and fairings designed for aerodynamic efficiency. The riding posture on a sports motorcycle has a forward lean with the foot pegs further rearward than on other motorcycle types. Examples of typical sports bikes include the Honda CBR series, the Yamaha YZF series, and the Kawasaki Ninja.

Scooter is a classification of motorbikes above 50cc that are operated differently from a typical motorcycle. For instance, the shifting is typically automatic, and does not require clutch operation. Also, both front and rear brakes may be operated by the hands, rather than a foot pedal for the rear brake. Figure 2.5 depicts a typical scooter. Examples of scooters include the Honda Ruckus and Metropolitan, the Vespa Primavera and Sprint, and the Genuine Buddy 125.

Three-wheeled motorcycles first appeared in the early to mid-1900s. One of the earliest on-road, three-wheeled motorcycles was manufactured by Harley-Davidson from 1932 to 1975 and was called the Servi-Car. The Servi-Car was designed to transport customer's cars from a maintenance garage to the customer's house. Harley-Davidson discontinued the Servi-Car in 1975 and did not begin making another three-wheeled motorcycle until the Tri Glide Ultra Classic in 2009. Three-wheeled motorcycles have experienced a surge in popularity recently. These motorcycles perform and handle differently from their two-wheeled counterparts in steering, braking, and acceleration. The analysis of crashes involving these motorcycles may need to account for these differences. Typical three-wheeled motorcycles are shown in Figure 2.6 and Figure 2.7.

Off-road motorcycles are generally referred to as dirt bikes (Figure 2.8). They are designed to handle rough, natural outdoor terrain. Off-road motorcycles, compared to the on-road versions, are typically simpler and lighter, with high clearance for obstacles, ruts, and longer suspension travel. They are often not outfitted to be street legal. *Dual-purpose* or *dual-sport*

FIGURE 2.5 Typical scooter.

FIGURE 2.6 Typical three-wheeled motorcycle.

FIGURE 2.7 Another type of three-wheeled motorcycle.

FIGURE 2.8 Typical off-road motorcycle.

motorcycles provide the benefits of being street legal with the ability to travel off-road as well. As expected from their name, the dual-purpose (or dual-sport or adventure) has some of the characteristics of on-road motorcycles, but with a higher ground clearance, greater suspension travel, and often knobby tires. Some dual sports are sized similarly to off-road motorcycles. *Adventure motorcycles*, on the other hand, which are also designed for both on-road and off-road uses, are typically larger. A typical dual-purpose adventure motorcycle

FIGURE 2.9 Typical dual-purpose motorcycle.

is depicted in Figure 2.9. Examples of typical dual-purpose motorcycles include the Honda XR650L, the KTM 500 EXC F, and the Kawasaki KLR650.[3] Typical adventure motorcycles include the KTM Adventure 790, the Honda Africa Twin, and the Harley-Davidson Pan America.

Motorcycle Dimensions and Inertial Properties

Geometric dimensions of the motorcycle are often needed when reconstructing a motorcycle crash. These dimensions can be obtained from manufacturer specifications or from inspecting an exemplar motorcycle. Another valuable source is Lightpoint Data (http://lightpointdata.com/motorcycle-specs). The motorcycle weight, its CG location, and its moments of inertia (MOI) may be needed. Typically, the wet weight of a motorcycle will be available from manufacturer specifications, and the weight distribution might also be available. This weight distribution data could be used to calculate the longitudinal position of the CG. Alternatively, an exemplar motorcycle could be weighed, or perhaps, the accident-involved motorcycle if the damage is not extensive enough to alter the CG position. The following equation will yield the longitudinal CG position [1]:

$$x_{cg,mc} = \frac{W_r}{W_f + W_r} \cdot WB \qquad (2.1)$$

[3.] https://www.chapmoto.com/blog/2010/09/14/top-5-dual-sport-motorcycles/

In this equation, $x_{cg,mc}$ is the distance from the center of the front axle to the CG, W_f is the weight carried by the front wheel, W_r is the weight carried by the rear wheel, and WB is the motorcycle wheelbase. If the weight of the rider and position of the rider's CG can be determined, then the combined longitudinal position of the CG can also be determined with the following equation:

$$x_{cg,mc+rider} = \frac{W_{mc} x_{cg,mc} + W_{rider} x_{cg,rider}}{W_{mc} + W_{rider}} \qquad (2.2)$$

In this equation, $x_{cg,mc+rider}$ is the distance the combined CG is behind the front axle, W_{mc} is the total weight of the motorcycle, W_{rider} is the weight of the motorcycle, and $x_{cg,rider}$ is the distance the rider's CG behind the front axle.

The CG height for a motorcycle is not typically reported in manufacturer specifications. The CG height can be measured or estimated based on available data. Measuring this parameter involves, first, measuring the weight supported by the wheels when the motorcycle is level, then elevating the rear wheel and remeasuring the wheel weights. Then, the CG height can be calculated with the following equation [1, 2]:

$$z_{cg,mc} = WB \cdot \frac{W_{r1} - W_{r2}}{W_t \tan\theta} + R_{avg} \qquad (2.3)$$

In this equation, $z_{cg,mc}$ is the height of the CG, W_{r1} is the weight supported by the rear wheel when the motorcycle is level, W_{r2} is the weight supported by the rear wheel when it is elevated, θ is the angle to which the motorcycle is lifted, and R_{avg} is the average radius of the front and rear wheel.

Once the motorcycle CG height is obtained, the combined CG height for the motorcycle and rider can be calculated with the following equation:

$$z_{cg,mc+rider} = \frac{W_{mc} z_{cg,mc} + W_{rider} z_{cg,rider}}{W_{mc} + W_{rider}} \qquad (2.4)$$

In this equation, $z_{cg,mc+rider}$ is the position of the combined CG above the ground and $z_{cg,rider}$ is the position of the rider's CG above the ground.

Cossalter et al. [3] tested two super-sport motorcycles and found that the CG height was equal to approximately 37% of the wheelbase. Foale [1] presented the CG heights for 39 motorcycles. His data are included in Table 2.1, organized by motorcycle type. As this table shows, the CG heights in Foale's data ranged from 11.6 to 24.7 in. (295 to 624 mm). Table 2.1 also includes the wheelbase for each motorcycle, along with the ratio of the CG height to the wheelbase. In Foale's dataset, the CG height was, on average, 34.2% of the wheelbase, with a standard deviation of 6.7% of the wheelbase. For the sport motorcycles, the CG height was, on average, 38.9% of the wheelbase, generally consistent with Cossalter's data. The standard deviation on this was approximately 4.6% of the wheelbase. For cruisers, the CG height was, on average, 27.2% of the wheelbase, with a standard deviation of 4.6% of the wheelbase. The average ratio for the touring motorcycles was 27.1% of the wheelbase, with a standard deviation of 5.8% of the wheelbase. For standard motorcycles, the CG height was, on average, 37.2% of the wheelbase, with a standard deviation of 5.2% of the

TABLE 2.1 Center of gravity (CG) data from Foale [1]

Motorcycle	Motorcycle type	Motorcycle wheelbase, WB (in.)	Center of gravity height, h (in.)	Ratio (h/WB)
Kawasaki Vulcan 1500 Drifter	Cruiser	65.2	11.6	0.18
Yamaha Road Star	Cruiser	66.5	15.7	0.24
Yamaha V-Star 1100	Cruiser	65.0	16.1	0.25
Yamaha V-Star 1100	Cruiser	65.0	17.4	0.27
Yamaha V-Star 1100 Classic	Cruiser	65.0	17.6	0.27
Harley-Davidson Ultra Classic FLHTCUI	Cruiser	62.7	18.2	0.29
Honda Shadow Sabre	Cruiser	65.0	19.5	0.30
Honda Magna	Cruiser	63.8	20.1	0.32
Harley-Davidson FXST Softail STD	Cruiser	66.9	23.2	0.35
BMW R1100GS	Dual-purpose	59.4	22.3	0.38
Kawasaki ZX-6R	Sport	55.1	20.1	0.36
Triumph Daytona	Sport	54.8	20.4	0.37
Suzuki Hayabusa	Sport	58.3	20.6	0.35
Honda CBR600F4i	Sport	55.0	20.8	0.38
Yamaha YZF-R6	Sport	54.3	21.6	0.40
Suzuki GSX-R600	Sport	55.0	21.6	0.39
Kawasaki ZX-6R	Sport	55.1	21.6	0.39
Honda CBR	Sport	54.7	21.6	0.39
Honda CBR1100XX	Sport	58.7	21.7	0.37
Triumph TT600	Sport	54.9	22.1	0.40
Yamaha YZF-R6	Sport	54.3	22.3	0.41
Triumph Sprint	Sport	57.9	22.5	0.39
Suzuki GSX-R600	Sport	55.0	22.7	0.41
Suzuki Bandit 600	Sport	56.0	23.2	0.41
Honda Nighthawk 750	Standard	59.1	18.4	0.31
Suzuki SV650	Standard	56.7	18.7	0.33
BMW F650GS	Standard	62.0	18.7	0.30
Triumph T-Bird Sport	Standard	62.2	20.3	0.33
Suzuki GZ250	Standard	57.1	20.5	0.36
Ducati Monster City	Standard	57.0	21.4	0.38
Buell X1	Standard	55.5	22.4	0.40
Buell Blast	Standard	55.0	22.9	0.42
Triumph Tiger 90	Standard	53.5	23.9	0.45
Buell X1	Standard	55.5	24.7	0.45
Honda Valkyrie Interstate	Touring	66.5	14.1	0.21
Honda GL 1800	Touring	66.6	15.6	0.23
Yamaha Royal Star Venture	Touring	67.1	16.3	0.24
Honda G1500 SE	Touring	66.5	19.2	0.29
BMW K1200LT	Touring	62.3	23.4	0.38

TABLE 2.2 Center of gravity (CG) heights for motorcycles from DiTallo et al. [4]

Motorcycle	Motorcycle type	Weight (lb)	Motorcycle wheelbase, WB (in.)	Center of gravity height, h (in.)	Ratio (h/WB)
2006 Harley-Davidson XL1200C	Cruiser	579	64.0	21.8	0.34
2010 Harley-Davidson Dyna Fat Bob FXDF	Cruiser	694	63.7	17.9	0.28
2011 Harley-Davidson Heritage Softail Classic FLSTCI	Cruiser	748	64.6	16.3	0.25
2011 Harley-Davidson Road King Classic FLHRCI	Cruiser	794	63.5	14.6	0.23
2009 Yamaha Varago 250 XV250	Cruiser	343	58.8	17.8	0.30
1989 Kawasaki Vulcan 1500 VN1500-A	Cruiser	636	63.0	19.7	0.31
1996 Kawasaki Vulcan 800 VN800-A	Cruiser	534	63.2	17.4	0.28
1998 Honda Shadow VT1100C2	Cruiser	616	65.0	20.9	0.32
1993 Honda XR250L	Dual-purpose	283	55.6	26.9	0.48
2012 Jmstar YY50QT-6	Scooter	252	51.0	16.4	0.32
2008 Taizhou Zhongneng GTR150	Scooter	229	50.5	18.1	0.36
1998 Buell S3 Thunderbolt	Sport	493	54.6	21.7	0.40
2001 Kawasaki Ninja 500R EX500-D	Sport	446	56.0	14.4	0.26
2001 Honda CBR600	Sport	416	55.0	21.6	0.39
1997 Kawasaki Ninja ZX-9R ZX900-B	Sport	355	56.0	20.4	0.36
1997 Kawasaki Ninja ZX-9R ZX900-B	Sport	501	57.0	22.4	0.39
1998 Honda CBR900RR	Sport	421	55.2	26.2	0.47
2000 Yamaha YZF-R6	Sport	408	54.0	20.3	0.38
2009 Yamaha FZ6R	Sport	447	56.8	20	0.35
1993 Yamaha FZR600	Sport	469	56.0	19.9	0.36
1992 Yamaha FZR600	Sport	421	56.0	17.2	0.31
2005 Suzuki GS500E	Sport	413	55.0	21.2	0.39
2011 Honda	Standard	304	55.6	20.2	0.36
2004 Kawasaki EN500C	Standard	480	62.8	20.8	0.33
2011 Harley-Davidson Ultra Classic Electra Glide FLHTCUI	Touring	881	63.5	16.9	0.27

© SAE International

wheelbase. Based on these numbers, it appears reasonable to lump cruisers with touring motorcycles and standard with sports motorcycle for developing equations to estimate the CG height of a motorcycle.

DiTallo et al. [4] presented the CG heights for 25 additional motorcycles and his data are included in Table 2.2. In DiTallo et al.'s dataset, the CG height was, on average, 34.0% of the wheelbase, with a standard deviation of 6.2% of the wheelbase. These numbers are consistent with Foale's data. In DiTallo et al.'s data, the CG height for the sport motorcycles was, on average, 36.9% of the wheelbase. The standard deviation on this was approximately 5.2% of the wheelbase. For cruisers, on the other hand, the CG height was, on average, 29.0% of the wheelbase, with a standard deviation of 3.5% of the wheelbase. Again, these numbers are consistent with Foale's data (and Cossalter's).

To develop equations to estimate the CG height of a motorcycle, the Foale and DiTallo et al. datasets were combined. The data were then partitioned with cruisers and touring motorcycles in one category and standard and sports motorcycles in another. These combined datasets were used to calculate an average and standard deviation for each population. This resulted in the following equation for *cruisers and touring* motorcycles. In this equation, h is the CG height and WB is the wheelbase of the motorcycle.

$$h = 0.2777 \times WB \pm 0.0454 \times WB \tag{2.5}$$

The following equation resulted for the *standard and sports* motorcycles:

$$h = 0.3759 \times WB \pm 0.0425 \times WB \tag{2.6}$$

The plus/minus on these equations is one standard deviation on each side of the mean. These equations can be used to estimate the CG height for a motorcycle when a measured CG height is not available. The dataset used to generate these equations included 23 cruiser/touring motorcycles and 37 standard/sport motorcycles. These data are plotted in Figure 2.10. Wheelbase is plotted on the horizontal axis and the ratio of CG height to wheelbase is plotted on the vertical axis. From this graph, it is apparent that standard/sport motorcycles generally have shorter wheelbases than cruiser/touring motorcycles and that the CG height for the standard/sport motorcycles is generally a higher percentage of the wheelbase than for the cruiser/touring motorcycles (though there is overlap of the datasets).

FIGURE 2.10 Ratio of the center of gravity (CG) height to wheelbase as a function of wheelbase.

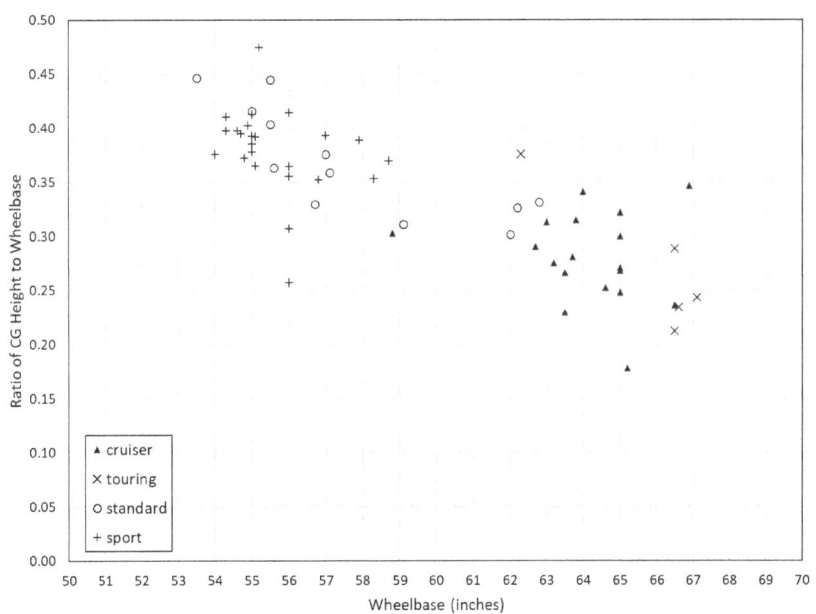

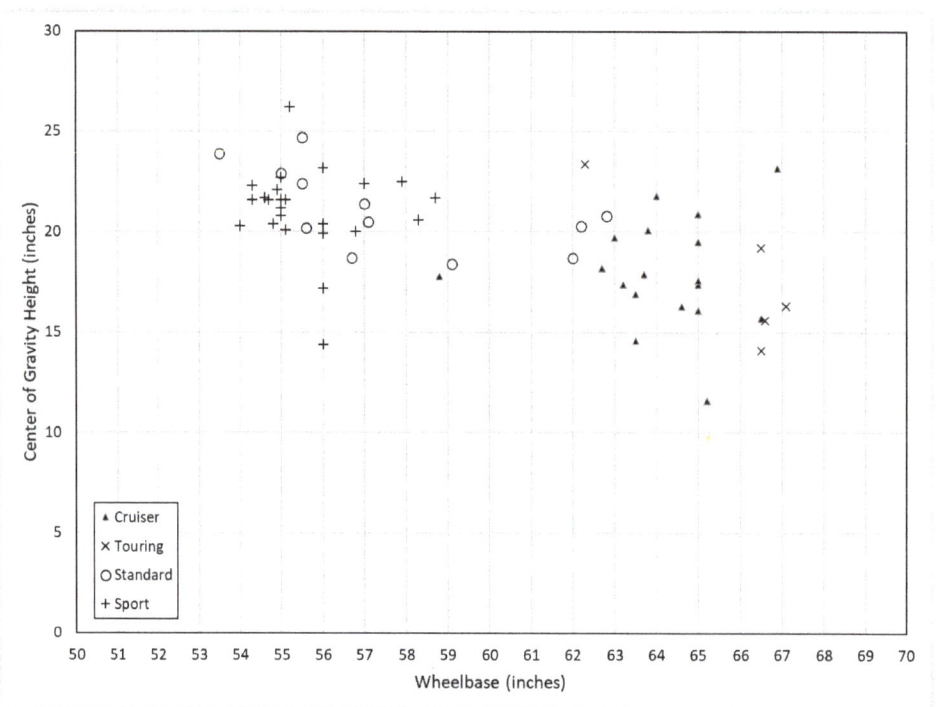

FIGURE 2.11 Center of gravity (CG) height as a function of wheelbase.

Figure 2.11 is a similar graph, but in this graph the actual CG height is plotted on the vertical axis. From this graph, it is apparent that the standard and sport motorcycles generally have higher CG than the cruiser and touring motorcycles, though there is overlap of the two datasets.

The MOI characterizes the resistance of a vehicle to rotation about its principal axes—roll, pitch, and yaw. Cossalter et al. [3] reported physical testing of two racing motorcycles to determine their inertial properties. These authors noted that motorcycle MOIs would be necessary inputs into a simulation of dynamic behavior and handling of a motorcycle. In the context of crash reconstruction, there will not be many instances where the analyst needs to evaluate the motorcycle MOIs. However, if they are needed, an estimate can be obtained with the prism method. The prism method assumes that the vehicle is a solid, homogeneous box. With this assumption, the MOIs about the principal axis are given by the following equations:

$$I_{roll} = \frac{1}{2} m \cdot (f_w^2 + h_{overall}^2) \tag{2.7}$$

$$I_{pitch} = \frac{1}{2} m \cdot (l_{overall}^2 + h_{overall}^2) \tag{2.8}$$

$$I_{yaw} = \frac{1}{2} m \cdot (l_{overall}^2 + f_w^2) \tag{2.9}$$

In these equations, I_{roll}, I_{pitch}, and I_{yaw} are the MOIs about the roll, pitch, and yaw axes. The vehicle mass is indicated with the letter m and $l_{overall}$, f_w, and $h_{overall}$ are the vehicle length, frame width, and height, respectively. If precise values are needed, testing could be conducted. Frank et al. [5] reported a measured value for the yaw moment of inertia of a Kawasaki Ninja ZX-10R of 320 in.-lb-sec².

Motorcycle Controls

Typically, the right side of a motorcycle contains controls for the ignition, braking, and accelerating, while the left side contains controls for shifting gears and signaling. The front and rear brakes are controlled independently on motorcycles with standard braking systems. The front brake is controlled with a lever near the right handgrip and the rear brake is controlled with a foot pedal on the right side (Figure 2.12). The throttle is on the right handgrip, the clutch is actuated using a lever near the left handgrip, and gear changes are commanded using a foot pedal on the left side of the motorcycle. The throttle is increased by rolling the grip toward the rider and decreased by rolling the throttle away from the rider. When released, the throttle will spring back to the idle position. Power from the engine to the wheels is disengaged by squeezing the clutch lever. Power is reengaged by releasing the clutch lever. The shifting pattern for most motorcycles is all the way down to first gear and all the way up for fifth or sixth gear (depending on how many gears there are). Neutral is between first and second gears and can be accessed from going up or down in gear. This gearing pattern is often referred to as, "one down, five up." Motorcycle transmissions are sequential, meaning that gears cannot be skipped. However, the gears will only engage if the clutch is released.

FIGURE 2.12 Motorcycle controls.

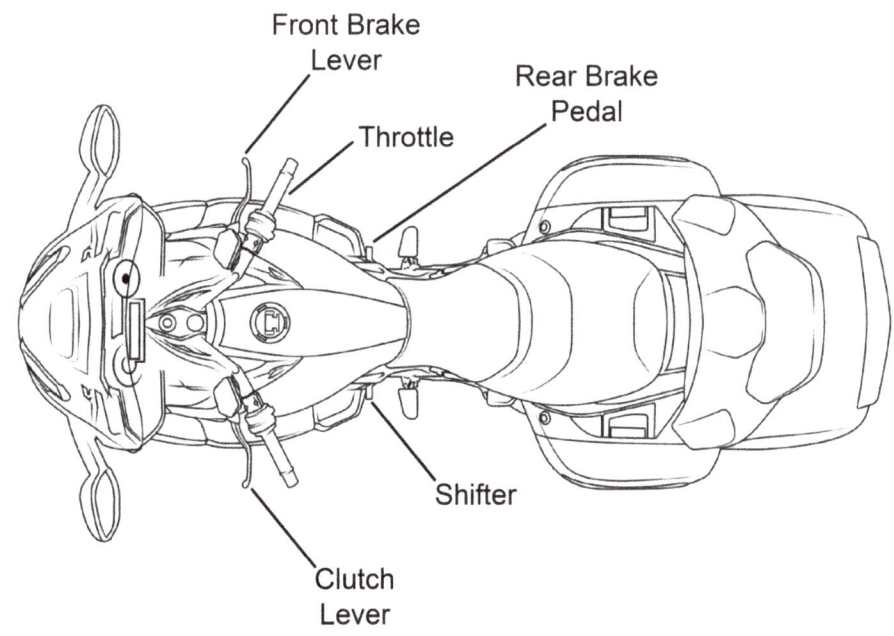

©2022, SAE International

Motorcycle Tires

Motorcycle tires differ from passenger car tires, particularly in terms of their cross section. Because motorcycle steering and cornering involve the operator leaning the motorcycle, motorcycle tires have a circular or U-shaped profile rather than the flatter profile of a passenger car tire. Because of this, the size and shape of the tire contact patches change as a motorcycle leans.

Motorcycle Tire Markings

Figure 2.13 depicts typical markings on the two sides of a motorcycle tire—in this case, a Metzeler ME Z2 sport touring radial tire. The markings on this tire include the following:

- Tire Manufacturer—Metzeler
- Tire Model—MEZ2
- North American Department of Transportation Compliance Symbol—DOT
- DOT is followed by the Tire Serial Number, which ends with a four-digit date code
- Tire Construction Details
- Size Designation—130/80R17
- Nominal Section Width (mm)—130
- Aspect Ratio—80
- Rim Diameter (inches)—17
- Motorcycle Tire Designation—M/C
- Load Index (Maximum Capacity at Maximum Pressure)—65
- Speed Symbol—H
- Maximum Load @ Maximum Pressure—290 kg (639 lb) @ 290 kPa (42 psi)

FIGURE 2.13 Typical markings on a motorcycle tire.

- Tire Construction—Steel Radial
- An arrow indicating the correct rotation direction for the tire

The date code, which is the last four digits of the tire serial number, specifies the week and year that the tire was manufactured. For example, the code on the tire in Figure 2.13 is 4116, indicating that the tire was manufactured in the 41st week of 2016. Table 2.3 lists the meaning of several load index numbers for motorcycle tires. Table 2.4 lists the meaning of the speed ratings, given in km/h. The owner's manual for a motorcycle will specify the tire size, construction, load range, and speed index intended to be installed on both the front and rear of the motorcycle.

As the Motorcycle Industry Council Tire Guide [6] notes, "proper air pressure is critical for tire performance and tire life. Under-inflation or overloading can cause sluggish handling, heavy steering, and internal damage due to over-flexing, and can cause the tire to separate from the rim. Over-inflation can reduce the contact area (and therefore available traction), and can make the motorcycle react harshly to bumps."

TABLE 2.3 Load indexes for motorcycle tires

Load index	Weight rating (lb)	Load index	Weight rating (lb)	Load index	Weight rating (lb)	Load index	Weight rating (lb)
50	419	56	494	62	584	68	694
51	430	57	507	63	600	69	716
52	441	58	520	64	617	70	739
53	454	59	536	65	639	71	761
54	467	60	551	66	661	72	783
55	481	61	567	67	677	73	805

TABLE 2.4 Speed ratings for motorcycle tires

Rating	Speed (km/h)
J	100
K	110
L	120
M	130
N	140
P	150
Q	160
R	170
S	180
T	190
U	200
H	210
V or VB	240
Z or ZR	240+
W	270
Y	300

Motorcycle Tire Friction Coefficients

Motorcycle tires are softer and stickier than passenger car tires [7]. Lambourn and Wesley [8] used a two-wheeled trailer (designed as a highway friction measuring device) to test three motorcycle tires designed for sports motorcycles to determine their peak and locked-wheel friction coefficients on asphalt (Figure 2.14). They tested two different asphalt surfaces (hot rolled asphalt and stone mastic asphalt) in both a dry and a wet condition (1-mm water depth). The tires were tested at nominal speeds of approximately 32, 64, and 100 km/h (approximately 20, 40, and 60 mph). On the dry, hot rolled asphalt, the motorcycle tires produced average peak friction coefficients between 1.1 and 1.3. These values generally increased with increasing speed. The average locked-wheel coefficients for the dry, hot rolled asphalt ranged between 0.7 and 0.9. There was slight speed dependence in these values, with the friction coefficient declining slightly with increasing speed. On the dry, stone mastic asphalt, the motorcycle tires produced average peak friction coefficients between 1.1 and 1.25. There was no speed dependence in these values. The locked-wheel coefficients on the dry, stone mastic asphalt fell between 0.9 and 0.65. These values exhibited significant speed dependence, with the range at 32 km/h (20 mph) falling between 0.8 and 0.9 and the range at 100 km/h (60 mph) falling between 0.65 and 0.76.

On the wet, hot rolled asphalt, the motorcycle tires produced average peak friction coefficients between 0.99 and 1.36. There was no obvious speed dependence in these values. The average locked-wheel coefficients on the wet, hot rolled asphalt showed significant speed dependence, falling between 0.8 and 0.9 at 32 km/h (20 mph) and 0.52 to 0.67 at 100 km/h (60 mph). On the wet, stone mastic asphalt, the motorcycle tires produced average peak friction coefficients between 1.0 and 1.15. The locked-wheel coefficients exhibited significant

FIGURE 2.14 Pavement friction tester used by Lambourn and Wesley [8].

speed dependence, falling between 0.68 and 0.76 at 32 km/h (20 mph) and between 0.37 and 0.4 at 100 km/h (60 mph).

In some situations, the tire temperature could also be an important consideration in assessing the available friction for a motorcycle tire—perhaps, for instance, when determining the friction limits for a motorcycle that has been on a long mountainous ride on a hot day. As Parks [9] observes, "as a tire gets hotter, the rubber becomes more compliant and has a greater ability to interlock with the tarmac, providing greater traction. This increased traction continues until the rubber exceeds its design temperature and begins to degrade … conversely, when tires are cold, the rubber becomes hard and doesn't conform to the peaks and valleys of the road surface as well as when the tires are warm, significantly reducing grip. This is especially true of race compound tires."

Tests like those reported in reference [8] can begin to define an upper limit on the performance capabilities of motorcycles. In many cases, the performance limits of the motorcycle-rider system will be defined by the rider, not by the tires or the motorcycle. Most riders will not be capable of fully utilizing the peak friction of their tires. In addition to that, the values reported by Lambourn and Wesley are applicable to dry and wet roadway surfaces that are generally free of debris and contaminants. Some motorcycle crashes involve debris or contaminants on the road. In such cases, the analyst may need to consider the effect of the debris or contaminants on the friction limits of the motorcycle tires [10, 11].

References

[1] Foale, T., Motorcycle Handling and Chassis Design – The Art and Science ed., 2006, self-published by Tony Foale and available for purchase at https://tonyfoale.com/book.htm.

[2] Shapiro, S., Dickerson, C., Arndt, S., Arndt, M. et al., "Error Analysis of Center-of-Gravity Measurement Techniques," SAE Technical Paper 950027, 1995, doi:10.4271/950027.

[3] Cossalter, V., Doria, A., and Mitolo, L., "Inertial and Modal Properties of Racing Motorcycles," SAE Technical Paper 2002-01-3347, 2002, doi:10.4271/2002-01-3347.

[4] DiTallo, M., Paul, E., Adamson, K., Green, T. et al., "Motorcycle Center of Gravity Data: Methodology and Reference," Collision: The International Compendium for Crash Research, 12(1): 50–60, 2017, ISSN 1934-8681.

[5] Frank, T., Smith, J., Fowler, G., Carter, J., et al., "Simulating Moving Motorcycle to Moving Car Crashes," SAE Technical Paper 2012-01-0621, 2012, doi:10.4271/2012-01-0621.

[6] Motorcycle Industry Council, "Tire Guide: All You Need to Know About Street Motorcycle Tires," https://www.msf-usa.org/downloads/MIC_Tire_Guide_2012V1.pdf, accessed on April 6, 2018.

[7] Bartlett, W., "Interpretation of Motorcycle Rear-Wheel Skidmarks for Accident Reconstruction," Proceedings, 4th International Conference on Accident Investigation, Reconstruction, Interpretation and the Law, Vancouver, BC, Canada, August 13–16, 2001, ISBN 088865796X.

[8] Lambourn, R.F. and Wesley, A., "Motorcycle Tire/Roadway Friction," SAE Technical Paper 2010-01-0054, 2010, doi:10.4271/2010-01-0054.

[9] Parks, L., Total Control: High Performance Street Riding Techniques, MBI Publishing, Minneapolis, MN, 2003, ISBN 978-0-7603-1403-6.

[10] Hall, G. and Painter, J., "Pavement Friction Reduction Due to Fine-Grained Earth Contaminants," SAE Technical Paper 2007-01-0736, 2007, doi:10.4271/2007-01-0736.

[11] Meyers, D. and Austin, T., "Dry Pavement Friction Reductions Due to Sanding Applications," SAE International Journal of Commercial Vehicles, 5(1): 239–250, 2012, doi:10.4271/2012-01-0603.

3

Braking and Acceleration

Braking Dynamics

A motorcycle's deceleration is maximized when both brakes are applied to the maximum extent possible without locking either wheel. During heavy braking, a significant portion of the weight of the motorcycle and rider shift to the front wheel, resulting in the available traction at the front tire being significantly greater than the available traction at the rear tire. Thus, the overall contribution of the front brake to deceleration is more than the contribution of the rear brake. As weight shift occurs during braking, the optimal braking pressure at the front and rear brakes varies, and thus, a rider's ability to achieve maximum deceleration depends on their skill level (assuming a braking system without rider aids such as antilock brakes or linked front and rear braking).

While expert riders may be able to achieve decelerations approaching or exceeding 1 g on many modern motorcycles, most riders will not be able to produce this level of deceleration. This is the result of the complex brake actuation required by a rider and the dynamic change in pressure between the front and rear brakes that needs to occur to maximize the deceleration. The rider needs to roll off the throttle and simultaneously apply both the front and rear brakes. As they begin to brake, weight shifts to the front of the motorcycle. When this occurs, the rider needs to apply more front-brake pressure and less rear-brake pressure. The rider needs to modulate the front brake to prevent lockup and to control the tendency of the rear tire to lift off the ground. They need to modulate the rear brake to prevent the rear wheel from locking as less weight on the rear tire makes this tire easier to lock. All the while, the rider will be gripping the tank with their knees to maintain body position and keep their butt as far back in the seat as they can to optimize the weight distribution and prevent the rear wheel from lifting. If the rider does lock the front wheel during braking, they will need to react quickly, releasing the brake and starting their brake modulation again. If they lock the rear wheel, the best strategy is typically for the rider to keep it locked until they come to a stop [1].

Cossalter et al. [2] developed a simple rigid body model for analyzing motorcycle braking dynamics. This model is shown graphically in Figure 3.1. In this figure, the following variables are introduced:

m = motorcycle mass
g = gravitational constant
h = center of mass height
a = distance from center of mass to center of front wheel
b = distance from center of mass to center of rear wheel
N_f = normal force at front wheel
N_r = normal force at rear wheel
S_f = braking force at front wheel
S_r = braking force at rear wheel

The equations of motion for this model are as follows:

$$ma_x = S_f + S_r \tag{3.1}$$

$$ma_z = mg - N_f - N_r = 0 \tag{3.2}$$

$$I_p \alpha_p = N_f a - N_r b - h(S_f + S_r) = 0 \tag{3.3}$$

Note that the coordinate system in Figure 3.1 is set up so that a positive acceleration in the x-direction slows the motorcycle down.

FIGURE 3.1 Simple motorcycle braking model [1].

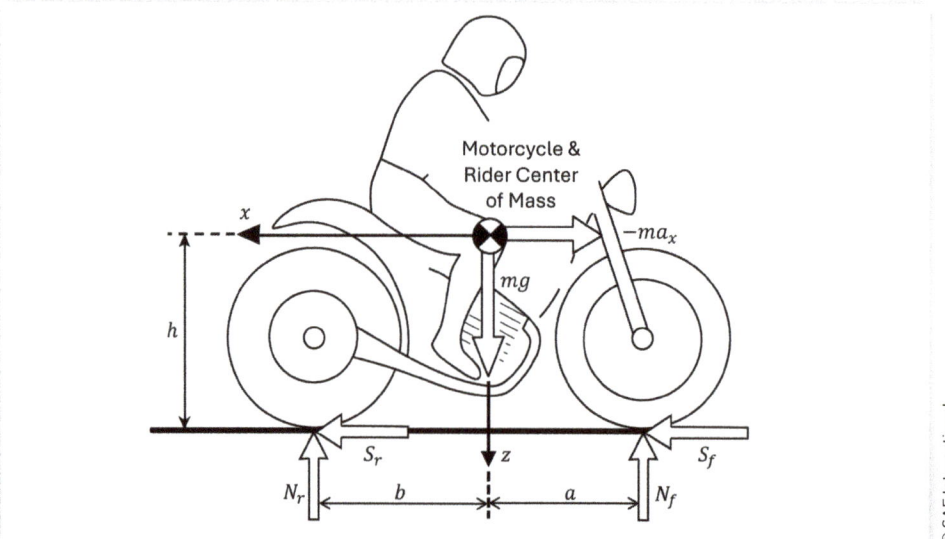

Algebraic manipulation of Equations (3.1) through (3.3) yields the following equations for the front and rear normal loads:

$$N_f = \frac{b}{a+b}mg + \frac{h}{a+b}ma_x \tag{3.4}$$

$$N_r = \frac{a}{a+b}mg - \frac{h}{a+b}ma_x \tag{3.5}$$

Equations (3.4) and (3.5) yield the following observations relevant to motorcycle braking dynamics: (1) Before braking begins, the distribution of weight between the front and rear wheels can be calculated based on the position of the center of gravity relative to the axles (the dimensions a and b). (2) When braking begins, a_x will begin increasing and weight will transfer from the rear axle to the front axle. (3) The higher the deceleration that develops, the more weight will transfer to the front wheel. (4) The higher the center of gravity, the more weight will transfer to the front of the motorcycle.

The rear-wheel load will go to zero and the rear wheel will lift off the ground when:

$$a_x \geq \frac{a}{h}g \tag{3.6}$$

Thus, the deceleration required to cause rear-wheel lift-off varies from motorcycle to motorcycle and depends on the longitudinal and vertical positions of the center of gravity. DiTallo et al. [3] presented a and h dimensions for 25 motorcycles and his data are included in Table 3.1. The value in the third column is a theoretical value for the longitudinal deceleration in g's that would be necessary for the rear wheel of these motorcycles to lift off the ground. These values illustrate that the sports bikes would generally require lower decelerations than the cruisers to cause rear-wheel lift-off. The dimensions in this table are for the motorcycle only, without a rider. When a rider is added, the center of gravity will generally move rearward and upward, increasing both a and h. Suspension effects are also neglected in Table 3.1. With a rider on the motorcycle and suspension effects accounted for, the decelerations needed to cause rear-wheel lift-off will generally be less than the values listed in Table 3.1.

The braking forces at the front and rear wheels are limited by the coefficients of friction and the normal loads at these wheels. As weight transfers from the rear to the front as the deceleration develops, more braking force will become available at the front wheel, and the available braking force at the rear wheel will decrease. To show this effect, Cossalter et al. introduced the following equation to describe the front-to-rear balance of braking:

$$\rho = \frac{S_r}{S_f + S_r} \tag{3.7}$$

Thus, ρ expresses the proportion of the total braking force that is produced at the rear wheel. Cossalter et al. derived the following expression for the ideal brake balance:

$$\rho_{ideal} = \frac{a}{a+b} - \frac{h}{a+b}\frac{a_x}{g} \tag{3.8}$$

©2022, SAE International

TABLE 3.1 Motorcycle characteristics from DiTallo et al. [3]

Motorcycle	Motorcycle type	a (in.)	h (in.)	Ratio (a/h)
2006 Harley-Davidson XL1200C (Sportster 1200 Custom)	Cruiser	34.9	21.8	1.6
2010 Harley-Davidson FXDF (Dyna Fat Bob)	Cruiser	33.8	17.9	1.9
2011 Harley-Davidson FLSTCI (Heritage Softail Classic)	Cruiser	34.6	16.3	2.1
2011 Harley-Davidson FLHRCI (Road King Classic)	Cruiser	34.9	14.6	2.4
2009 Yamaha XV250	Cruiser	31.1	17.8	1.8
1989 Kawasaki VN1500-A	Cruiser	34.3	19.7	1.7
1996 Kawasaki VN800-A	Cruiser	34.6	17.4	2.0
1998 Honda VT1100C2	Cruiser	34.7	20.9	1.7
1993 Honda XR250L	Dual-purpose	29.6	26.9	1.1
2012 Jmstar YY50QT-6	Scooter	31.3	16.4	1.9
2008 Taizhou Zhongneng GTR150	Scooter	32.2	18.1	1.8
1998 Buell S3 Thunderbolt	Sport	27.7	21.7	1.3
2001 Kawasaki EX500-D	Sport	28.9	14.4	2.0
2001 Honda CBR600	Sport	25.9	21.6	1.2
1997 Kawasaki ZX900-B	Sport	29.8	20.4	1.5
1997 Kawasaki ZX900-B	Sport	27.4	22.4	1.2
1998 Honda CBR900RR	Sport	26.5	26.2	1.0
2000 Yamaha YZF-R6	Sport	26.3	20.3	1.3
2009 Yamaha FZ6R	Sport	26.9	20.0	1.3
1993 Yamaha FZR600	Sport	28.8	19.9	1.4
1992 Yamaha FZR600	Sport	28.8	17.2	1.7
2005 Suzuki GS500E	Sport	27.4	21.2	1.3
2011 Honda	Standard	32.3	20.2	1.6
2004 Kawasaki EN500C	Standard	34.7	20.8	1.7
2011 Harley-Davidson FLHTCUI (Ultra Classic Electra Glide)	Touring	36.2	16.9	2.1

© SAE International

For two-wheeled motorcycles, the front brake will produce a significant portion of the available braking force. However, Equation (3.8) shows that rear-brake application will increase in effectiveness as the distance between the front axle and the center of gravity (a) increases. Lowering the center of gravity also increases the effectiveness of rear-wheel braking. From Equation (3.8), we can observe that rear-wheel braking becomes entirely ineffective when $\rho_{ideal} = 0$. This occurs when the rear-wheel lift-off (stoppie) condition of Equation (3.6) is met.

Braking Dynamics Example

DiTallo et al. [3] reported a center of gravity height of 20.3 in. for a 2000 Yamaha YZF-R6. This motorcycle had a weight of 408 lb, a wheelbase of 54 in., and approximately 51.2% of the weight was on the front axle. Thus, without a rider, the load on the front wheel would be 209 lb and the load on the rear wheel would be 199 lb. For this example, assume a 170-lb rider, approximately 66% of whose weight is carried by the rear wheel when the motorcycle is at a steady speed with no braking or acceleration. With these assumptions, the total weight of the motorcycle-rider combination would be 578 lb, with 267 lb on the front axle and 311 lb on the rear axle. Using Equation (3.5), the dimensions a and b can be calculated as follows:

$$a = (a+b)\frac{N_r}{mg} = 54 \text{ in.} \times \frac{311 \text{ lb}}{578 \text{ lb}} = 29.05 \text{ in.}$$
$$b = (a+b) - a = 54 \text{ in.} - 29.05 \text{ in.} = 24.94 \text{ in.}$$

The combined center of gravity height for the motorcycle-rider combination can be calculated with the following formula:

$$h_{combined} = \frac{W_{mc} h + W_{rider} h_{rider}}{W_{mc} + W_{rider}}$$

In this formula, $h_{combined}$ is the combined center of gravity height and h_{rider} is the height of the rider's center of gravity above the ground in a normal riding posture. The seat height for a 2000 Yamaha YZF-R6 is approximately 32.3 in. If we assume for the sake of this example, that the rider's center of gravity would be in the area of their hips, we could reasonably assume that their center of gravity height would be approximately 36 in. Thus:

$$h_{combined} = \frac{408}{578} 20.3 + \frac{170}{578} 36 = 24.9 \text{ in.}$$

Thus, under the static loads, the ratio a/h is 1.17. Equation (3.5) can be used to track the normal force at the rear wheel as follows:

$$\frac{N_r}{mg} = \frac{a}{a+b} - \frac{h}{a+b}\frac{a_x}{g}$$
$$\frac{N_r}{mg} = \frac{29.05}{54} - \frac{24.9}{54}\frac{a_x}{g}$$
$$\frac{N_r}{mg} = 0.54 - 0.46\frac{a_x}{g}$$

Thus, when braking begins, 54% of the weight would be on the rear axle. As the deceleration increases, weight would shift to the front of the motorcycle, and when a deceleration of 1.17 g is reached the rear wheel would lift off. As this process unfolds, the front tire would develop increasing traction for braking, because of increased normal load, whereas the available traction of the rear tire would be diminished. This is a deceleration that most riders would not be capable of generating. However, this calculation neglects suspension effects and forward movement by the rider during the braking, so the actual deceleration to produce rear-wheel lift-off would be less. This is discussed in the following section, along with the deceleration capabilities of typical riders.

Motorcyclist Braking and Deceleration Capabilities

This section reviews published studies related to the deceleration that motorcyclists can produce during braking. Several distinctions are important in considering these studies. First, there is a difference between the capabilities of the motorcycle and the deceleration an average rider can achieve. On a dry paved surface, most riders will not be able to fully utilize the braking capabilities of their motorcycle (rider aids on newer motorcycles such as antilock brakes and linking of the front and rear brakes can help them get closer to achieving this). Second, it is important to note that many of these studies were conducted with riders who were not confronted with emergency situations. They were conducted on test tracks, roads without hazards, or parking lots. Thus, these studies capture what riders are capable of under ideal circumstances. When confronted with actual emergencies, some riders may not achieve the decelerations that riders in these studies did.[1]

An important point is made by Ayres and Kubose [5], who noted that "it has long been understood that faster responses tend to be less accurate … Clearly there is no single value that can be said to represent the reaction time of even one person for one task, without considering the accuracy of task performance." This is relevant to the braking deceleration achieved by motorcyclists because "a driver would tend to respond quickly although perhaps not very accurately when a collision is imminent, but would take a more careful approach (accurate, appropriate) with more time available." Thus, "one should not expect … that drivers will efficiently use all time available to make an optimum avoidance maneuver … it is unreasonable to expect optimal response timing and maneuver performance by drivers faced with emergencies."

It is also worth stating that riding experience does not necessarily equate to the ability to achieve higher decelerations during emergency braking (see, e.g., the Bartlett and Greear [16] study discussed in what follows). When we speak of riding experience, it is important to ask: experienced at what? Even riders who have ridden many miles are not frequently confronted with the need to brake in an emergency fashion, and so, even experience riders may not be experienced at emergency braking. Experience does not necessarily equate to skill, and skill in one area may not equate to skill in another area. On the other hand, the data do appear to show that antilock brakes (particularly with linked or integrated braking systems) help riders achieve higher levels of deceleration more consistently with less risk of capsizing, and much of this benefit (at least the consistency part) is likely to carry over to emergency braking in the face of actual emergencies.

Tolhurst and McKnight [6] tested and compared five methods of braking in a straight line and three methods of braking in a curve. For all eight methods, the rider applied the

---

[1.] As Ouellet [4] has observed: "… laboratory studies [of motorcycle braking] share common shortcomings … The first is that the test riders are not subjected to unexpected, real collision situations; their lives are never threatened with immediate termination. To do so would grossly violate the ethics of experimentation involving humans. Test riders usually don't panic and stomp on the rear brake only, scream scatological monosyllables, or freeze at the controls when an evasive maneuver is called for. But riders in real accidents often make various combinations of these ineffective responses. Thus, test conditions of necessity differ fundamentally from real world collision situations."

front brake to the maximum extent possible without locking the wheel. For the straight-line braking tests, which were run from a nominal speed of 40 mph, the method of rear-wheel braking varied as follows: no rear-wheel braking, light rear-wheel braking, locked rear-wheel braking, pumping of the rear brake, and heavy rear-wheel braking just below the level necessary to lock the wheel. For the tests related to braking in a curve, which were run from a nominal speed of 30 mph, the method of rear-wheel braking varied as follows: no rear-wheel braking while keeping the motorcycle leaned, heavy rear-wheel braking while keeping the motorcycle leaned, and heavy rear-wheel braking while righting the motorcycle.

This study utilized three expert riders operating three different motorcycles—a Yamaha FJ1100 (a sport touring motorcycle), a Yamaha 550 Vision (a standard motorcycle), and a Suzuki GS 550 (a standard motorcycle). For the straight-line braking, adding the heavy rear-wheel braking to the front-wheel braking (below the level necessary to lock the wheel) produced the highest deceleration and the shortest stopping distance. The lowest deceleration and longest stopping distances resulted when only the front brake was applied. This study did not examine rear wheel only braking. For braking in a curve, the highest deceleration and shortest stopping distances were achieved by righting the motorcycle while applying heavy braking to both brakes (without locking the wheels). The lowest deceleration and longest stopping distances were generated by continuing to lean in the curve and not applying any rear brake.

Tolhurst and McKnight [6] noted that there were "highly significant differences among the three [riders] … [these] differences are more easily attributed to differences in the design of motorcycle, particularly the tire 'footprint,' than to the skill of the riders." These authors only reported a single average stopping distance for each braking method, and so their article does not enable deeper analysis of motorcycle-to-motorcycle or rider-to-rider differences. These authors also noted that "applying both brakes just short of lock-up can demand a high degree of braking skill. Less proficient riders might find one of the alternative methods to be more effective for them." Given that this study utilized expert riders, the decelerations reported are unlikely to be achieved by the average rider.

Prem [7, 8] conducted "emergency straight-path braking" tests with 59 volunteer riders. He used the Motorcycle Operator Skill Test (MOST) to provide a quantitative assessment of the riders' skill level. The MOST takes the riders through a series of tasks designed to test their steering and braking performance. The braking maneuver from this test required the riders to brake aggressively to a stop from a speed of 32 km/h (20 mph). A red signal light was activated to indicate to the riders when they should begin braking. The motorcycle used by the volunteers, a Honda CB400T, was instrumented to record the rider's front and rear brake-lever force inputs and motorcycle speed.

Prem analyzed differences in braking technique between skilled and less-skilled riders. He found that skilled riders applied higher levels of front-brake force than the less-skilled riders. Less-skilled riders preferred the use of the rear brake. The skilled riders also modulated the level of front- and rear-wheel braking to maintain optimum braking as weight shifted toward the front of the motorcycle during heavy braking. The less-skilled riders maintained a generally constant level of pedal pressure independent of the weight shift. More skilled riders also exhibited shorter braking reaction times, though it should be kept in mind that this reaction was to an illuminating light that the riders knew would illuminate, not to an actual emergency.

Fries et al. [9] performed testing with five different motorcycles to determine the deceleration of the motorcycles when the rider employed the rear brake only and when a combination of front and rear braking was employed. They tested a 1968 Harley-Davidson FLH (a touring motorcycle), a 1978 BMW R90 (a standard motorcycle), a 1982 Honda XR500R (a dirt bike), a 1972 Honda SL350 (a standard motorcycle), and a 1972 Honda SL125S (a standard motorcycle). These motorcycles are depicted in Figure 3.2. Each motorcycle was tested at nominal speeds of 20, 30, and 40 mph (32.2, 48.3, and 64.4 km/h) on worn asphalt. The experience level of the riders was not reported in the study. Overall, the deceleration from rear only braking was less than when heavy front braking was also used. The range of deceleration for rear only braking was 0.31 to 0.52 g. The range of decelerations when heavy front-wheel braking was also employed was between 0.54 and 0.88 g.

These authors observed that "when faced with an emergency stopping situation, or avoidance situation, a motorcycle [rider] has the decision of whether to stop using the rear brake only, front and rear brakes combined, or by laying the motorcycle down. There are several common misconceptions about motorcycles. One is that they will stop faster if they are laid down on their side … When a motorcycle is stopped by laying it on its side there is a delay in implementing the deceleration … The test results show that laying a motorcycle over and rear-wheel braking have very similar deceleration factors. However, when laying a motorcycle over there is an impact and risk of injury when the motorcycle hits the pavement. Also, all control is lost. If the motorcycle is kept upright, it is possible to reduce braking and steer. Front- and rear-wheel braking provides the best deceleration factors. Our testing also demonstrates that even during hard braking with front and rear brakes, the experienced driver consistently maintained a straight path without causing the motorcycle to fall." On the other hand, it should be acknowledged that the

FIGURE 3.2 Motorcycles used in the testing by Fries et al. [9].

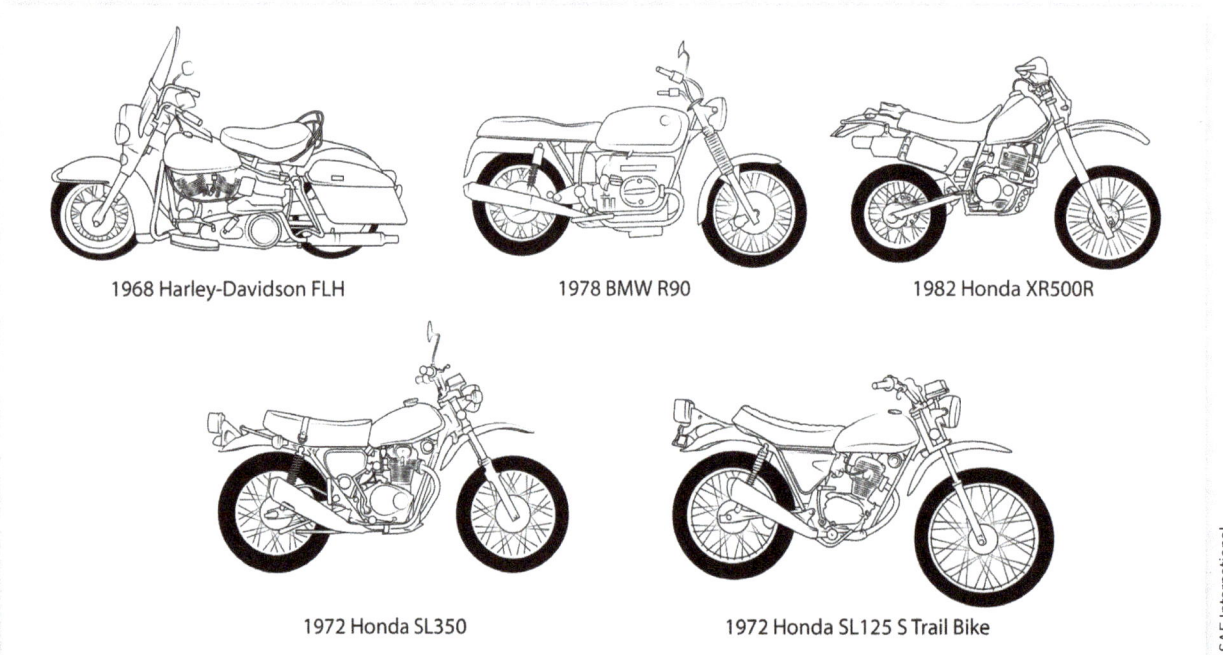

FIGURE 3.3 Kawasaki 1000 police motorcycle used in the testing by Hunter [10].

© SAE International

riders in this study were not confronted with an emergency, and thus, had a different level of urgency than that which riders in real crash scenarios might be confronted. In some scenarios, riders will not be able to avoid a collision regardless of the deceleration they are able to achieve. In these situations, it is not unusual for riders to unintentionally lay the motorcycle down or cause the motorcycle to pitch-over by over-braking the front brake. Of course, this can also occur when such a level of urgency was not warranted. Thus, the relevance of these braking errors to causation is something that will have to be evaluated on a case-by-case basis.

Hunter [10] reported acceleration and braking tests conducted by the Washington State Patrol on a dry, level roadway with a 1983 and a 1985 Kawasaki 1000 police motorcycles. This motorcycle is depicted in Figure 3.3. For deceleration tests with rear braking only, Hunter reported decelerations between 0.35 and 0.36 g. For deceleration tests with front braking only, Hunter reported decelerations between 0.64 and 0.74 g. For deceleration tests with heavy front and rear braking, Hunter reported decelerations between 0.63 and 0.96 g. For the rapid acceleration tests, Hunter reported accelerations between 0.48 and 0.73 g. He did not specify the experience level of the riders who conducted these tests, although he indicates that they both worked for the Washington State Patrol (presumably they were experienced motor officers). In the discussion, this paper observes that "overall operator skill has a great influence on the deceleration ability of the motorcycle."

Hugemann and Lange [11] conducted 74 instrumented braking tests with 18 different riders, 15 of whom were riding their own motorcycle. Motorcycle types were not specified. The riders had varying levels of experience (less than 12,500 mi and up to 80,000 mi.) and were instructed to brake from 50 km/h (31 mph) to a standstill "within the shortest possible distance." The tests were conducted on dry asphalt. Riders characterized as "skilled" exhibited mean decelerations between 0.70 and 0.81 g. Riders characterized as "novice" exhibited decelerations between 0.44 and 0.52 g. Individual test results were not reported in this article.

©2022, SAE International

FIGURE 3.4 Motorcycles tested by Bartlett [12].

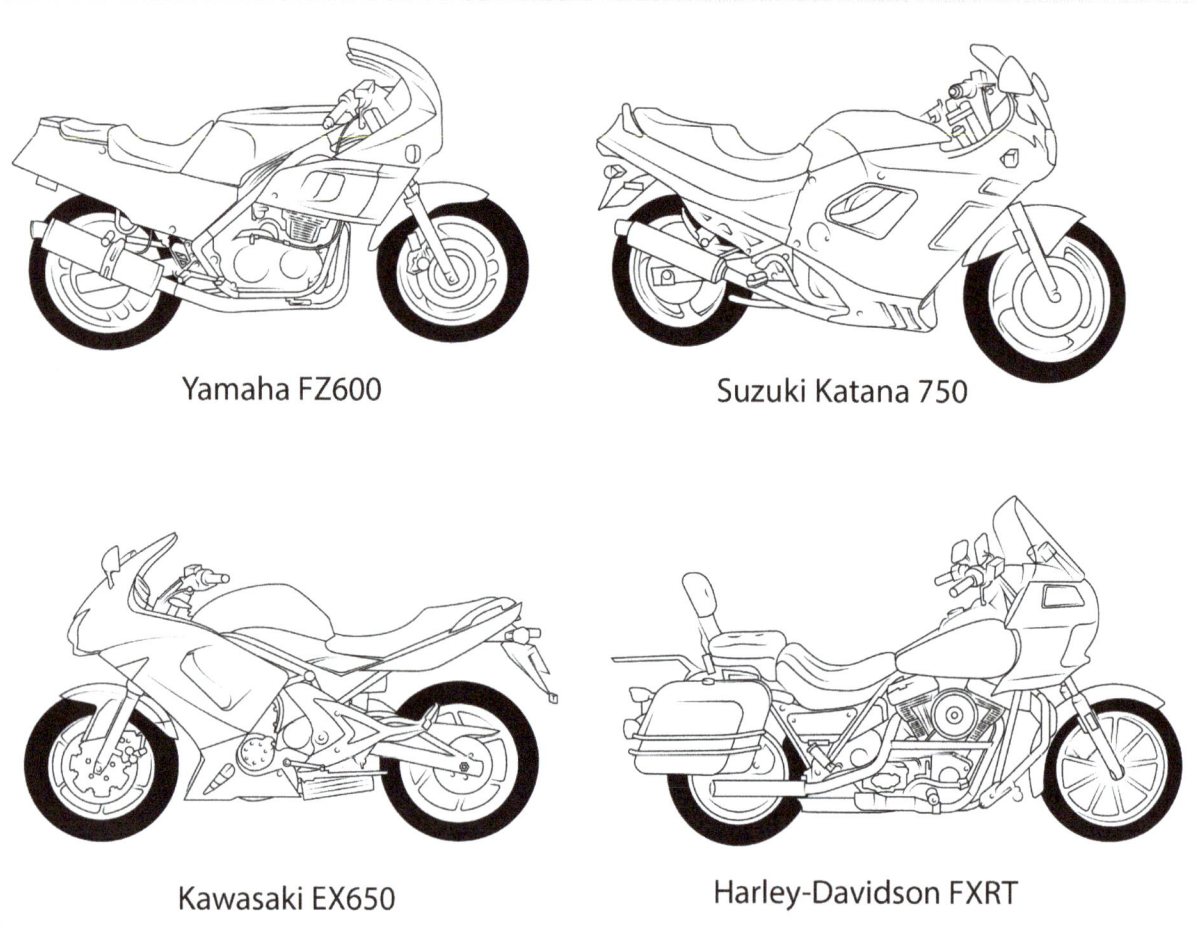

Bartlett [12] reported testing with four motorcycles—a Harley-Davidson FXRT, a Yamaha FZ600, a Suzuki Katana 750, and a Kawasaki EX650. These motorcycles are depicted in Figure 3.4. For tests that utilized only the rear brake, the maximum decelerations between these four motorcycles varied between 0.38 and 0.46 g. For tests that utilized only the front brake, the maximum decelerations varied between 0.88 and 0.89 g. Bartlett reported testing with combined front and rear braking for the Harley-Davidson. This produced a maximum deceleration of 0.96 g. In this testing, the Yamaha brake pads were deteriorated, resulting in metal-to-metal contact. The maximum deceleration produced with the Yamaha with these deteriorated brake pads was 0.75 g. The experience and skill levels of the test rider were not reported in this study.

Ecker and his colleagues [13] conducted a study composed of approximately 600 tests performed by more than 300 riders of varying levels of experience (novice to 40+ years) operating an instrumented Honda CB500 (Figure 3.5). Most of the riders were participants in motorcycle safety courses. As the riders were operating the Honda around a training

FIGURE 3.5 Honda CB500 used in Ecker's testing [13].

facility, the test coordinator would trigger a red light mounted to the instrument cluster, signaling the rider to "make a full stop emergency braking maneuver." The authors noted that "the test persons were aware of the imminent signal to start the maneuver. However, the test coordinator could vary the instant of triggering the maneuver via remote control within several seconds so that there was some uncertainty involved for the test persons." These tests were conducted on dry asphalt from a speed of approximately 60 km/h (37 mph). The average deceleration for all 600 runs was 0.63 g with a standard deviation of 0.12 g. One conclusion of this study was that "a correlation between experience and deceleration is hardly recognizable, especially for more than 1 year of riding experience." Another conclusion was that half of the tested individuals utilized 56% or less of the deceleration that could be achieved with the motorcycle.

Vavryn and Winkelbauer [14] examined the influence of rider experience level and the effectiveness of antilock braking systems (ABS). He reported the results of 800 tests performed with 181 subjects on two different motorcycles. The riders were asked to "come to a complete stop as soon as possible without falling off the vehicle." Initial speeds were either 50 or 60 km/h (31 or 37 mph). The subjects performed two tests on their own motorcycles and then two runs on a motorcycle equipped with ABS. One of the ABS-equipped motorcycles was a standard-style BMW and the other was a scooter equipped with linked ABS. The average deceleration for experienced motorcyclists on their own motorcycles was 0.67 g (SD = 0.14 g). When riding the motorcycles equipped with ABS, that number jumped up to 0.80 g (SD = 0.11 g). Eighty five percent of the subjects exhibited improved braking with the ABS-equipped motorcycle and the novice riders achieved almost equal braking decelerations to the experienced riders when operating the ABS-equipped motorcycles. Vavryn also noted that "the deceleration the novice drivers achieved with ABS almost equals the experienced drivers' deceleration. All of the novices improved their deceleration with ABS." Without ABS, the novice riders achieved an average deceleration of 0.57 g.

Bartlett et al. [15] reported hundreds of brake tests from reconstruction classes conducted at the Institute of Police Technology and Management (IPTM) between 1987 and 2006. These tests were conducted at various locations around the country with 112 different motorcycles and riders. They were conducted on dry asphalt or concrete. Initial speeds in the tests were nominally 20, 30, and 40 mph (32.2, 48.3, and 64.4 km/h). The riders in these tests were typically motorcycle unit officers or instructors from a police agency. Thus, this study was of experienced riders operating in parking lots. The data in this study included 275 rear-brake only tests, 239 front-brake only tests, and 221 tests with combined front and rear braking. This data yielded the conclusion that the decelerations were normally distributed with a mean and standard deviation for the rear only braking of 0.37 g ± 0.06 g, with front only braking of 0.60 g ± 0.16 g, and with combined front- and rear-wheel braking of 0.74 g ± 0.15 g.

Bartlett and Greear [16] presented brake test data from students in a motorcycle training program (*S*kills *T*raining *A*dvantage for *R*iders from the State of Idaho) with three skill levels—Basic I, Basic II, and Experienced. The authors noted that "the Basic I program is for riders who are new to motorcycling, with virtually no experience, and is conducted on STAR training motorcycles. These bikes are typically 250cc or smaller, with front disc and rear drum brakes. The Basic II program is for riders who are returning to motorcycling or those who have ridden on dirt, but not on the street, i.e., riders with some experience but not much on street cycles. These riders also use the program's training motorcycles. The Experienced program is for riders who have been riding for more than one year and is conducted using the riders' own motorcycles."

The culmination of each program was a riding skills test, which included a stopping test conducted in a parking lot. Riders were instructed to approach the stopping area at a steady speed between 15 and 20 mph (24.1 and 32.2 km/h). Once in the stopping area, they were to stop the motorcycle as quickly as they could with maximum braking. Bartlett reported the results of 288 tests, close to 100 tests for each experience level. The results of these tests "were almost indistinguishable" for three skill levels. The Basic I group produced decelerations with a mean and standard deviation of 0.60 g ± 0.16 g; the Basic II group produced decelerations of 0.64 g ± 0.14 g; and the experienced group produced decelerations of 0.61 g ± 0.14 g. When the experience levels were combined into one dataset, the decelerations were 0.62 g ± 0.15 g. In this study, no information was reported about the front-to-back braking split used by each rider or about the braking systems on the motorcycles for the tests where the students use their own motorcycles. The summary of braking data contained in Table 3.2 assumes that the riders in this study generally used both brakes and that the motorcycles did not have antilock brakes.

Dunn et al. [17] reported brake test data and tire mark characteristics for the three motorcycles depicted in Figure 3.6—a 1995 BMW R1100RS (sport-touring with antilock brakes), a 2003 Buell XB9R (sport), and a 2005 Harley-Davidson XL 1200 Sportster Custom (cruising/touring). They tested three different braking strategies—best effort front braking only, best effort rear braking only, and best effort front and rear combined braking. Initial speeds for the tests were nominally 25, 45, and 60 mph (40.2, 72.4, and 96.6 km/h) and most of the tests were conducted on a flat, dry asphalt surface. One set of tests was conducted on wet asphalt with the BMW, a motorcycle equipped with antilock brakes. The riders used in this study varied in years of experience, from 2 to 35 years. Nothing was reported regarding the annual mileage covered by the riders.

For the BMW, the rear-braking-only strategy produced decelerations between 0.364 and 0.416. The front-braking-only strategy produced decelerations between 0.671 and 0.828. The combined front and rear braking strategy produced decelerations between 0.642 and

FIGURE 3.6 Motorcycles used in the testing reported by Dunn et al. [17].

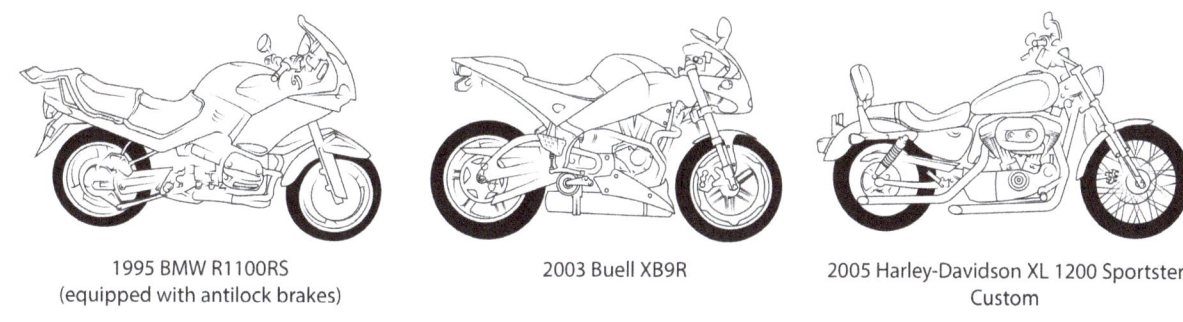

1995 BMW R1100RS (equipped with antilock brakes)

2003 Buell XB9R

2005 Harley-Davidson XL 1200 Sportster Custom

© SAE International

0.842. For all three strategies, the decelerations increased with increasing speed. On the wet asphalt surface, the BMW produced decelerations with both brakes between 0.637 and 0.827. For the Buell, the rear-braking-only strategy produced decelerations between 0.345 and 0.380. The front-braking-only strategy produced decelerations between 0.548 and 0.709. The combined front and rear braking strategy produced decelerations between 0.612 and 0.708. Again, for all three strategies, the decelerations increased with increasing speed. For the Harley-Davidson, the rear-braking-only strategy produced a deceleration of 0.386 (this strategy was only tested at 45 mph). The front-braking-only strategy produced a deceleration of 0.518 (this strategy was only tested at 45 mph). The combined front and rear braking strategy (tested at 45 and 60 mph) produced decelerations between 0.658 and 0.674.

Dunn et al. [17] found that "at the extreme, the rear tire of the Buell lifted off the ground in some tests." Frank et al. [18] noted that "pitch-over events are common in motorcycle accidents and can be caused by impact to the front wheel and occasionally by hard brake application … Provided there is sufficient tire/road friction, at the limit of the braking capacity of the motorcycle the weight on the rear tire is zero. Though not inevitable, this is the point at which the motorcycle can pitch-over." Frank et al. conducted 18 sled tests to evaluate the trajectory and velocity of riders and passengers on motorcycles that pitched over because of braking. This testing used target decelerations of 1.0, 1.15, and 1.3 g. Target speeds for the testing were 20, 30, and 33.5 mph (32.2, 48.3, and 53.9 km/h). The lowest braking deceleration that produced a pitch-over in Frank's testing was 1.0 g with a test speed of 30.2 mph (48.6 km/h).

Fatzinger et al. [19] reported a study of motorcycle deceleration for sport motorcycles during over-braking of the front wheel. Testing was conducted with a 2002 Kawasaki ZRX1200R, a 2006 Yamaha YZF-R6, and a 2013 Ninja EX300. Thirteen tests were completed, with initial speeds ranging from 50 to 60 mph. All three motorcycles had independently actuated front and rear brakes without antilock brakes. Testing was conducted on a flat asphalt surface. Brake pressure was applied to the front-brake lever or the rear-brake pedal with elastic straps. Electronically controlled valves installed in each brake line prevented this pressure from being applied to the calipers until the motorcycle was up to speed. Of the 13 tests, 3 were performed with a 6 ft, 1 in. and 175 lb dummy on the motorcycle. In some of the tests, rear-wheel braking was applied in addition to the front braking, and in some of the tests, no rear braking was applied. Front-wheel lockup was achieved in nine of the tests.

Fatzinger et al. reported that the deceleration achieved by the motorcycle with front-wheel lockup depended on the lean angle of the motorcycle at the beginning of the braking. They reported that the average deceleration when the front wheel locked up and the initial lean was approximately 2° or less was in the range of 0.69 and 0.8 g. The average deceleration when the initial lean was around 3° or 4° was between 0.51 and 0.67 g. The average deceleration when the initial lean angle was between 8° and 9° was 0.32 to 0.39 g. A test where the front-wheel braking resulted in a pitch-up had a deceleration of approximately 0.8 g. A test that resulted in complete pitch-over had a deceleration of 0.98 g. Rear-brake application did not significantly increase the deceleration of the motorcycles when front-wheel lock had been achieved. Also, there were "no significant differences noted in the peak and average decelerations between the tests" with and without the dummies.

Peck et al. [20] examined the effect of tire pressure on the deceleration achieved with full application of the rear brake only. This testing utilized a 2003 Suzuki GSF1200 equipped with Michelin Pilot Road radial tires. The tests were run from a nominal speed of 30 mph (48.3 km/h)—three tests with the rear tire at 40 psi and three tests with the rear tire at 20 psi. The front tire was inflated to the manufacturer recommended tire pressure of 36 psi for all tests. These authors documented the size of the tire contact patch by using a rear swingarm stand to suspend the rear tire above a piece of brown paper, putting paint on the tire, and then lowering the tire onto the paper. The size of the rear tire contact patch was 46% larger at 20 psi than at 40 psi, and the average deceleration was 5% greater at 20 psi than at 40 psi. For the tests at 40 psi, the three tests yielded the following decelerations (g): 0.324, 0.321, and 0.327 (average = 0.324). For the tests at 20 psi, the three tests yielded the following decelerations (g): 0.341, 0.339, and 0.338 (average = 0.339). These findings related to the influence of tire pressure are consistent with results reported by others for passenger cars [21, 22].

In addition to these studies, the author of this book conducted testing with a 2003 BMW F650GS Dakar R13 under rear-brake only deceleration. This motorcycle, which is depicted in Figure 3.7, was an adventure motorcycle without ABS, without linking between the front

FIGURE 3.7 2003 BMW F650 GS Dakar R13.

and rear brakes, and with knobby tires. Testing was conducted on dry asphalt and utilized an experienced rider. In this testing, the rear brake was applied with sufficient severity to cause rear tire skidding. In three runs, rear-brake application only produced decelerations of 0.31 g, 0.37 g, and 0.35 g.

Table 3.2 summarizes the decelerations from the studies reviewed in this section. These decelerations can potentially be applied for calculating speed loss of a motorcycle because of maximal braking by the operator or for assessing a motorcyclist's ability to avoid a crash. The reconstructionist will need to consider conditions relevant to each crash. For example, what evidence is there related to the braking strategy rider utilized (rear only, front only, or front and rear combined).

TABLE 3.2 Summary of braking decelerations from various studies (dry pavement)

Study	Best effort braking decelerations on dry roadway (g)			
	Rear brake only (no antilock brakes, ABS)	Front brake only (no antilock brakes, ABS)	Front and rear combined (no antilock brakes, ABS)	Front and rear with antilock brakes (ABS)
Tolhurst and McKnight [6] (expert riders)		0.77	0.97	
Fries et al. [9]	0.31–0.52		0.54–0.88	
Hunter [10] (experienced riders)	0.35–0.36	0.64–0.74	0.63–0.96	
Hugemann and Lange [11]				
Bartlett [12]	0.38–0.46	0.88–0.89	0.96	
Ecker et al. [13] (mixed experience levels)			0.63 ± 0.12	
Vavryn and Winkelbauer [14]			0.67	0.8
Bartlett et al. [15] (motor officers)	0.37 ± 0.06	0.60 ± 0.16	0.74 ± 0.15	
Bartlett and Greear [16] (students)			0.62 ± 0.15	
Dunn et al [17]	0.345–0.386	0.518–0.709	0.612–0.708	0.642–0.842
Fatzinger et al [19]		0.34–0.87	0.39–0.98	
Peck et al. [20]	0.321–0.341			
Rose (adventure motorcycle, experienced rider, knobby tires)	0.31–0.37			

Motorcycles with Integrated, Linked, and Antilock Braking Systems

This section reviews additional studies that focused on motorcycle braking systems equipped with integrated front and rear brakes or antilock brakes. These studies provided illustrative results of the influence of these systems, but these systems vary in how various manufacturers implement them. Reconstructionists can refer to the owner's manual for specific motorcycles for information specific to individual motorcycles. Individual motorcycles can also be tested.

In 1988, BMW became the first manufacturer to offer a motorcycle with an ABS [23]. Yamaha followed in 1991 and Honda in 1996. Starting in 2003, Ducati offered an ABS with an on-off switch, giving the operator control over whether the ABS system was active or not. BMW also offered an on-off switch on some of their models. This is commonly a feature on motorcycles intended for both on and off-road uses. In 2008, Harley-Davidson began offering an optional ABS system with independently controlled front and back brakes. For an independent ABS, the front and rear brakes are controlled independently, but the antilock system prevents either wheel from locking. For an integrated ABS, both brakes are applied regardless of which lever is used and the antilock system prevents either wheel from locking. For an integrated braking system, the degree to which application of the front-brake lever results in rear braking, and the degree to which application of the rear-brake foot pedal results in front braking, varies. The owner's manual for particular makes and models can be consulted for model-specific details (or an exemplar motorcycle could be tested).

Mortimer [24] examined the effectiveness of integrated brakes without ABS. His testing utilized a 1979 Yamaha XS-400 with standard brakes as the original equipment and a 1982 Yamaha XS-1100 with integrated brakes as the original equipment. Mortimer modified both motorcycles so that they could be operated in either a standard braking mode or an integrated braking mode. The integrated mode on the XS-400 could only be operated with the right-side rear-brake foot pedal. On the XS-1100, the integrated braking would be activated with either the right-side foot pedal or hand lever. Five experienced riders were used. Tests were run from a nominal speed of 25 mph (40.3 km/h), and the riders attempted to stop the motorcycle in as short a distance as possible. Each rider performed testing on each motorcycle, and they made five stops in each test condition. The tests were run with the hand brake only, the foot brake only, and then both.

Mortimer noted that "the stopping distances were directly measured at the point where the motorcycle came to a stop in terms of the distance from the cones marking the entrance to the braking course. The stopping distance was translated into the mean deceleration during the stop, assuming an initial speed of 40.3 km/h." This manner of measuring the stopping distance and deceleration is prone to error because there is no way to know, in any given instance, if the riders began braking at the cone or to know that the rider started braking from a speed of precisely 40.3 km/h (25 mph). Mortimer found the greatest benefit from integrated brakes for the condition of braking with the foot pedal only. He noted that "use of the foot brake alone of the XS-400 motorcycle produced a 72% greater mean deceleration in the integrated than the separated mode. Similarly, use of the foot brake of the XS-1100 motorcycle in the integrated mode produced a 50% increase in mean deceleration compared with the separated mode … In addition, when both brakes were used on the larger motorcycle there were significant and consistent increases in deceleration obtained on both the dry and wet pavements in the integrated mode compared with the separated mode, but the increments were not as large as those found for the operation of the foot brake alone."

Vavryn and Winkelbauer [14] examined the effectiveness of ABS, reporting the results of 800 tests performed with 181 subjects on two different motorcycles. The riders were asked to "come to a complete stop as soon as possible without falling off the vehicle." Initial speeds were either 50 or 60 km/h (31 or 37 mph), and the subjects performed two tests on their own motorcycle, and then two runs on a motorcycle equipped with ABS. One of the ABS bikes was a standard BMW, while the other was a scooter equipped with linked ABS. The average braking deceleration for motorcyclists on their own motorcycle was 0.67 g (SD = 0.14 g). However, when riding the motorcycles equipped with ABS, that number jumped up to 0.80 g (SD = 0.11 g). Eighty five percent of the subjects exhibited improved

braking with the ABS-equipped motorcycle and the novice riders achieved almost equal braking decelerations to the experienced riders when operating the ABS-equipped motorcycles.

Green [25] reported a test program conducted by NHTSA, in cooperation with Transport Canada (TC), "to assess the effectiveness of anti-lock braking systems (ABS) and combined braking systems (CBS) on motorcycles." Six motorcycles were tested on both dry and wet asphalt—a 2002 Honda VFR 800 with ABS and CBS, a 2002 BMW F650 with ABS, a 2002 BMW R 1150R with ABS and CBS, a 2002 BMW R 1150R without ABS or CBS, a 2004 Yamaha FJR1300 with ABS, and a 2004 Yamaha FJR1300 without ABS. Green observed that, with ABS, "the stopping distances were very consistent from one run to another." Without ABS, "the stopping distances were less consistent because the rider while modulating the brake force, had to deal with many additional variables at the same time … Test results from non-ABS were noticeably more sensitive to rider performance variability." On average, ABS reduced the stopping distances by approximately 5%.

Anderson et al. [26] reported deceleration testing of motorcycles with the following different braking systems—standard brakes (1990 Harley Davidson Road King FLHTPI), integrated brakes without ABS (1986 Yamaha Venture Royale XVZ13), independent ABS brakes (1999 BMW R1100RPT), integrated ABS brakes, and linked brakes (2003 Honda VFR800 Interceptor). The authors tested each of these systems on an asphalt surface (automobile $\mu = 0.83$) with application of the rear pedal only, the front lever only, and with both levers applied. The initial speed for the tests was approximately 56 km/h. All the tests utilized the same operator with many years of riding experience. The authors noted that "there was no wheel lockup or skidding during any of the tests runs."

Table 3.3 summarizes the results of the testing for each braking system. The values in this table are primarily of use for showing the comparison between the different braking systems. This study used only a single experienced rider and the variability in decelerations was not reported. Thus, the values in this table should not be blindly applied to a reconstruction, without consideration of how a typical rider would perform with each system. In addition, the authors noted that "this testing only analyzed motorcycle braking during the period of maximum, and near-constant, deceleration. Operator reaction time and brake system lag time were not addressed in this study, although such investigation may be worthwhile as extensions of the work presented in this paper. The systems that utilize linkages to actuate both front and rear brakes, such as the integrated and linked brakes of the BMW and Honda motorcycles herein, may have lag times and mechanical behavior that affects the resultant onset of deceleration." The trend in these values is, however, consistent with the benefits that would be expected from the various braking systems.

TABLE 3.3 Summary of average braking decelerations for the various systems tested by Anderson et al. [26]

	Average deceleration (g)				
	Standard	Integrated	Antilock brakes (ABS)	Integrated antilock brakes (ABS)	Linked
Pedal only	0.42	0.58	0.40	0.98	0.62
Hand lever only	0.65	0.74	0.89	0.92	0.86
Both levers	0.71	0.88	0.93	1.00	0.93

©2022, SAE International

Anderson et al. concluded that "motorcycle braking systems that actuate both front and rear brakes with the application of only one control lever produce more effective braking than independent front and rear brakes on a standard system. When combined with antilock control the benefits of the combined system are increased. Perhaps more importantly, however, is that the motorcycle is also more stable during the braking maneuver. The increased stability along with the simplified brake application combine to reduce the load on the operator during the stressful moment of hard braking to avoid a crash. The operator does not have to concentrate on modulating pressure between two separate controls and simultaneously keep the motorcycle stable and prevent the wheels from locking, as the system performs these functions and permits the operator to focus on avoiding the crash."

Dinges and Hoover [27] reported a series of maximal braking tests on a dry, asphalt surface with and without the antilock brakes active on a 2011 BMW S1000RR (a super sport motorcycle). This motorcycle was tested in three modes related to the ABS—sport mode, race mode, and ABS disabled. The BMW was equipped with partially integrated braking when the ABS was active. When the ABS was deactivated, the integral braking was also deactivated. These authors reported a coefficient of friction for the test surface of 0.7, measured using a Ford Expedition with the ABS disabled. Their testing yielded 420 braking runs, with target speeds ranging from 40 to 60 mph. Table 3.4 lists the average decelerations reported by Dinges and Hoover for each mode with three braking conditions—rear brake only, front brake only, and front and rear braking combined. In addition to these decelerations, the authors reported hydraulic pressure build times, noting that "the average time it takes to build pressure in the front brake system is between 0.2 and 0.3 seconds … The rear brake system is similar, but a range of 0.2 to 0.4 seconds is shown from the data."

TABLE 3.4 Average decelerations reported by Dinges and Hoover [27]

	Rear brake Only	Front brake only	Front and rear combined
ABS off	0.37	0.79	0.79
Race mode antilock brakes (ABS)	0.40	0.80	0.80
Sport mode antilock brakes (ABS)	0.38	0.76	0.72

© SAE International

Speed and Avoidance Calculations

For many crashes, braking precedes an impact and there may be zero, one, or two tire marks deposited. The following equation can be used to calculate the speed of the motorcycle at the onset of maximal braking or skidding (v_{skid}).

$$v_{skid} = \sqrt{2 f_{skid} g d_{skid} + v_{impact}^2} \qquad (3.9)$$

In this equation, f_{skid} is the deceleration of the motorcycle during the skidding or maximal braking in gravitational units, g is the gravitational acceleration, d_{skid} is the length of the skid mark(s), and v_{impact} is the speed of the motorcycle at impact. The impact speed would be determined using methods discussed in a later chapter.

Accident reconstructionists are also often asked to determine how a crash could have been avoided, and there will be instances in which an assumption will be made about the level of deceleration a motorcyclist should have been able to achieve. Examining the data in Table 3.2, the studies that stand out as representative of a mix of experience levels are the Ecker et al. [13] and Bartlett and Greear [16] studies. Based on these studies, motorcyclists riding motorcycles with conventional braking systems will generally be able to achieve a deceleration of 0.62 g ± 0.15 g on a dry road under ideal circumstances. With antilock brakes, particularly if there is integration between the front and rear brakes, motorcyclists are likely to achieve higher decelerations with greater consistency than they would have with a motorcycle with a conventional braking system. These systems are present on many newer motorcycles, and the level of deceleration that the average rider can achieve is likely to increase in the future. Again, these braking studies were conducted under ideal conditions. The level of deceleration that can be achieved during a specific emergency must consider conditions present that may have affected a rider's ability to achieve these expected levels. External factors such as roadway conditions, other traffic, the presence of cargo or passengers, or what the specific avoidance decision a rider makes may need to be considered when assigning an expected braking level to a specific crash scenario.

Engine Braking

Jansen et al. [28] measured and reported closed-throttle engine braking in each forward gear for a cross section of eight typical street motorcycles. Table 3.5 lists the average decelerations for each motorcycle in each gear. These decelerations could be used to calculate precollision speed loss experienced by a motorcycle prior to the onset of braking. Illustratively, if we were to assume a 0.25 sec lag between throttle release and brake application, the engine drag would account for approximately ½ to 1 mph of speed loss during this lag. The values in Table 3.5 could also be used for calculating postimpact speed loss in the rare instances when a motorcycle stays upright and rolls out, rather than simply capsizing.

TABLE 3.5 Average deceleration in normalized gravitational constant (g) by forward gear for each motorcycle

	N	1	2	3	4	5	6
2015 BMW R nineT (standard)	0.10	0.24	0.22	0.18	0.13	0.15	0.15
2003 Buell XB9S Lightning (standard/naked)	0.08	0.18	0.17	0.13	0.13	0.12	n/a
2015 Harley-Davidson XL1200X Forty-Eight (cruiser)	0.09	0.15	0.15	0.15	0.15	0.15	n/a
2006 Honda CBR1000RR (sport)	0.09	0.16	0.15	0.16	0.16	0.14	0.12
2013 Kawasaki KLR650 (dual-sport)	0.12	0.21	0.19	0.15	0.14	0.15	n/a
2019 Yamaha R3 (sport)	0.10	0.18	0.18	0.18	0.17	0.15	0.13
2017 Yamaha XSR900 (standard)	0.11	0.22	0.21	0.17	0.16	0.14	0.13
2013 Yamaha FZ6R (sport)	0.07	0.18	0.17	0.15	0.13	0.13	0.12
Average	0.09	0.19	0.18	0.16	0.15	0.14	0.13

© 2022, SAE International

This analysis could be made more nuanced by recognizing that the engine braking "deceleration is highly sensitive to the motorcycles' speed and RPM. The engines are at higher RPM for a given speed in numerically lower gears, while the RPMs are convergent for a given speed in higher gears ... Because RPM and ground speed are directly related through the overall reduction ratio for each gear, a reasonable conclusion is the resulting deceleration appears to be more correlated to the engine RPM than to the motorcycle speed." As an example, at 2000 rpm, the BMW R nineT exhibited an engine braking deceleration around 0.1 g, but at 7000 rpm, the deceleration was above 0.3 g. Often, the reconstructionist will not know the rpm at which a motorcyclist was operating, but if the speed and gear position are known, the rpm can be inferred. Most reconstructions of motorcycle crashes will not exhibit significant sensitivity to the engine braking, but for instances where the engine braking is being considered, the data in reference [28] provide guidance. In some instances, testing of a specific motorcycle (beyond those tested in this reference) could be warranted.

Motorcycle Acceleration Capabilities

Bartlett [29] studied motorcycle accelerations and reported that "maximum motorcycle acceleration rates are highly nonlinear ... In a perfect world, the investigator will have access to the accident motorcycle or an identical unit for testing with reliable and accurate measurement equipment. In the real world, though, this is often not the case. This leaves the investigator to review published data on their motorcycle capabilities." Bartlett discussed two techniques reconstructionists could use to estimate accelerations for a motorcycle. However, these techniques only provide a maximum rate.

The first technique is applicable to instances when the reconstructionists have access to some time/speed data, typically from an enthusiast magazine. In this instance, Bartlett proposed that the analyst should (1) plot the data points available as speed versus time; (2) determine the best-fit curve for the speed versus time data; (3) calculate the predicted speed for every 0.1 sec using the equation for the best-fit curve; (4) calculate the distance covered in each 0.1 sec; and 5) calculate the cumulative distance across the data.

The second technique is applicable to instances when no time/speed data are available. In this instance, Bartlett suggests that the analyst should determine the combined weight of the motorcycle (wet) and rider and the horsepower of the motorcycle. Bartlett noted that "the horsepower measured at the rear wheel using the most common type of dynamometer is commonly expected to be about 85% of that reported at the crankshaft due to losses in the driveline. In the current analysis, comparing 30 such pairs of claimed/measured data

TABLE 3.6 Bartlett's motorcycle acceleration equations

Parameter	Equation	R^2
Quarter mile time (lb/hp < 30)	$t = 6.5729 \cdot x^{0.278}$	0.97
Quarter mile speed (lb/hp < 30)	$S = 239.90 \cdot x^{-0.351}$	0.97
Quarter mile time (lb/hp > 30)	$t = 5.9504 \cdot x^{0.322}$	0.90
Quarter mile speed (lb/hp < 30)	$S = 172.78 \cdot x^{-0.267}$	0.70
0-30 mph	$t = 0.0023x^2 - 0.01x + 1.23$	0.78
0-60 mph	$t = 0.0067x^2 + 0.1x + 2.13$	0.97
0-90 mph	$t = 0.0342x^2 + 0.1x + 3.47$	0.89
0-100 mph	$t = 0.0312x^2 + 0.45x + 2.55$	0.92

© SAE International

from published sources reveals the average loss from the claimed value to be approximately 13%, which matches the expectation quite well. If only published 'crankshaft' or 'claimed' horsepower values are available, they should be degraded by 15% to approximate rear wheel horsepower." Bartlett reported the following table of equations. In these equations, x is the weight-to-horsepower ratio. These equations yield points for use with the first technique.

These equations can be used to place an upper limit on the speed a motorcycle could have been traveling in scenarios where a crash was preceded by the motorcycle accelerating.

Some newer motorcycles are equipped with traction control. This is another factor that could influence the acceleration achieved by a motorcycle, because the traction control will limit the power supplied to the rear wheel to avoid a loss of traction at the rear wheel. This is most likely to come into play on lower traction surfaces, or when a rider applies the throttle in a curve. These systems typically utilize the same wheel speed sensors utilized by the ABS on the motorcycle. As with the ABS present on many newer motorcycles, the motorcycle owner's manual will typically describe some of the details of the traction control system present on a motorcycle. In some instances, testing of the behavior of these systems could be necessary for a reconstruction.

Determining Speed Based on Gear

In some instances, on-scene investigating officers will document the gear the motorcycle was immediately following the crash. If they do not, this determination can sometimes be made during a later inspection. If the reconstructionist can determine the gear the motorcycle was in at the time of a collision or loss of control, this information can be used to place a lower and upper limit on the speed of the motorcycle [30]. The range of possible speeds is likely to be large, if a full range of rpm is used. Keep in mind that a motorcyclist could downshift during braking prior to a collision, although that may not be likely when the rider is engaged in an urgent emergency response. Also, and if the motorcycle capsizes onto its left side during the crash, the shifter could engage with the ground in a way that alters the gear position. When this occurs, damage to the shift lever will be present, and potentially damage to surrounding components (like the crankcase) that interact with the shift lever during the crash and exhibit evidence of the shift levers movement. This author has documented such shifting occurring in two motorcycle drop and slides tests. This can be ruled out if there is no damage to the shift lever evidencing that it engaged with the ground.

To make a determination of speed based on gear position, the analyst will need to determine the rolling radius of the rear tire of the motorcycle (r_{rear}), the maximum engine speed in rotations per minute ($S_{rpm,max}$), the primary reduction ratio (r_{pr}), the final reduction ratio (r_{fr}), and the gear ratio for the gear of interest (r_i). The primary reduction ratio is the reduction that occurs from the engine to the transmission. The final reduction ratio is the ratio of the crank sprocket to the wheel sprocket. Assuming the rolling radius of the rear tire of the motorcycle has been obtained in units of in., Equation (3.10) will yield the distance this tire rolls for each revolution, in units of ft (c_{rear}). Then, Equation (3.11) will yield the maximum road speed ($S_{road,max}$) for the gear of interest, in units of mph.

$$c_{rear} = \frac{2\pi \cdot r_{rear}}{12} \qquad (3.10)$$

$$S_{road,max} = \frac{S_{rpm,max}}{r_{pr} \cdot r_{fr} \cdot r_i} \cdot c_{rear} \cdot \frac{60}{5280} \qquad (3.11)$$

In applying the results of these calculations, the reconstructionist should keep in mind that a motorcyclist may downshift during precollision braking.

Case Study: Determining Speed Based on Gear

To test the accuracy of Equation (3.11), physical testing was conducted to determine the maximum speed of a 2008 Lifan 200GY-5 dual-sport motorcycle (Figure 3.8) in each of its five gears. This motorcycle had a 15 horsepower, 200cc single cylinder gasoline engine, a five-speed manual transmission, and original equipment tires. The engine on this motorcycle reaches the redline at 10,500 rpm. Manufacturer specifications were obtained to determine the gearing parameters for this motorcycle. The manufacturer specifications also reported that this motorcycle had a maximum speed of 100 km/h (62 mph). The maximum engine speed was controlled by an ignition limiter.

Additional parameters used in the gear train calculations are included in Table 3.7. Static testing was conducted, and it was determined that the rear tire would compress approximately 0.7 in. with a rider sitting on the motorcycle. In the calculations, this value was reduced slightly—to 0.5 in.—in recognition of tire recovery that would occur at speed. The calculations based on Equation (3.11) are included in Table 3.8. These calculations provided the road speed for each gear and rpm combination over the range of 7500 to 10,500 rpm (in 250 rpm increments). These calculations predicted that the motorcycle would reach a maximum speed of approximately 29.9 mph in first gear, 44.0 mph in second gear, 59.1 mph in third gear, 62 mph in fourth gear, and 62 mph in fifth gear.

Prior to testing, the Lifan motorcycle was warmed to operating temperature and the tires were checked to verify if they were inflated to manufacturer specifications. Testing was performed on a nearly flat roadway at Front Range Airport in Watkins, CO, on August 1, 2017. Each test run consisted of accelerating the motorcycle up to its maximum speed for a given gear with the operator attempting to achieve the greatest acceleration possible. Each gear was tested in both directions on the roadway because the road had a slight grade. This

FIGURE 3.8 Lifan motorcycle utilized for physical testing.

TABLE 3.7 Gearing parameters for the Lifan motorcycle

Theoretical rear tire diameter (in.)	24.9
Rear tire compression (in.)	0.5
Rear tire diameter (ft)	2.03
Tire rolling distance/Rev (ft)	6.39
Engine idle speed (rpm)	1500
Maximum engine speed (rpm)	10500
Maximum road speed (km/h)	100.0
Maximum road speed (mph)	62.1
Primary reduction ratio	3.333
Final reduction ratio	2.706
Gear ratio, 1st gear	2.769
Gear ratio, 2nd gear	1.882
Gear ratio, 3rd gear	1.400
Gear ratio, 4th gear	1.130
Gear ratio, 5th gear	0.960

TABLE 3.8 Gear train calculations for the Lifan motorcycle

Gear	Engine speed (rpm)												
	7500	7750	8000	8250	8500	8750	9000	9250	9500	9750	10,000	10,250	10,500
1	21.35	22.06	22.78	23.49	24.20	24.91	25.62	26.34	27.05	27.76	28.47	29.18	29.89
2	31.42	32.46	33.51	34.56	35.61	36.65	37.70	38.75	39.79	40.84	41.89	42.94	43.98
3	42.23	43.64	45.05	46.46	47.86	49.27	50.68	52.09	53.50	54.90	56.31	57.72	59.13
4	52.32	54.07	55.81	57.56	59.30	61.05	62.79	64.53	66.28	68.02	69.77	71.51	73.25
5	61.59	63.64	65.70	67.75	69.80	71.86	73.91	75.96	78.01	80.07	82.12	84.17	86.23

allowed for any roadway grade effects on the speed test results to be eliminated. A VBOX Sport data acquisition system was attached to the motorcycle to document vehicle speed during each run. The motorcycle operator also wore a chest-mounted Go-Pro camera to document the engine speed as reported by an aftermarket digital tachometer. Five separate tests were completed by the operator—one for each gear.

The data from each of the five test runs are depicted in Figure 3.9. Time is plotted on the horizontal axis and speed on the vertical axis. In first gear, the motorcycle achieved an average maximum speed of approximately 30.3 mph, 0.3 mph greater than what Equation (3.11) predicted. In second gear, the motorcycle achieved an average maximum speed of approximately 43.6 mph, 0.4 mph lower than what Equation (3.11) predicted. In third gear, the motorcycle achieved an average maximum speed of 54.4 mph, approximately 4.7 mph lower than what Equation (3.3) would predict for a redline engine speed of 10,500 rpm. However, in third gear, the video revealed that the maximum engine speed was limited to approximately 9800 rpm. At this engine speed, Equation (3.11) predicted a speed of 55.2 mph, 0.8 mph greater than what was achieved. In fourth gear, the motorcycle achieved an average maximum speed of 58.4 mph. However, in fourth gear, the video revealed that the maximum engine speed was limited to approximately 8300 rpm. At this engine speed, Equation (3.11) predicts a road speed of 57.9 mph, 0.5 mph less than what was achieved.

In fifth gear, the motorcycle achieved an average maximum speed of 59.9 mph. This was the run with the greatest discrepancy between the downhill and uphill directions. In the downhill direction, the motorcycle achieved a speed of 62.4 mph and in the uphill direction a speed of 57.4 mph. However, in fifth gear, the video revealed that the maximum engine speed was limited to approximately 7600 rpm. At this engine speed, Equation (3.11) predicts a road speed should be able to reach the maximum of 62 mph, 2.1 mph greater than the average of what was achieved, but in line with the downhill maximum. Thus, Equation (3.11) provided a reasonable estimate of the maximum speed achievable in each gear if the analyst recognized the rpm limits in each gear. Unfortunately, in this instance, this required testing of the motorcycle, which would eliminate the need for Equation (3.11). This likely would not be an issue for a more powerful motorcycle because the rpm limits would be higher. The maximum error was the 2.1 mph in fifth gear. This is a 3.4% overestimate. These results are summarized in Table 3.9.

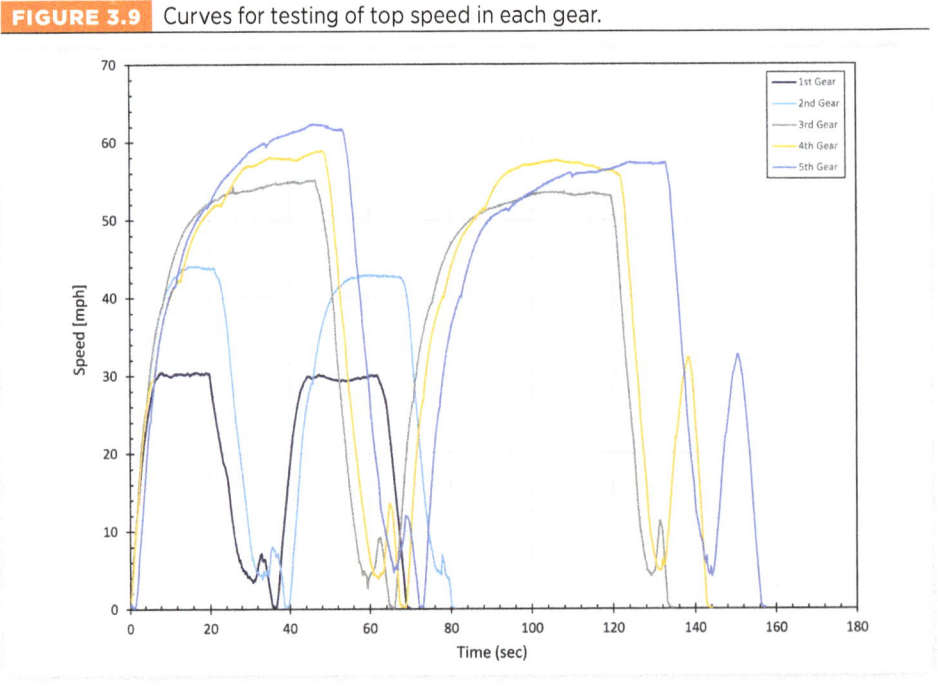

FIGURE 3.9 Curves for testing of top speed in each gear.

TABLE 3.9 Summary of the results from testing of the Lifan motorcycle

Gear	Calculated maximum speed (mph)	As-tested maximum speed (mph)	Difference, calculated versus actual (mph)	Percent difference
1	29.9	30.3	−0.4	−1.3%
2	44.0	43.6	+0.4	+0.9%
3	55.2	54.4	+0.8	+1.5%
4	57.9	58.4	−0.5	−0.9%
5	62	59.9	+2.1	+3.4%

Motorcycle Autonomous Emergency Braking (MAEB)

Rider assistance and autonomous systems are under development and beginning to be implemented on some motorcycles, including motorcycle autonomous emergency braking (MAEB). Specifically, Harley-Davidson, Honda, and BMW are developing MAEB. These systems are also being developed and evaluated in a research setting. These systems are designed to closing the throttle, apply the brakes, or amplify a rider's braking input in the case of an inevitable collision. Initially, the development of these systems focused on autonomous brake applications that would occur if the motorcycle was traveling straight (not leaning). Some recent studies, though, have explored the implementation of AEB during cornering or swerving, via an automatic braking control (ABC) to control the vehicle along a curved path [31]. MAEB systems include an obstacle detection device (radar or lidar), a triggering algorithm to define and detect inevitable collisions, and the automatic braking device.

Initially, MAEB systems will be tuned to respond only when a collision is inevitable. A critical aspect of these systems is the definition of inevitable collision states [32]. An inevitable collision state occurs when, no matter what future actions the involved drivers or riders take, a collision will eventually take place. To keep these systems from being over-responsive, the algorithms must assume that riders are more capable in both braking and swerving than they likely are [33]. According to reference [31], an inevitable collision state can usually be identified up to 0.5 sec prior to a collision. If the rider is not braking, when the inevitable collision is identified, the system can apply moderate braking (perhaps generating a deceleration around 0.3 g) to reduce the severity of the collision. If the rider is braking, the system can amplify that braking. It seems apparent that, if these systems apply braking at 0.3 g as much as 0.5 sec prior to the collision, not many collisions are likely to become avoidable simply by the addition of these systems. More likely, they will simply reduce the impact speed of the motorcycle. Reference [31] noted an important limitation of MAEB systems for motorcycle collisions: "… the potential benefits of the proposed system expressed in terms of impact speed reduction or avoidance of fall events cannot be directly correlated with actual benefits for the rider in terms of injury mitigation. In fact, risk curves expressing the level of injury for the rider as a function of kinematic quantities (such as impact speed) are not currently available for riders."

References

[1] Peck, L., "Motorcycle Braking," www.louispeck.com/motorcycle-braking, accessed on November 5, 2021.

[2] Cossalter, V., Lot, R., and Maggio, F., "On the Braking Behavior of Motorcycles," SAE Technical Paper Number 2004-32-0018, 2004, doi:10.4271/2004-32-0018.

[3] DiTallo, M., Paul, E., Adamson, K., Green, T., et al., "Motorcycle Center of Gravity Data: Methodology and Reference," *Collision: The International Compendium for Crash Research*, 12(1): 50–60, 2017, ISSN 1934-8681.

[4] Ouellet, J.V., "Lane Positioning for Collision Avoidance: A Hypothesis," *Proceedings of the 1990 International Motorcycle Safety Conference: The Human Element*, vol. 2, Orlando, FL, 1990, 9.58–9.80.

[5] Ayres, T. and Kubose, T., "Speed and Accuracy in Driver Emergency Avoidance," *56th Annual Meeting of the Human Factors and Ergonomics Society*, Boston, MA, 2012, ISBN 978-0-945289-41-8.

[6] Tolhurst, N. and McKnight, A., "Motorcycle Braking Methods," SAE Technical Paper 860020, 1986, doi:10.4271/860020.

[7] Prem, H., "The Emergency Straight-Path Braking Behaviour of Skilled versus Less-skilled Motorcycle Riders," SAE Technical Paper 871228, 1987, doi:10.4271/871228.

[8] Prem, H., "Motorcycle Rider Skill Assessment," PhD Thesis, University of Melbourne, Parkville, VIC, Australia, 1983.

[9] Fries, T.R., Smith, J.R., and Cronrath, K.M., "Stopping Characteristics for Motorcycles in Accident Situations," SAE Technical Paper 890734, 1989, doi:10.4271/890734.

[10] Hunter, J.E., "The Application of the G-Analyst to Motorcycle Acceleration and Deceleration," SAE Technical Paper 901525, 1990, doi:10.4271/901525.

[11] Hugemann, W. and Lange, F., "Braking Performance of Motorcyclists," 1993, https://static1.squarespace.com/static/57394f537da24fc27bc554bf/t/5e8e42db0998ad77056cd312/1586381532397/%281993%29+Hugemann+-+Braking+Performance+of+Motorcyclists.pdf, accessed on November 5, 2021.

[12] Bartlett, W., "Motorcycle Braking and Skidmarks," *Mechanical Forensics Engineering Services*, LLC, 2000, unpublished article, https://www.mfes.com/motorcyclebraking.html, accessed on June 1, 2017.

[13] Ecker, H., Wasserman, J., Hauer, G., Ruspekhofer, R. et al., "Braking Deceleration of Motorcycle Riders," *International Motorcycle Safety Conference*, Orlando, FL, March 1-4, 2001.

[14] Vavryn, K. and Winkelbauer, M., "Braking Performance of Experienced and Novice Motorcycle Riders – Results of a Field Study," *International Conference on Traffic & Transport Psychology*, Nottingham, England, 2004.

[15] Bartlett, W., Baxter, A., and Robar, N., "Motorcycle Braking Tests: IPTM Data Through 2006," *Accident Reconstruction Journal*, 17(4): 19–21, 2007, ISSN: 1057-8153.

[16] Bartlett, W. and Greear, C., "Braking Rates for Students in a Motorcycle Training Program," *Accident Reconstruction Journal*, 20(6): 19–29, 2010, ISSN: 1057-8153.

[17] Dunn, A.L., Dorohoff, M., Bayan, F., Cornetto, A. et al., "Analysis of Motorcycle Braking Performance and Associated Braking Marks," SAE Technical Paper 2012-01-0610, 2012, doi:10.4271/2012-01-0610.

[18] Frank, T., Smith, J., Hansen, D., and Werner, S., "Motorcycle Rider Trajectory in Pitch-Over Brake Applications and Impacts," *SAE International Journal of Passenger Cars—Mechanical Systems*, 1(1): 31–42, 2009, doi:10.4271/2008-01-0164.

[19] Fatzinger, E., Landerville, J., Bonsall, J., and Simacek, D., "An Analysis of Sport Bike Motorcycle Dynamics during Front Wheel Over-Braking," SAE Technical Paper 2019-01-0426, 2019, doi:10.4271/2019-01-0426.

[20] Peck, L., Deyerl, E., and Rose, N., "The Effect of Tire Pressure on the Deceleration Rate of a Motorcycle Under Application of the Rear Brake Only," *Accident Reconstruction Journal*, 27(4): 19–21, 2017, ISSN: 1057-8153.

[21] Baumann, F.W., Schreier, H., and Simmermacher, D., "Tire Mark Analysis of a Modern Passenger Vehicle with Respect to Tire Variation, Tire Pressure and Chassis Control Systems," SAE Technical Paper 2009-01-0100, 2009, doi:10.4271/2009-01-0100.

[22] Rievaj, V., Vrabel, J., and Hudak, A., "Tire Inflation Pressure Influence on a Vehicle Stopping Distances," *International Journal of Traffic and Transportation Engineering*, 2(2): 9–13, 2013, doi:10.5923/j.ijtte.20130202.01.

[23] Baxter, A. and Robar, N., "An Examination of the Performance of Motorcycle Brake Systems," *Accident Investigation Quarterly*, 47: 28–31, 2007, ISSN 1082-6521.

[24] Mortimer, R., "Braking Performance of Motorcyclists with Integrated Brake Systems," SAE Technical Paper 861384, 1986, doi:10.4271/861384.

[25] Green, D., "A Comparison of Stopping Distance Performance for Motorcycles Equipped with ABS, CBS and Conventional Hydraulic Brake Systems," *USDOT, NHTSA, International Motorcycle Safety Conference*, Long Beach, CA, March 2006.

[26] Anderson, B., Baxter, A., and Robar, N., "Comparison of Motorcycle Braking System Effectiveness," SAE Technical Paper 2010-01-0072, 2010, doi:10.4271/2010-01-0072.

[27] Dinges, J. and Hoover, T., "A Comparison of Motorcycle Braking Performance with and without Anti-Lock Braking on Dry Surfaces," SAE Technical Paper 2018-01-0520, 2018, doi:10.4271/2018-01-0520.

[28] Jansen, H., LeBlanc, B., Wilhelm, C., Shaw, T. et al., "Quantifying Engine Braking for Various Common Street Motorcycles," SAE Technical Paper 2020-01-0880, 2020, doi:10.4271/2020-01-0880.

[29] Bartlett, W., "Estimating Maximum Motorcycle Acceleration Rates," *Collision*, 6(1): 30–37, 2011, ISSN 1934-8681.

[30] Baxter, A.T., *Motorcycle Crash Investigation*, Institute of Police Technology and Management, Jacksonville, FL, 2017, ISBN 978-1-934807-18-7.

[31] Savino, G., Giovannini, F., Piantini, S., Baldanzini, N. et al., "Autonomous Emergency Braking for Cornering Motorcycle," *24th ESV Conference Proceedings*, Paper Number 15-0220, Gothenburg, Sweden, June 8–11, 2015.

[32] Savino, G., Giovannini, F., Fitzharris, M., and Pierini, M., "Inevitable Collision States for Motorcycle-to-Car Collisions," *IEEE Transactions on Intelligent Transportation Systems*, 17(9): 2563–2573, 2016.

[33] Giovannini, F., Savino, G., Pierini, M., and Baldanzini, N., "Analysis of the Minimum Swerving Distance for the Development of a Motorcycle Autonomous Braking System," *Accident Analysis & Prevention*, 59: 170–184, 2013.

4

Cornering, Swerving, and Changing Lanes

A reconstructionist may need to calculate the lean angle that a motorcyclist would achieve when traversing a curve. In some instances, this lean angle can reveal if the rider was operating at the motorcycle's limits or at their own psychological (or willingness) limits. This analysis may reveal factors that contributed to a crash. Similarly, in evaluating a motorcyclist's ability to avoid a crash, a reconstructionist may need to analyze the distance necessary for a motorcyclist to swerve laterally or make a lane change. This chapter introduces equations that can be used to analyze cornering, swerving, and lane change maneuvers.

Analysis of a Motorcycle Traversing a Curve

The lean angle required for a motorcycle traversing a curved path is the angle that brings the overturning moment generated by the tire frictional forces into balance with the opposing moment generated by the tire forces perpendicular to the road surface. The required lean angle increases with increasing speed and decreasing path radius. Fricke [1] and Cossalter [2] report that the lean angle of a motorcycle for a given path and speed can be calculated with the following equation:

$$\theta = tan^{-1}\left(\frac{v_{mc}^2}{g \cdot r}\right) \quad (4.1)$$

In this equation, θ is the lean angle of the motorcycle, v_{mc} is the forward velocity of the motorcycle, g is the gravitational acceleration, and r is the path radius. This equation assumes that: (1) the motorcycle is traveling a steady speed high enough that the rider would have used countersteering to initiate the lean; (2) the rider is leaning at the same angle as the motorcycle; and (3) the part of the tire contacting the road does not change as the motorcycle leans. For a real motorcycle tire, as the motorcycle leans, the portion of the tire contacting the road will change.

Incorporating Roadway Superelevation

Rose et al. [3] derived a form of Equation (4.1) that included the roadway superelevation and yielded the lean angle relative to the road surface. In deriving this equation, they used the nomenclature identified in Figure 4.1, which is a diagram of a motorcyclist traversing a leftward curve with a superelevation. The superelevation angle has been given the symbol ϕ, and θ designates the lean angle of the motorcycle relative to the road surface. Thus, the lean angle of the motorcycle and rider relative to the vertical direction is $\phi + \theta$. In this derivation, counterclockwise rotations are positive, and, therefore, the superelevation and lean angles for a leftward curve are positive. Figure 4.1 also depicts the forces applied to the cornering motorcycle—the combined weight of the motorcycle and rider applied at their effective center of mass (W), the lateral friction force ($F\_friction$) applied at the tire contact patch, and the normal force (F_{normal}) also applied at the tire contact patch.

To remain in equilibrium as it traverses the curve, the resultant of the normal and friction forces must act along the line connecting the tire contact patch to the center of mass of the motorcycle and rider. For this to be the case, the friction and normal forces will be related to the lean angle as defined by the following equation:

$$F_{friction} = F_{normal} \cdot \tan\theta \tag{4.2}$$

Newton's second law dictates that the sum of the forces in the lateral direction is equal to the mass multiplied by the lateral acceleration:

$$\sum F_{lat} = m \cdot a_{lat} \tag{4.3}$$

By examination of Figure 4.1, it can be seen that:

$$\sum F_{lat} = F_{lat} \cdot \sin\phi + F_{friction} \cdot \cos\phi \tag{4.4}$$

Further, for a motorcycle traversing a curved path:

$$a_{lat} = \frac{v_{mc}^2}{r} \tag{4.5}$$

Therefore, Equation (4.3) can be rewritten as follows:

$$F_{normal} \cdot \sin\phi + F_{friction} \cdot \cos\phi = \frac{W}{g} \cdot \frac{v_{mc}^2}{r} \tag{4.6}$$

FIGURE 4.1 Forces applied to a cornering motorcycle.

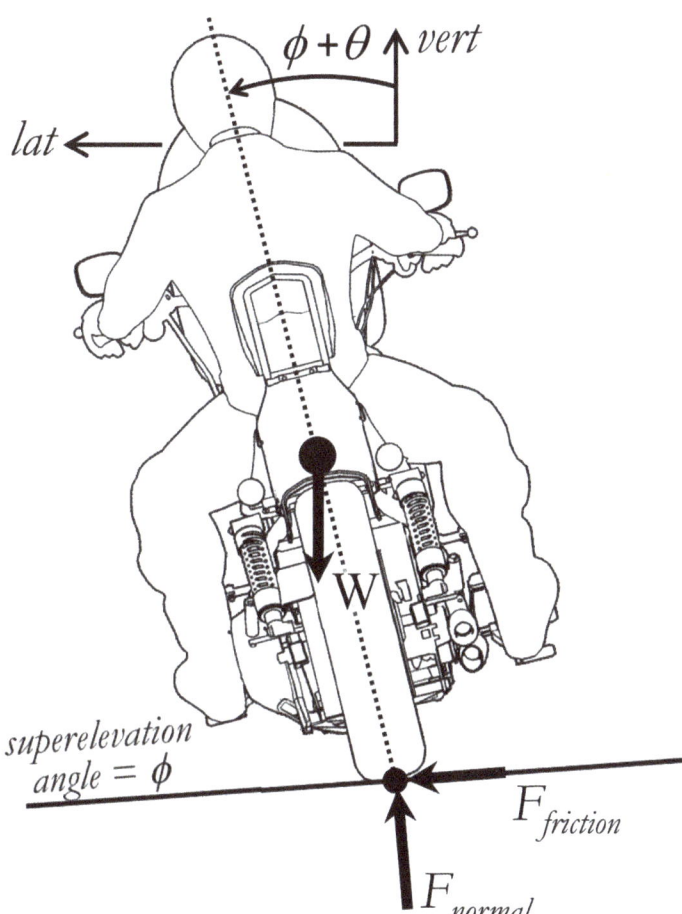

Substituting Equation (4.2) into Equation (4.6) yields:

$$F_{normal} \cdot (\sin\phi + \tan\theta \cos\phi) = \frac{W}{g} \cdot \frac{v_{mc}^2}{r} \qquad (4.7)$$

As it traverses the curve, the motorcycle depicted in Figure 4.1 is in static equilibrium in the vertical direction. Therefore, the sum of the forces in the vertical direction is equal to zero. This can be written as follows:

$$-W + F_{normal} \cos\phi - F_{friction} \sin\phi = 0 \qquad (4.8)$$

Substituting Equation (4.2) into Equation (4.7) and solving for F_{normal} yields:

$$F_{normal} = \frac{W}{\cos\phi - \tan\theta \sin\phi} \tag{4.9}$$

Substituting Equation (4.9) into (4.7) yields the following equation:

$$\frac{1}{g} \cdot \frac{v_{mc}^2}{r} = \frac{\sin\phi + \tan\theta \cos\phi}{\cos\phi - \tan\theta \sin\phi} \tag{4.10}$$

Roadway superelevation typically does not exceed 6° or 7° [4]. Given this, we can employ small-angle assumptions to simplify Equation (4.1). Specifically:

$$\cos\phi \approx 1 \tag{4.11}$$

$$\sin\phi \approx \tan\phi \tag{4.12}$$

Substituting Equations (4.11) and (4.12) into (4.10) yields the following equation:

$$\frac{1}{g} \cdot \frac{v_{mc}^2}{r} = \frac{\tan\phi + \tan\theta}{1 - \tan\phi \tan\theta} \tag{4.13}$$

Using trigonometric identities, it can be show that:

$$\frac{\tan\phi + \tan\theta}{1 - \tan\phi \tan\theta} = \tan(\phi + \theta) \tag{4.14}$$

Therefore:

$$\theta = \tan^{-1}\left(\frac{v_{mc}^2}{g \cdot r}\right) - \phi \tag{4.15}$$

Equation (4.15) shows that the superelevation angle simply reduces the required lean angle relative to the road surface by 1° for each degree of superelevation. From Equations (4.1) and (4.15), the following relationships can be observed. (1) as a motorcyclist's speed around a curve increases, their lean angle will also increase to maintain the same path of travel; (2) superelevation in a curve reduces the magnitude of lean relative to the roadway; and (3) the larger the radius of a curve, the less lean angle required for a motorcyclist to traverse that curve.

Example: Lean Angle Calculation

As an example, consider a motorcycle following a curve with a radius of 300 ft and a superelevation of 5° at a speed of 55 mph. Equation (4.15) would be applied as follows.

Within the equation, the 55-mph speed is multiplied by 1.4667 to convert it to units of ft/s. The calculated lean angle relative to a line normal to the road surface is 29°. The lean angle relative to vertical is 34°.

$$\theta = tan^{-1}\left(\frac{(55 \cdot 1.4667)^2}{32.2 \cdot 300}\right) - 5° = 29°$$

Assumptions

Equations (4.1) and (4.15) were developed with several assumptions. First, these equations assume the motorcycle is traveling a constant velocity of great enough magnitude that the rider would have initiated the lean through countersteering. Second, they assume that the motorcycle and its rider have the same lean angle. This is often an accurate assumption, but clearly a rider has the option of leaning either more than or less than the motorcycle, and reconstructionists should be attentive to situations where it might not apply. As Cossalter [2] has noted:

> *The motorcycle roll angle on a turn is influenced, in a significant way, by the rider's driving style. By leaning with respect to the vehicle, the rider changes the position of his center of gravity with respect to the motorcycle … if the rider remains immobile with respect to the chassis, the center of gravity of the motorcycle-rider system remains in the motorcycle plane … if the rider leans towards the exterior of the turn, the center of gravity is also moved to the exterior of the turn with respect to the motorcycle. As a result, he needs to incline the motorcycle further so that the tires, being more inclined than necessary, operate under less favorable conditions … If the rider leans his torso towards the interior of the turn and at the same time rotates his leg so as to nearly touch the ground with his knee, he manages to reduce the roll angle of the motorcycle plane. When racing, the riders move their entire bodies to the interior of the turn, both to reduce the roll angle of the motorcycle and to better control the vehicle on the turn.*

Along these same lines, when there is a passenger on the motorcycle, the passenger may lean more than or less than the operator, and this influences the trajectory of the motorcycle (see Figure 4.2). As Baxter [5] has noted, "a passenger who is unfamiliar with a motorcycle and will not lean with the operator, or leans in the opposite direction during a turn, makes steering and handling much more difficult. If the passenger leans in the wrong direction, it has a negative effect on the operator's lean and steering inputs, canceling them out, so that the motorcycle continues in a straight line." In crashes involving a passenger on the motorcycle in addition to the operator, this is a factor worth considering.

Finally, Equations (4.1) and (4.15) assume that the motorcycle tires have no width, such that the part of the tire contacting the roadway does not change as the motorcycle leans. In reality, as the motorcycle leans, the portion of the tire contacting the road changes, and the contact patch moves in the direction of the lean. This reduces the effective lean angle of the motorcycle. Thus, because of this assumption, the actual lean angle required for a curved path will be higher than that predicted by Equations (4.1) or (4.15).

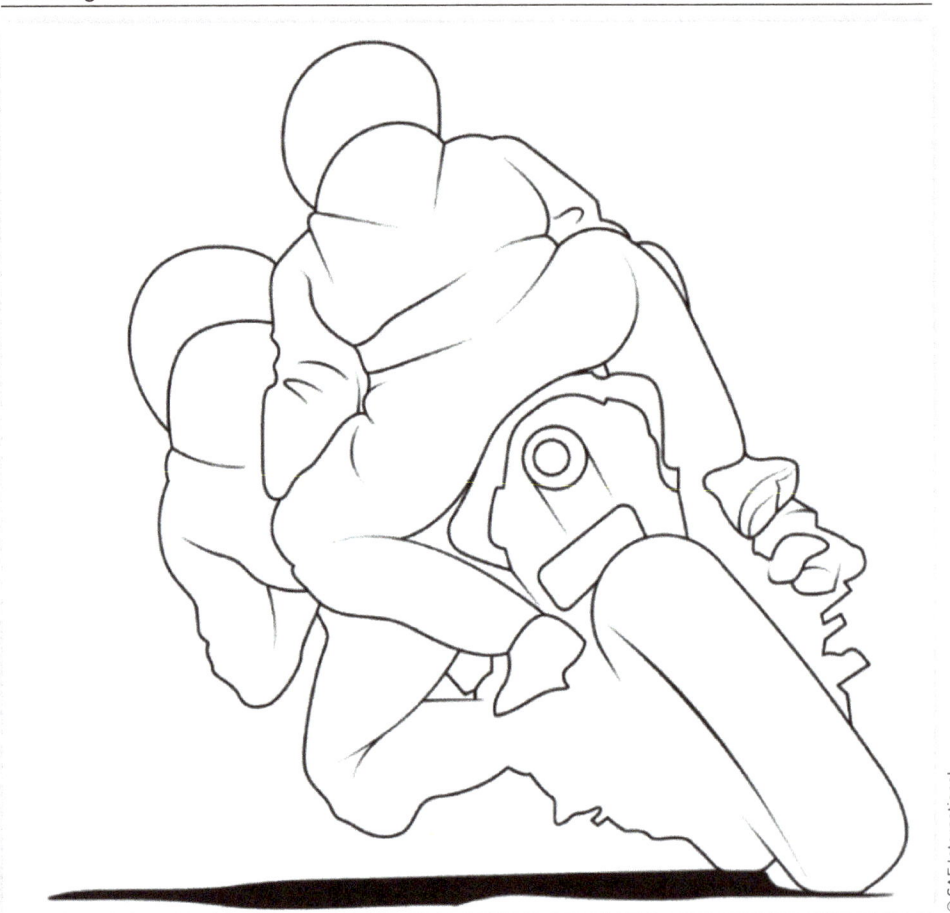

FIGURE 4.2 Motorcycle, operator, and passenger with different lean angles while cornering.

Case Study: Application of the Lean Angle Equations at the Apex of a Curve

To test the accuracy of Equations (4.1) and (4.15), physical testing was conducted with two motorcycles—a 2012 Suzuki DR650SE Enduro and a 2007 Kawasaki VN900-D (see Figure 4.3) [6]. The photographs of Figure 4.4 show cross-sectional views of the rear tires of these motorcycles. Figure 4.5 is an aerial photograph showing the area where the testing was conducted. Because this lot was used for teaching the Motorcycle Safety Foundation (MSF)'s Basic and Advanced Rider Courses, the range was premarked for the various exercises in those courses. This testing utilized the two semicircle paths that form a large oval marked with blue paint, visible near the center of Figure 4.5.

Three riders were utilized for this testing—one novice and two expert riders (these are self-characterizations). The riders were instructed to make turns at different speeds, while trying to maintain a constant velocity during the turns. The riders also tried to keep their body at the same lean angle as the motorcycle. Each rider rode the motorcycle around the

CHAPTER 4 Cornering, Swerving, and Changing Lanes

FIGURE 4.3 Motorcycles used for testing.

FIGURE 4.4 Contact between the Suzuki's tire and the test surface.

FIGURE 4.5 Parking lot and range used for testing.

©2022, SAE International

FIGURE 4.6 Motorcycle and rider at apex of curve during one test.

oval path many times, varying their speed and direction of travel. The path, speed, and lean angle of the motorcycle were continuously documented using a Racelogic VBOX that measured speed, position, and roll angle at 20 Hz. The VBOX system utilized two GPS antennas. A metal crossbar was strapped to the rear of the motorcycle and the GPS sensors were magnetically attached to this crossbar (Figure 4.3 and Figure 4.6).

One sensor was attached at the motorcycle centerline and another near the left extent of the crossbar. The metal crossbar was damped with a piece of Styrofoam placed between the crossbar and the rack on the back of the bike. The VBOX data logger was placed in the left saddlebag of the motorcycle. This testing was also captured with two video cameras recording at 30 fps. Testing was conducted on two separate days. On both days, the range was clear and dry, and the wind was below a level that required the riders to make any adjustment to their riding. The VBOX data from each cornering maneuver were analyzed to determine the path radius of the motorcycle, speed of the motorcycle, and lean angle relative to gravity (rather than relative to the road surface) of the motorcycle. These values were tabulated for the time during each cornering maneuver when the lean angle reached a maximum. The path radius was determined by fitting a curve to the VBOX positional data for each curve. To normalize the results from the testing and plot them all on one graph, the path radius and speed were used to calculate a lateral acceleration for each maneuver.

Figure 4.7 is a graph that depicts the lateral acceleration and resulting lean angle from each cornering maneuver from this testing compared to the lean angle Equation (4.1) would yield. Lateral acceleration in g's is plotted on the horizontal axis, and the lean angle in degrees is plotted on the vertical axis. This is the lean angle of the motorcycle itself because the instrumentation was attached to the motorcycle. Points plotted in gray are

FIGURE 4.7 Test results compared to lean angle equation.

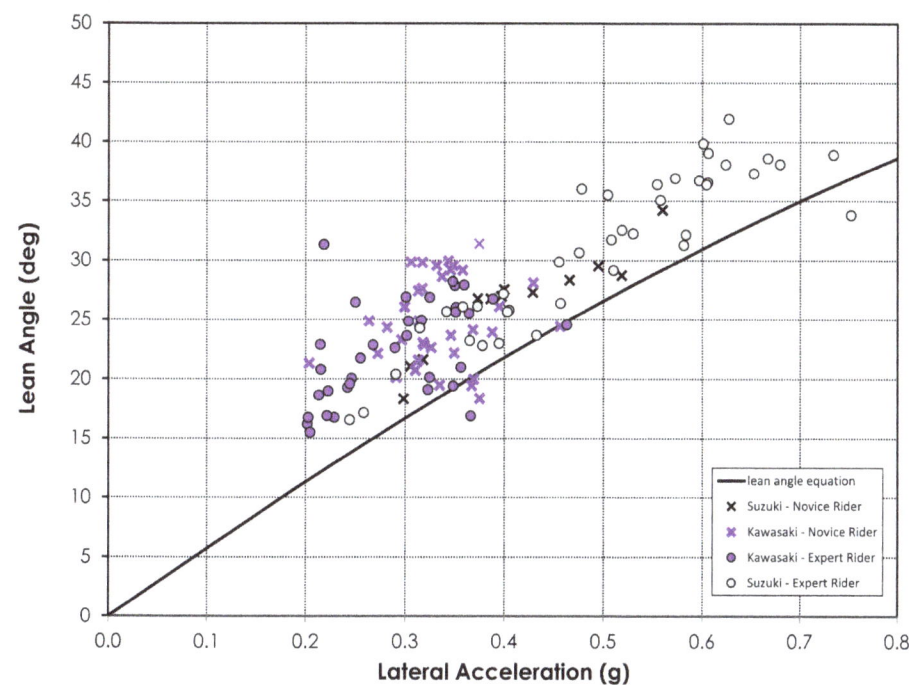

for the Suzuki motorcycle and points plotted in purple are for the Kawasaki motorcycle. Points plotted with an "x" are for the novice rider and points plotted with a circle are for the expert riders.

Several trends emerge in Figure 4.7. First, while Equations (4.1) and (4.15) prescribe a particular lean angle for a given lateral acceleration, the test data show scatter in the lean angle for any given lateral acceleration. Second, the lean angle is nearly always greater than the lean angle predicted by Equations (4.1) or (4.15). Wahba et al. [7] reported similar results. To some degree, this trend is expected because Equations (4.1) and (4.15) neglect the effects of the motorcycle tires having width and any effects from the rider's body positioning.

Cossalter [2] showed that the additional lean angle required because of the tire width could be calculated using Equations (4.16) and (4.17). In these equations, t is the tire width and h is the combined motorcycle and rider center of gravity (CG) height.

$$\theta = \theta_{Equation\,(1)} + \Delta\theta_{tire\ width} \tag{4.16}$$

$$\Delta\theta_{tire\ width} = \sin^{-1}\left(\frac{\frac{t}{2} \times \sin(\theta_{Equation\,(1)})}{h - \frac{t}{2}}\right) \tag{4.17}$$

Figure 4.8 and Figure 4.9 are similar to Figure 4.7. Figure 4.8 plots the test points for the Suzuki motorcycle and Figure 4.9 plots the points for the Kawasaki motorcycle. These figures also plot Equation (4.16) for each motorcycle with a dashed line. In calculating the curve for Equation (4.16), the average of the front and rear tire widths of each motorcycle was used. The CG height of the Suzuki was estimated with Equation (2.3) and with Equation (2.2) for the Kawasaki. Each rider's seated CG height was estimated to be at their navel. Each motorcycle and rider were then weighed, and a combined CG height for the motorcycle and rider was calculated using the formula of Equation (4.18):

$$CG_{combined} = \frac{CG_{mc} \times W_{mc} + CG_{r} \times W_{r}}{W_{total}} \quad (4.18)$$

The results displayed on these graphs demonstrate that incorporating the tire width into the calculation of lean angle reduces the average error in the calculation. However, it did not fully explain the error. There are other factors that likely explain the difference between the theoretical and actual lean angles. This research did not measure or quantify the rider's position during the maneuvers. A discrepancy between the motorcycle and rider lean angles could explain why the motorcycle lean angle was consistently under-predicted by Equations (4.1) and (4.15). If this was the case, it implies that both the novice and expert riders consistently leaned less than they leaned their motorcycles. Review of the testing video revealed that the novice rider did tend to lean less than the motorcycle and had a tendency when riding the Kawasaki to turn the front wheel during low-speed cornering, which reduces the motorcycle lean angle required for a given maneuver. The expert riders had similar tendencies, but to a lesser degree.

Two additional areas deserve comment. First, the Kawasaki motorcycle had a geometric limit of around 30° of lean. Based on Equation (4.16), this geometric limit should have allowed

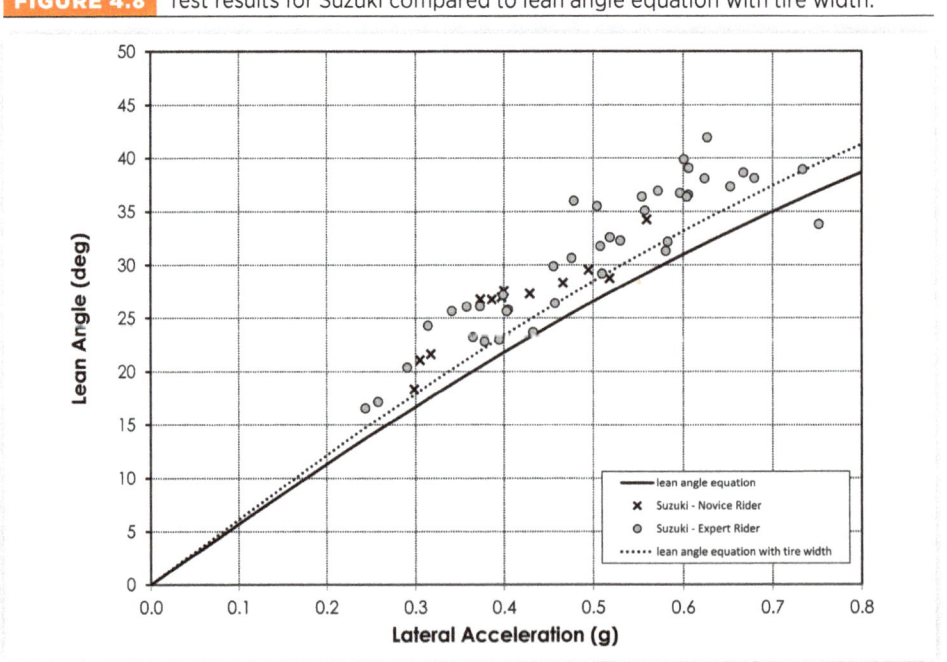

FIGURE 4.8 Test results for Suzuki compared to lean angle equation with tire width.

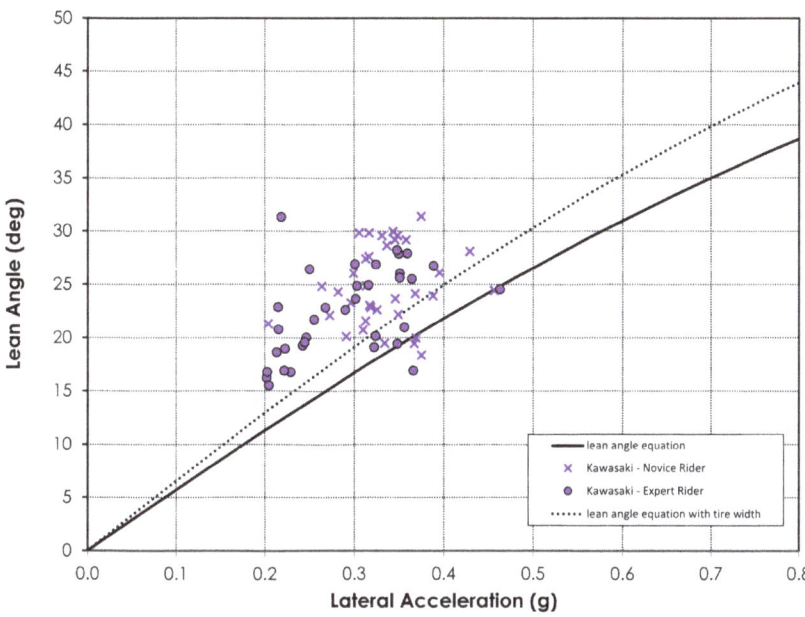

FIGURE 4.9 Test results for Kawasaki compared to lean angle equation with tire width.

the riders to reach a lateral acceleration level of 0.5 g. While the expert rider achieved a lateral acceleration of 0.46 g on this motorcycle and the novice rider achieved 0.45 g, the bulk of the data points for this motorcycle fell below 0.4 g. Both riders reported repeatedly scraping components of this motorcycle during the testing. The Suzuki, on the other hand, had a geometric limit exceeding 40° of lean. Neither rider reported scraping components of the motorcycle during the testing with this motorcycle. The expert rider achieved a lateral acceleration level of 0.75 g with this motorcycle, whereas the novice only achieved 0.56 g. In the test results with the Kawasaki, it is difficult to discern any difference in the lean angle behavior of the novice and expert riders. With the Suzuki, the novice rider, for the most part, kept the lean angle below 30°. The expert rider achieved lean angles with this motorcycle that approached and exceeded 40°. These results indicate that the observed limits for lean angle on the Kawasaki are being driven by the geometric limit defined by component of this motorcycle contacting the road surface.

Case Study: Application of the Lean Angle Equations in the Time Domain

To test the accuracy of Equations (4.1) and (4.15) in the time domain, physical testing was conducted using a 2007 Suzuki GSX-R750 motorcycle (Figure 4.10) [8]. This testing utilized a single experienced rider who was a certified safety instructor through the MSF. The rider traversed the route with the goal of maintaining safety and varying his speed in accordance with the characteristics of the roadway. No special instructions were given to the rider in terms of how he should lean his body relative to the motorcycle. The risks of this ride were discussed with the rider, and he was aware of and consented to these risks.

FIGURE 4.10 Suzuki motorcycle used for testing.

This testing involved the rider driving the motorcycle westbound along County Road 95 between Frazier Park, California, and State Highway 33. Seven curves were identified for analysis on this section of roadway. The rider also drove southbound along Highway 33 and three additional curves were identified for analysis on this section. Three of the ten curves were digitally mapped with a Faro laser scanner. One of these three was also mapped with a total station. The remaining seven were mapped with aerial imagery. Figure 4.11 is an aerial photograph showing the curve that was mapped with both a Faro scanner and a total station. This photograph is oriented such that north is up on the page. Riders traveling southbound through this curve would first traverse from the top left to the top right of this photograph and then exit the curve traveling from right to left across the image. Figure 4.12 is a photograph showing the overall geometry of the curve from a ground level perspective. Riders traveling southbound through this curve would travel toward the viewer of this image.

Throughout the entire route, the path, speed, and lean angle of the motorcycle were continuously recorded using a Racelogic VBOX that measured speed, position, and roll angle at 20 Hz (a similar setup to that used in the previous section). The VBOX system utilized two GPS antennas. A metal crossbar was strapped to the rear of the motorcycle and the GPS sensors were magnetically attached to this crossbar, near its outer extents (Figure 4.13). The motorcycle was scanned with a Faro laser scanner both before and after the ride, so that any displacement of this bar that might occur over the course of the ride could be quantified. The VBOX data logger was carried in the rider's backpack. This testing was also captured with a video camera and a GoPro camera attached to a chase vehicle and with two GoPro cameras attached to the rider (helmet and chest). These cameras were recording at a rate of 30 fps. At the time of the testing, the road surface was dry.

FIGURE 4.11 Aerial photograph showing one of the curves.

FIGURE 4.12 Photograph showing the same curve.

FIGURE 4.13 VBOX sensors attached to motorcycle.

The VBOX data from each of the ten curves were analyzed to determine the actual path radius, speed, and lean angle relative to gravity (rather than relative to the road surface) of the motorcycle. These values were tabulated for the motorcycle's entire traversal of each curve. The path radius was calculated using positional data from the VBOX. When the motorcycle was instrumented, the authors attempted to position the antenna bracket perpendicular to the vertical axis of the motorcycle. The motorcycle was then scanned with a Faro laser scanner in its instrumented state. The scan data were examined, and it was found that the bar had a 2.2° angle relative to the roll axis of the motorcycle. The measured lean angles were adjusted to reflect this offset. Adjustments were also made to the radius to compensate for the primary antenna placement relative to the CG of the motorcycle. The positional data from the VBOX were analyzed to determine the instantaneous radius of the path of the motorcycle at each point along each curve. To eliminate excessive noise because of the sensitivity of the radius calculation to small changes in positional data, points selected for this calculation were selected ½ sec apart. Thus, the "instantaneous" radius was calculated over 1-sec intervals.

Figure 4.14 is a graph that compares the VBOX-measured lean angles (dashed black) to the lean angles calculated with Equation (4.1) (green) and with Equation (4.16) (red) for turn 4. Positive values for lean angle indicate a leftward lean, while negative values indicate a rightward lean. Examination of this graph reveals that Equation (4.1) tended to underestimate the fully developed lean for each curve. Equation (4.16), on the other hand, closely predicted the lean angle throughout the course of this curve.

The results depicted here were typical for the curves tested. Table 4.1 lists the average differences between the measured lean angle and the lean angle calculated with Equation (4.16). This difference was calculated in two ways. First, it was calculated simply as the

FIGURE 4.14 Comparison of measured and calculated lean angles for turn 4.

TABLE 4.1 Average differences between measured lean angle and calculated lean angle for each curve

Turn number	Average difference (deg)	Average of absolute value of difference (deg)
1	−0.25	0.84
2	−0.05	1.56
3	−0.58	0.90
4	0.05	0.73
5	0.10	0.89
6	0.10	0.89
7	−0.02	0.82
8	−0.28	1.18
9	0.32	0.77
10	−0.42	1.14
All Data	−0.10	0.94

average of the point-to-point differences, with consideration for the positive and negative signs associated with left and right turns. Second, it was calculated as the average of the absolute value of the difference. The average difference between the predicted lean angle of Equation (4.16) and the actual lean angle for all the data was −0.10°. Thus, the average error was near zero and the differences that existed were sometimes smaller and sometimes larger than the actual lean angle. The average of the absolute differences between the predicted and actual lean angles for all the data was 0.94°. Equation (4.16) typically predicted the actual lean angle within 3° (sometimes underestimating and sometimes overestimating). The difference between calculated and actual was seldom greater than 5°. This equation also reasonably models the buildup of lean angle through the progression of a curve (in the time domain). Thus, it could be used to compare the rate at which a motorcyclist would need to lean when traversing a curve at various speeds.

Friction-Limited Speed

Three factors limit the speed at which a motorcycle can traverse a curve. The first of these is the limit of the available friction between the motorcycle tires and the roadway. The second is a geometric limit that is defined by the lean angle at which the foot peg of the motorcycle—or some other component—contacts the roadway. The third is the limit imposed by the rider's willingness limits—their willingness to approach either the geometric or friction limits of their motorcycle.

On a curve with no superelevation, the friction limit is reached when the lateral acceleration (in gravitational units) of the motorcycle/rider combination is equal to the coefficient of friction between the roadway and the motorcycle tires. Equation (4.19) expresses this limit.

$$v_{mc,max} = \sqrt{\mu g r} \qquad (4.19)$$

Roadway superelevation decreases both the necessary friction force and the angle a motorcyclist needs to lean to traverse a curve. In determining the limits for a motorcycle or rider on a curve, the superelevation needs to be considered, because most curves are banked in a way that reduces the lean angle the motorcycle and rider will achieve relative to the roadway. To develop a comparable equation for a curve with superelevation, the lateral forces depicted in Figure 4.1 can again be equated to the lateral inertial force, as written in Equation (4.6). When the motorcycle reaches its friction limits:

$$F_{friction} = \mu F_{normal} \qquad (4.20)$$

Substituting Equation (4.20) into (4.6) yields the following equation.

$$F_{normal}(\sin\phi + \mu\cos\phi) = \frac{W}{g} \cdot \frac{v_{mc}^2}{r} \qquad (4.21)$$

Returning now to the sum of forces in the vertical direction, Equation (4.8), and substituting Equation (4.20) into this equation yields:

$$F_{normal} = \frac{W}{\cos\phi - \mu\sin\phi} \qquad (4.22)$$

Substituting Equation (4.22) into (4.21) yields the following equation:

$$\frac{\sin\phi + \mu\cos\phi}{\cos\phi - \mu\sin\phi} = \frac{1}{g} \cdot \frac{v_{mc}^2}{r} \qquad (4.23)$$

©2022, SAE International

Again, employing small angle assumptions, and then solving for the maximum, friction-limited speed in Equation (4.23) results in the following equation:

$$v_{mc,max} = \sqrt{\frac{\mu + \tan\phi}{1 - \mu\tan\phi} gr} \qquad (4.24)$$

Lambourn [9] examined friction coefficients between motorcycle tires and dry asphalt roadways and found peak friction coefficients of 1.2. With this level of friction, a motorcycle traversing a flat, 250-ft radius curve would have a friction-limited speed of 67 mph. On a flat, 500-ft radius curve, the same motorcycle would have a friction-limited speed of 95 mph. If superelevation is present, these speeds would increase. As the following section discusses, many motorcycles do not have the geometric clearances necessary for the lean angle that these speeds would require. Thus, physical limits of a motorcycle are likely to be determined by its maximum lean angle, not the friction limits of its tires.

Geometric Limit on Speed

The second factor that limits the speed of a motorcycle around a curve is the geometric limit that is defined by the lean angle at which motorcycle components other than the tires—a foot peg or crash bar, for instance—contact the roadway. This limit is defined by the geometry of each motorcycle but is generally in the range of 25° to 50° [10]. On a flat, 250-ft curve, a motorcycle that can lean 25° can achieve a speed of 42 mph before components begin to contact the roadway. A motorcycle that can lean 50° can achieve the friction-limited speed of 67 mph. Thus, for many motorcycles, the geometric limit is more limiting than the friction limit.

Suspension loading and compression have a small effect on the geometric limit for a motorcycle on a curve. As the load on the suspension increases, the springs compress more, and the ground clearance of components on the motorcycle decreases. To account for this in evaluating the geometry-limited speed of a motorcycle around a curve, an equation can be developed to calculate the suspension load for a rider operating a motorcycle through a curve at a given speed. Developing this equation begins by parsing the weight of the motorcycle/rider combination into individual components as follows:

$$W = W_{rider} + W_{mc-sprung} + W_{mc-unsprung} \qquad (4.25)$$

In this equation, the weight of the motorcycle has been divided into sprung and unsprung components. The sprung weight of the motorcycle is the portion of the motorcycle weight that is supported by the suspension, and the unsprung weight is the weight of the motorcycle components that support the suspension. The unsprung portion of the motorcycle weight does not contribute to compressing the suspension springs. Thus, when the motorcycle is upright, the force compressing the suspension is as follows:

$$F_{suspension} = W_{rider} + W_{mc-sprung} \qquad (4.26)$$

The force compressing the suspension increases during cornering. To quantify this increase, the lateral component of the force on the suspension can be equated to the inertial force of the rider and sprung weight of the motorcycle, as follows:

©2022, SAE International

$$F_{suspension}\sin(\theta + \phi) = \frac{W_{rider} + W_{mc-sprung}}{g} \cdot \frac{v_{mc}^2}{r} \quad (4.27)$$

Substitution into Equation (4.15) and solving for $F\_suspension$ yields the following equation:

$$F_{suspension} = (W_{rider} + W_{mc-sprung})\frac{1}{\cos(\theta + \phi)} \quad (4.28)$$

Equation (4.28) demonstrates that the force on the suspension of a motorcycle depends on the superelevation of the roadway and the motorcycle lean angle. Because the lean angle depends on the motorcycle speed and the radius of the curve being traversed, so does the force on the suspension, and thus, the compression of the suspension. This means that the maximum lean angle of a motorcycle depends, to some degree, on the speed and curve radius.

To explore and illustrate the significance of these suspension effects, the suspension on a 2003 Harley-Davidson FXD motorcycle was tested, and the ground clearance of the motorcycle under various loading conditions was documented (Figure 4.15). For each loading configuration, a Faro Laser Scanner Focus³ᴰ was used to document the position of the motorcycle undercarriage relative to the test platform and to measure the ground clearance. With the fuel tank approximately five-eighths full, this motorcycle weighed 671 lb. Without a rider, the motorcycle had 5.34 in. of ground clearance, as measured at the sidestand stop.

The following four additional configurations were tested: (1) the motorcycle with a 201-lb rider, including his gear (Figure 4.16); (2) the motorcycle with a 201-lb rider and 51 lb of ballast placed on the fuel tank in front of the rider; (3) the motorcycle with a 201-lb

FIGURE 4.15 2003 Harley-Davidson FXD.

rider and 105 lb of ballast placed on the fuel tank in front of the rider; and (4) the motorcycle with a 201-lb rider and 155 lb of ballast placed on the fuel tank in front of the rider.

Table 4.2 lists the ground clearance for the motorcycle undercarriage for the various loading conditions. With the ground clearance that resulted from loading the suspension of this motorcycle with a rider, this motorcycle would have a maximum lean angle of 33°. With the ground clearance that resulted from loading the suspension of this motorcycle with a rider and 155 lb of ballast, this motorcycle would have a maximum lean angle of 32°. These values show that the suspension load has an effect, though relatively minor, on the maximum lean angle.

TABLE 4.2 Harley-Davidson FXD ground clearance under various loading conditions

Loading condition	Ground clearance (in.)
No Rider	5.34
Rider (201 lb)	4.90
Rider + 51 lb	4.77
Rider + 105 lb	4.67
Rider + 155 lb	4.48

FIGURE 4.16 2003 Harley-Davidson FXD with rider.

The maximum lean angle for a motorcycle can often be reasonably estimated from what is reported in manufacturer specifications, without the need for any physical testing. The maximum lean angle reported in these specifications will typically have been determined according to the procedure described in the Society of Automotive Engineers (SAE) Recommended Practice J1168 [11]. This recommended practice specifies that the front and rear suspension systems on the motorcycle be compressed to 75% of their maximum travel. The motorcycle is then leaned until a component contacts the test surface and the lean angle is measured. Because this procedure specifies the motorcycle suspension being compressed, the resulting value will be a reasonable approximately of the geometric limit in most cases.

Willingness to Lean

The amount of lean required by a rider as they traverse a curve is determined by their speed in the curve. Many riders reach a limit on their willingness to continue to lean their motorcycle before they reach the maximum lean angle of their motorcycle. Watanabe and Yoshida [12] found that the maximum lean angles utilized by novice riders were typically in the range of 15° to 25° and those used by experienced riders were in the range of 34° to 40°. These results imply that the experienced riders in the study by Watanabe and Yoshida would approach the lean angle limits of many motorcycles, whereas novice riders stopped well short of the motorcycle limits. Using the middle values of these ranges, these results further imply that on a flat, 250-ft curve, an experienced rider would be willing to lean far enough to traverse the curve at a speed of 53 mph, whereas a novice rider would only be willing to lean far enough to traverse the curve at a speed of 37 mph.

Thus, speed can contribute to causing a motorcycle crash in a curve even if the motorcyclist's speed has not reached either the friction-limited or geometry-limited speeds. If a motorcyclist enters a curve at a speed that requires them to lean beyond the roll angle they are willing to achieve, then the motorcycle will drift toward the outside of the curve (assuming no other action is taken by the rider) and may cross out of its lane and off the shoulder or into oncoming traffic, depending on whether the subject curve is toward the left or right.

Lane Change, Swerve, and Turn-Away

In some situations, a motorcyclist will attempt a lane change, swerve, or turn-away to avoid a hazard. A *lane change* is defined as a complete lateral move by a vehicle from one lane to another. For a lane change to the left, this would involve an initial rightward countersteer and lean to the left, followed by a subsequent return to an upright orientation, and then a leftward countersteer and lean to the right, followed by a return to an upright position. The term *swerve* refers to a maneuver that is like a lane change in the sense that it involves a complete lateral movement, but it is not referenced to any marking on the roadway and will often be smaller laterally and quicker than a lane change. The term *turn-away* is defined as half of a swerve.

Consider the characteristics of a leftward turn-away with a triangular- or sinusoidal-shaped lateral acceleration profile that occur across the turn-away distance. Initially, the motorcycle will be upright with zero lateral acceleration and heading straight down the road. The rider will press on the left handlebar to countersteer and the motorcycle will lean left. As the motorcycle leans, the lateral acceleration will develop, reaching the peak lateral acceleration halfway through the turn-away. As the lateral acceleration develops, the motorcycle will yaw in a counterclockwise direction. The maximum lean angle will be reached at the same time the maximum lateral acceleration occurs, halfway through the turn-away. Then the rider will press on the right handlebar and the motorcycle will begin returning to an upright position. The lateral acceleration will return to zero. The maximum change in heading of the motorcycle will occur at the end of the turn-away. To complete a lane change or swerve, this same process can then be repeated to the right.

Rice [13] examined the effect of skill level on the technique used by motorcyclists making lane changes. He studied the degree to which riders used countersteering and body lean to initiate and execute these maneuvers. The lane change tests involved a 12-ft lateral move to the left for which the riders were afforded 60 ft in which to complete the maneuver. All the tests utilized a 1974 Honda CB360G (a standard motorcycle weighing just less than 400 lb). Four riders were utilized, and they were instructed to continue increasing their speed through the course until they could no longer successfully complete the required lane change. Rice observed that "the rider has great flexibility in selecting a combination of leaning motions and steer torque applications [countersteering] for successfully performing a simple lane change maneuver. This result alone points strongly to the need for considering motorcycle performance in terms of the rider-vehicle combination—the handling qualities problem—rather than on vehicle dynamics alone (given, of course, reasonable response characteristics in the machine)." Rice noted that the most experienced rider used a large initial countersteer to initiate the lane change. For the least experienced rider, Rice found significant variability from run-to-run in the degree to which the rider utilized countersteering and body lean. One moderately experienced rider tended to emphasize control of the motorcycle through lean of his body rather than through large countersteer inputs.

In his discussion of motorcycle avoidance maneuvers, Limpert [14, 15] suggested a maximum lateral acceleration of 0.65 g for motorcyclists. However, Watanabe and Yoshida [12] found that novice riders utilized maximum lean angles in the range of 15° to 25°. This implies a range of maximum lateral accelerations for these riders in the range of 0.268 to 0.466 g. Thus, novice riders are unlikely to achieve a lateral acceleration of 0.65 g. Watanabe and Yoshida found that experienced riders utilized maximum lean angles in the range of 34° to 40°. This implies a range of maximum lateral accelerations for these riders in the range of 0.675 to 1.0 g. Thus, in evaluating a motorcyclist's ability to avoid a crash by swerving, the rider's experience and skill level may need to be considered. Also, if the rider employs any braking or acceleration during the maneuver, this will consume some of the available friction and this may limit the lateral acceleration the rider could utilize for lateral movement.

Daily et al. [16] presented Equation (4.29) for calculating turn-away distance. In this equation, S is the vehicle speed in mph, $L_{turn-away}$ is the lateral turn-away distance in ft, $a_{lat,avg}$ is the average lateral acceleration in gravitational units, and $d_{turn-away}$ is the longitudinal distance required for the maneuver. Equation (4.30) is an analogous expression for a complete lane change or swerve. In this equation, $L_{lane-change}$ is the lateral movement accomplished by the lane change.

$$d_{turn-away} = 0.366 \cdot S \sqrt{\frac{L_{turn-away}}{a_{lat,avg}}} \tag{4.29}$$

$$d_{lane-change} = 0.732 \cdot S \sqrt{\frac{L_{lane-change}}{2} \cdot \frac{1}{a_{lat,avg}}} \tag{4.30}$$

Bartlett and Meyers [17] reported testing conducted with four experienced motorcyclists on four motorcycles, swerving 2 m (6.5 ft) to their left after passing through a gate at speeds of 40 to 88 km/h (25 to 55 mph). The riders were instructed to swerve as rapidly as safely possible "to cross a line 2 m (6.5 ft) away from the left edge of the approach chute … A group of observers on the sidelines marked the point where the motorcycle's front tire crossed the line." The following motorcycles were tested: a 2010 Kawasaki Ninja 250, a 2005 Yamaha R6, a 2009 Harley-Davidson FLHTPI, and a 1990 Harley-Davidson FXRT. The testing was conducted on a dry roadway with a coefficient of friction, measured with a Ford Crown Victoria police cruiser, of 0.75. The authors reported the speed, distance, and time for each swerve maneuver. They did not measure the lean angles or the lateral accelerations utilized in the maneuvers. Instead, they assumed a lateral acceleration of 0.65 g and used that to refine the empirical coefficient for Equation (4.29). This is a strange assumption, because there must have been variability in the lateral accelerations used by the riders. The data reported by Bartlett and Meyers was reanalyzed here using Equation (4.29). Instead of assuming the same lateral acceleration for each maneuver, Equation (4.29) was used to calculate an apparent lateral acceleration for each test, assuming that the original coefficient for the equation is correct. The resulting lateral accelerations are reported in Table 4.3. These are not measured lateral accelerations, but calculated lateral accelerations assuming the validity of Equation (4.29). That being the case, they should be applied in conjunction with that equation. Based on the average acceleration reported in this table for each rider, there does appear to be dependence on the rider in this data.

Figure 4.17 is a graph that plots the lateral acceleration from these tests against the maneuver speed. Each motorcyclist's runs are depicted with a different point style (open circles, open squares, x's, and filled triangles). No relationship is apparent between speed and lateral acceleration, but rider-to-rider differences are apparent. The rider on motorcycle D (filled triangles), for instance, generated apparent lateral accelerations between 0.46 and 0.72, whereas the rider on motorcycle C (open circles) generated apparent lateral accelerations between 0.22 and 0.40. Because the riders were instructed to "swerve left as rapidly as possible," these differences likely reflect these riders' willingness limits on these motorcycles (which could reflect their familiarity with the motorcycle, the geometry of the motorcycle, their personality, and other factors).

Varat et al. [18] presented data from a series of lane change maneuvers conducted with instrumented motorcycles. They documented the rider control inputs (throttle, brake, and steering force) and the resulting motorcycle response (steering angle, roll angle, and lateral and longitudinal acceleration) with a 100-Hz sampling rate. Instrumentation included an optical speed sensor, a tri-axial accelerometer, a roll angular rate sensor, a steering angle sensor, and a steering torque cell. This testing utilized a range of motorcycles (off-road, sport, and touring). The tests consisted of 53 normal (nonevasive) lane change maneuvers on a public road with straight 12-ft lanes (3.7 m). The roadway was smooth and newly paved with asphalt. This reference only presented data from a single lane change maneuver for each motorcycle. The remainder of the data was not included in the article. When the speeds, lateral movements, and lane change distances reported in this reference are used in conjunction with the lane change form of Equation (4.29), they yield apparent lateral accelerations between 0.05 and 0.16 g. Lateral accelerations of this level indicate mild lane change maneuvers that most riders would have little difficulty completing.

Shuman and Husher [19] reported a series of swerve tests on a level, dry asphalt roadway with a single rider on his own 2004 Honda RC51 sport motorcycle. They conducted tests with lateral offsets of 6.5 ft and 13 ft (2 m and 4 m) at speeds between 25 and 40 mph (40 and 64 km/h). They stated that "the swerve type maneuver is investigated using active,

TABLE 4.3 Lateral acceleration for the Bartlett and Meyers testing calculated with Equation (3.31)

	MC ID & test #	Speed (mph)	Longitudinal turn-away distance (ft)	Lateral acceleration (g)	Average (g)
Police cruiser	A1	29	45.5	0.35	0.38
	A2	31	44.3	0.43	
	A3	40	65.8	0.32	
	A4	39	61.0	0.36	
	A5	51	78.0	0.37	
	A6	50	67.5	0.48	
Sport motorcycle	B1	32	45.5	0.43	0.49
	B2	28	41.5	0.40	
	B3	41	57.3	0.45	
	B4	50	70.3	0.44	
	B5	49	54.7	0.70	
	B6	55	69.2	0.55	
Sport motorcycle	C1	30	50.2	0.31	0.31
	C2	32	52.7	0.32	
	C3	40	64.6	0.33	
	C4	38	56.0	0.40	
	C5	44	77.6	0.28	
	C6	43	85.7	0.22	
Cruiser	D1	29	39.7	0.46	0.59
	D2	28	34.4	0.58	
	D3	38	41.8	0.72	
	D4	35	42.4	0.59	
	D5	43	56.6	0.50	
	D6	46	52.1	0.68	
	D7	54	72.1	0.49	
	D9	53	60.6	0.67	

© SAE International

purposeful countersteering with minimal rider body lean." This statement is contradicted, however, by the images in their report, which show the rider leaning his body. These tests were captured with video and documented with a Racelogic VBOX III 100Hz data acquisition system. Shuman and Husher observed that "the beginning and end points of the swerve have the motorcycle upright and proceeding down the roadway at laterally offset road positions. The total swerve distance involves multiple steering inputs by the rider and a sequence of responses by the motorcycle. This total swerve distance is contrasted with the shorter distance whereby the front and rear tires have successfully moved laterally the desired distance. This wheel clearance distance may be sufficient to clear short roadway hazards even if the total swerve is not yet complete. Because obstacle size varies in the real world, it is helpful to consider both the total distance for the swerve as well as the time for the wheels to clear the desired lateral offset … it takes less time and distance to avoid an object near ground level than to avoid larger obstacles."

©2022, SAE International

FIGURE 4.17 Lateral acceleration versus speed graph for the Bartlett and Meyers turn-away testing.

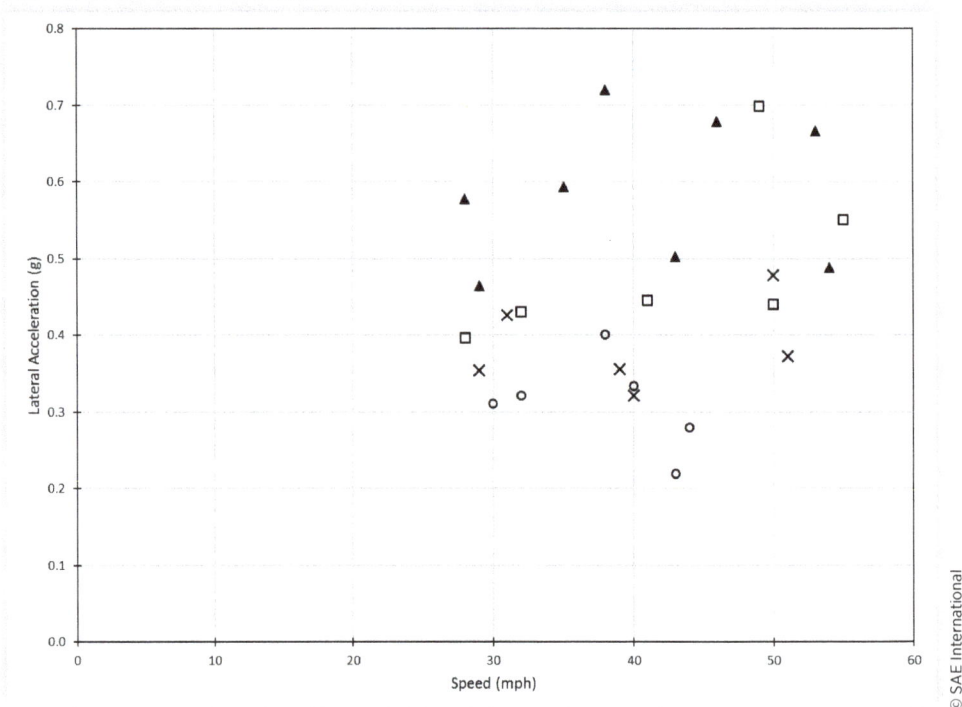

Table 4.4 lists data reported by Shuman and Husher. The lateral accelerations listed were not reported in the paper. The lateral accelerations for the turn-aways were calculated with Equation (4.29). The lateral accelerations for the full swerves were calculated with this same equation, with the exception that a coefficient of 0.732 was used and only half of the full lateral distance was input.

Araszewski et al. [20] presented additional equations for modeling lane changes or swerves. For example, for a sinusoidally developing steer input, the following equation will yield the lane change or swerve distance:

TABLE 4.4 Turn-away and swerve data from Shuman and Husher [19]

	Speed (mph)	Lateral movement (ft)	Longitudinal turn-away distance (ft)	Lateral acceleration (g)	Average (g)	Longitudinal swerve distance (ft)	Lateral acceleration (g)	Average (g)
Sport motorcycle	30	6.5	47	0.35	0.42	69	0.33	0.40
	30	13	56	0.50		78	0.52	
	35	6.5	50	0.43		73	0.40	
	35	13	65	0.50		100	0.43	
	40	6.5	60	0.39		95	0.31	
	40	13	86	0.38		115	0.42	

$$d_{sine} = v\sqrt{2\pi \frac{L_{lane-change}}{a_{lat,peak}}} \qquad (4.31)$$

In this equation, d_{sine} is the longitudinal lane change distance for a sinusoidal steering input in meters, v is the forward velocity of the vehicle during the maneuver (assumed constant) in m/s, $L_{lane-change}$ is the lateral lane change distance in meters, and a_{peak} is the peak lateral acceleration during the maneuver in m/s². The English unit version of Equation (4.31) is as follows:

$$d_{sine} = 0.732 \cdot S \sqrt{\frac{\pi}{2} \cdot \frac{L_{lane-change}}{2} \cdot \frac{1}{a_{lat,peak}}} \qquad (4.32)$$

In Equation (4.32), d_{sine} is the longitudinal lane change distance for a sinusoidal steering input in feet, the speed S is in mph, $L_{lane-change}$ is the lateral lane change distance in feet, and a_{peak} is the peak lateral acceleration during the maneuver in g.

For a steering input that develops in a triangular fashion, the corresponding equation is:

$$d_{tri} = v\sqrt{8\frac{L_{lane-change}}{a_{lat,peak}}} \qquad (4.33)$$

Equation (4.30) is the English unit equivalent to Equation (4.30), with $a_{lat,peak} = 2 \boxtimes a_{lat,avg}$. The time that elapses during the lane change can be calculated as follows:

$$t = \frac{d_{sine}}{v} \qquad (4.34)$$

Or:

$$t = \frac{d_{tri}}{v} \qquad (4.35)$$

Reference [20] also reported the following equations for the maximum heading angle achieved during the lane change:

$$\Phi_{sine,max} = \left(\frac{180}{\pi^2}\right)\frac{a_{lat,peak}t}{v} \qquad (4.36)$$

$$\Phi_{tri,max} = \left(\frac{45}{\pi}\right)\frac{a_{lat,peak}t}{v} \qquad (4.37)$$

In these equations, $\Phi_{sine,max}$ and $\Phi_{tri,max}$ are the maximum heading angles achieved during the lane changes with a sinusoidal or triangular steering input. These heading angles will occur halfway through the lane change. Maximum lean angles for the motorcycle will occur ¼ and ¾ through the lane change maneuvers when the maximum lateral acceleration is achieved. The maximum lean angle can be calculated with the following equation:

$$\theta_{max} = tan^{-1}(a_{lat,peak}) \qquad (4.38)$$

In this equation, θ_{max} is the maximum lean angle and $a_{lat,peak}$ is the peak lateral acceleration in g's. The lane change models presented here assume a symmetric lane change, and so this maximum lean angle will be achieved twice during the lane change or swerve.

Example: Turn-Away and Swerve Calculation

As an example, consider a motorcyclist traveling 45 mph that needs to move 6 ft laterally to avoid debris in their lane. Consider average lateral accelerations of 0.3, 0.4, and 0.5 g. Equation (4.29) would be applied as follows. These are distances for a turn-away.

$$d_{turn-away} = 0.366 \cdot 45 \text{ mph} \sqrt{\frac{6 \text{ ft}}{0.3g}} = 73.7 \text{ ft}$$

$$d_{turn-away} = 0.366 \cdot 45 \text{ mph} \sqrt{\frac{6 \text{ ft}}{0.4g}} = 63.8 \text{ ft}$$

$$d_{turn-away} = 0.366 \cdot 45 \text{ mph} \sqrt{\frac{6 \text{ ft}}{0.5g}} = 57.1 \text{ ft}$$

Once completing these maneuvers, the motorcyclist would still need to return their motorcycle to its original heading. The total distance for the maneuver would be double the values calculated below and the resulting complete maneuver would have a lateral distance of 12 ft associated with it. This would essentially amount to the motorcyclist making a lane change. These calculations would be carried out as follows:

$$d_{lane\ change} = 2 \cdot 0.366 \cdot 45 \text{ mph} \sqrt{\frac{6 \text{ ft}}{0.3g}} = 147.4 \text{ ft}$$

$$d_{lane\ change} = 2 \cdot 0.366 \cdot 45 \text{ mph} \sqrt{\frac{6 \text{ ft}}{0.4g}} = 127.6 \text{ ft}$$

$$d_{turn-away} = 2 \cdot 0.366 \cdot 45 \text{ mph} \sqrt{\frac{6 \text{ ft}}{0.5g}} = 114.2 \text{ ft}$$

If a motorcyclist was going to make the complete maneuver within a lateral distance of 6 ft, the calculations would be carried out as follows:

$$d_{swerve} = 2 \cdot 0.366 \cdot 45 \text{ mph} \sqrt{\frac{3 \text{ ft}}{0.3g}} = 104.2 \text{ ft}$$

$$d_{swerve} = 2 \cdot 0.366 \cdot 45 \text{ mph} \sqrt{\frac{3 \text{ ft}}{0.4g}} = 90.2 \text{ ft}$$

$$d_{swerve} = 2 \cdot 0.366 \cdot 45 \text{ mph} \sqrt{\frac{3 \text{ ft}}{0.5g}} = 80.7 \text{ ft}$$

References

[1] Fricke, L.B., *Traffic Crash Reconstruction*, 2nd ed., Northwestern University Center for Public Safety, Evanston, IL, 2010, ISBN 0-912642-03-3.

[2] Cossalter, V., *Motorcycle Dynamics*, 2nd ed., self-published, 2006, ISBN 978-1-4303-0861-4.

[3] Rose, N.A., Carter, N., and Pentecost, D., "Analysis of Motorcycle and Rider Limits on a Curve," *Collision: The International Compendium for Crash Research*, 9(1): 28–36, ISSN 1934-8681.

[4] *Policy on Geometric Design of Highways and Streets*, 1990 edition, American Association of State Highway and Transportation Officials (AASHTO), Washington, DC, ISBN 978-1560510017.

[5] Baxter, A.T., *Motorcycle Crash Investigation*, Institute of Police Technology and Management, Jacksonville, FL, 2017, ISBN 978-1-934807-18-7.

[6] Carter, N., Rose, N.A., and Pentecost, D., "Validation of Equations for Motorcycle and Rider Lean on a Curve," *SAE International Journal of Transportation Safety*, 3(2): 126–135, 2015, doi:10.4271/2015-01-1422.

[7] Wahba, R., Timbario, T., Nelson, J., Bayan, F. et al., "Motorcycle Lean Angle Variation around a Constant Radius Curve at Differing Speeds and Travel Paths with an Evaluation of Data Measurement Systems," SAE Technical Paper 2019-01-0437, 2019, doi:10.4271/2019-01-0437.

[8] Rose, N.A., Carter, N., and Smith, C., "Further Validation of Equations for Motorcycle Lean on a Curve," SAE Technical Paper Number 2018-01-0529, 2018, doi:10.4271/2018-01-0529.

[9] Lambourn, R.F. and Wesley, A., "Motorcycle Tire/Roadway Friction," SAE Technical Paper 2010-01-0054, 2010, doi:10.4271/2010-01-0054.

[10] Bartlett, W., "Lean Angle Selection by Motorcycle Riders," *Accident Reconstruction Journal*, 21(2): 29–30, 2011, ISSN 1057-8153.

[11] "Motorcycle Bank Angle Measurement Procedure," SAE Surface Vehicle Recommended Practice, J1168, March 2012.

[12] Watanabe, Y. and Yoshida, K. "Motorcycle Handling and Performance for Obstacle Avoidance," *International Congress on Auto Safety*, San Francisco, CA, July 1973.

[13] Rice, R., "Rider Skill Influences on Motorcycle Maneuvering," SAE Technical Paper 780312, 1978, doi:10.4271/780312.

[14] Limpert, R., *Motor Vehicle Accident Reconstruction Cause and Analysis*, 4th ed., Lexis Nexis, Michie, VI, 1994, ISBN 1-55834-207-9:679.

[15] Limpert, R., *Motor Vehicle Accident Reconstruction Cause and Analysis*, 7th ed., LexisNexis Matthew Bender, MA, 2013, ISBN 978-0-7698-5811-1:36-6.

[16] Daily, J., Shigemura, N., and Daily, J., *Fundamentals of Traffic Crash Reconstruction*, Institute of Police Technology and Management, FL, 2006, ISBN 1-884566-63-4:476-479.

[17] Bartlett, W. and Meyers, D., "Time and Distance Required for a Motorcycle to Turn Away from an Obstacle," SAE Technical Paper 2014-01-0478, 2014, doi:10.4271/2014-01-0478.

[18] Varat, M.S., Husher, S.E., Shuman, K.F., and Kerkhoff, J.F., "Rider Inputs and Powered Two Wheeler Response for Pre-Crash Maneuvers," *Safety, Environment, Future V: Proceedings of the 5th International Motorcycle Conference*, Essen, Germany, 2004.

[19] Shuman, K.F. and Husher, S.E., "Do I Brake or Do I Swerve – Motorcycle Crash Avoidance Maneuvering," *International Motorcycle Safety Conference*, Long Beach, CA, 2006.

[20] Araszewski, M., Toor, A., Overgaard, R., and Johal, R., "Lane Change Maneuver Modeling for Accident Reconstruction Applications," SAE Technical Paper 2002-01-0817, 2002, doi:10.4271/2002-02-0817.

5

Physical Evidence from Motorcycle Crashes

This chapter describes physical evidence that may be present on the ground, on the motorcycle, or on the struck or striking vehicle following a motorcycle crash. The first section focuses on evidence deposited at the scene, the second on damage sustained by the motorcycle, and the third on damage sustained by a struck vehicle. After these sections, this chapter covers techniques that can be utilized to document, measure, and diagram this evidence.

Scene Evidence

Scene evidence from motorcycle crashes often includes roadway markings such as tire marks, gouges, scrapes, fluid deposits, and debris. Additional evidence may be present on other scene objects that were struck, such as trees, poles, and guardrails. This evidence will typically need to be documented and preserved for later use in the reconstruction. This documentation process includes three phases: identification, documentation, and quantification. Identification involves recognizing the presence of a piece of evidence, either at the crash site or in photographs or video, and then determining if the evidence is from the subject crash. In the early stages of evidence preservation, it may not be apparent if a piece of evidence is from the subject crash, and it is typically better to capture the evidence and to rule it in or out during later analysis.

During the documentation phase, the investigator can record the evidence with still photography and video. The use of unmanned aerial vehicles (UAVs) or elevated cameras is becoming more common and these can provide perspectives that are useful in later analysis. Nearby bridges or hillsides can also be used to gain these elevated vantage points for taking photographs or video. During the quantification phase, the evidence is measured and located within its surroundings using tape measures, a measurement wheel, a total station, a laser scanner, or through photogrammetry using photographs or video captured with a UAV [1]. Evidence documentation typically culminates in creating a scaled evidence diagram that can be used for analysis of the vehicle motion. A sample evidence diagram for a single vehicle motorcycle crash is included in Figure 5.1. This diagram includes tire marks, scrapes, gouges, fluid, and police paint. This evidence diagram was utilized, along with the damage to the motorcycle, to determine the motion of the motorcycle as it capsized, impacted a guardrail system, and slid along the ground. The reconstructed motion for this motorcycle is shown in Figure 5.2.

FIGURE 5.1 Sample evidence diagram.

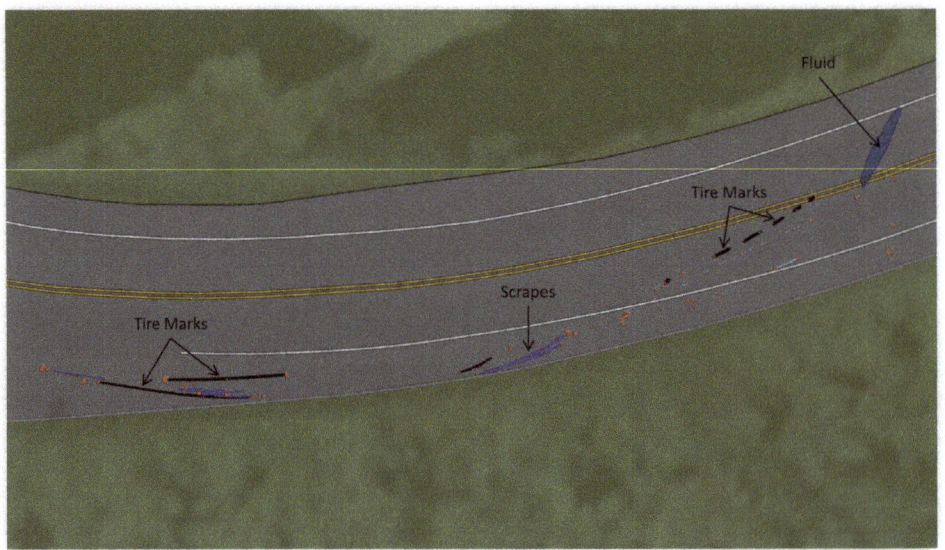

FIGURE 5.2 Evidence diagram with motorcycle positions.

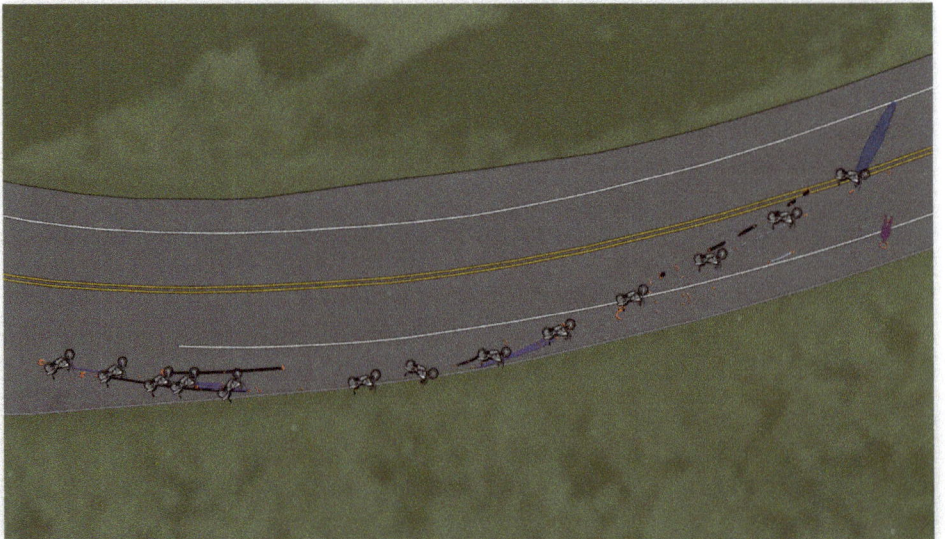

Skid Marks

A skid mark is a tire mark deposited by a tire that is locked because of heavy braking. Two questions that arise related to a single skid mark deposited by a motorcycle are: (1) Was the mark deposited by the front or rear tire? and (2) Was the rider utilizing both brakes or only one?

Physical inspection of the crash-involved motorcycle can be useful in determining if a skid is from the front or rear tire because skidding will often result in a flat spot, abrasions, or scuffing on the skidding tire. In addition, the length of a tire mark can sometimes reveal if it was deposited by the front or rear tire. Locking the front wheel of a motorcycle will quickly lead to capsizing. On the other hand, a motorcycle is inherently more stable with the rear wheel locked, and so a skid mark deposited by the rear tire can be significantly longer than one deposited by the front tire. Of course, a rider could lock the rear wheel and then release the rear brake. Thus, a tire mark deposited by a rear tire can be either short or long, whereas a skid mark deposited by a front tire will always be short. Fatzinger et al. [2] reported a study of motorcycle falls that result from front-wheel overbraking. They completed 13 tests, with initial speeds ranging from 50 to 60 mph. Front-wheel lockup was achieved immediately after brake application in eight of the tests and capsizing resulted. Based on these tests, Fatzinger et al. reported that the motorcycles deposited front-wheel skid marks that were between 26 and 57-ft long. Rear-wheel tire marks can sometimes reach lengths of 100 to 150 ft. Finally, the motion of the motorcycle and the rider after a skid mark is deposited can sometimes reveal if a skid mark was deposited by the front or rear wheel. A rear-wheel skid that leads to a low-side fall will often lead to the motorcycle and rider sliding along the same path, but with the motorcycle leading. On the other hand, a fall that results from front-wheel lock will often lead to the motorcycle and rider sliding along adjacent paths.

Fatzinger et al. reported that, for the motorcycles they tested, when front-wheel lockup occurred, application of the rear brake added little to the deceleration achieved by the motorcycle. This is consistent with the fact that, during heavy braking, a significant portion of the weight of the motorcycle and rider shifts to the front wheel of the motorcycle. This results in the available traction at the front wheel being significantly greater than that at the rear wheel. If the reconstructionist determines that a single motorcycle skid mark was deposited by the front tire, then it may not be essential to determine the degree to which the rear brake was in use. In such a case, the deceleration used in the speed calculations would usually not differ significantly based on the presence or absence of rear brake application.

On the other hand, if a single motorcycle skid mark is determined to be from the rear tire, the determination of whether the front brake was also utilized can make a significant difference in the speed calculations. Unfortunately, the mere presence of a skid mark from the rear wheel does not reveal the degree to which the rider was employing the front brake. But perhaps some characteristics of the rear tire mark could reveal something about front brake use. For example, Bartlett [3, 4] has observed that "there are many people who aver that one can determine if the front brake was in use by examining a rear wheel skid mark for a 'lazy-S' appearance. They claim that a straight rear skid is a clear indication of front brake use." Fricke [5] falls in this camp, stating that "the use of the front brake (but not locking the wheel) provides greater braking stability, so a straight skid is more likely to have been produced using both the front and rear brakes, while a weaved/hooked skid is an indication of the rear brake being used alone." Similarly, Baxter [6] states that "if only the rear brake was applied to the point of locking the wheel, the skid mark left by the rear tire will be a long lazy S curve, trailing off in the direction of the road slope … the skid mark left during front and rear brake application … will be straight …."

Bartlett noted that "when a rider applies the rear brake only, a rearward-directed force on the tire at the contact patch acts to slow the motorcycle. This condition is inherently stable, and the rear wheel will tend to track straight. If the rear wheel is locked, it will still follow the front wheel while on level pavement (tire-roadway forces act to align the vehicle), but if the roadway is sloped, the rear wheel will tend to drift downgrade in response to gravity and the wheel's significantly reduced lateral force generation capability. By applying torque to the handlebars, the rider generates lateral forces which act on the chassis, effectively pushing the chassis to one side. A series of reversing force applications can cause the rear wheel of the motorcycle to 'weave' back-and-forth … Thus, given a locked rear wheel with no front braking, though the system is stable, lateral motions of the rear wheel can result from roadway geometry as well as from rider inputs. These motions will not necessarily occur, but they can."

Further, "if a rider applies significant front brake while the rear wheel is locked, the system may become inherently unstable. As the longitudinal force at the front contact patch exceeds the longitudinal force of the locked rear wheel, the rear wheel will try to swap places with the front. While this is easily done in a car, a motorcycle will fall over if this situation is not countered by a corrective action by the operator. The dynamic forward load transfer resulting from the increased braking action will lighten the load on the rear tire, further reducing its lateral and longitudinal force capacity. Thus, in considering only the external forces acting on the motorcycle, the tendency of the rear wheel to track in a straight line directly behind the front wheel is lowest when the front brake is in heavy use and the rear wheel is locked. This is essentially the opposite of the commonly observed nature of motorcycle skidmarks … this suggests that forces internal to the rider/motorcycle system provide a dominant influence on the shape of the skidmarks."

In addition, Bartlett conducted testing related to this issue. He concluded that "a single long straight skid mark may well be caused by a rear wheel skid generated while the front brake was used (particularly if the motorcycle was still traveling at some speed at the end of the mark), but one can NOT be confident to the level of 'more likely than not', let alone 'beyond a reasonable doubt' that this is true. The 'lazy-S' shape clearly indicates a rear brake skid mark, but without additional supporting information does not indicate 'to a reasonable degree of certainty' that the front brake was not used. Similarly, a serpentine mark can be generated with front brake use, and is dependent on the operator's 'body language'. Without testing the actual tires in use on a particular bike, or in some cases examining the tires themselves prior to their being driven on much, it may not be possible to confidently identify which tire made a particular short mark, or the vehicle's braking condition. With representative exemplar marks made under known conditions (rear only and front with rear) and using the actual tires in use on the accident unit, it would probably be possible to identify the braking condition at the time of the accident."

Dunn et al. [7] reported braking tests and tire marks for three motorcycles—a 1995 BMW R1100RS (sport-touring with antilock brakes), a 2003 Buell XB9R (sport), and a 2005 Harley-Davidson XL 1200 Sportster Custom (cruising/touring). They tested three different braking strategies—best effort front braking only, best effort rear braking only, and best effort front and rear combined braking. Initial speeds for the testing were nominally 25, 45, and 60 mph and most of the tests were conducted on a flat, dry asphalt surface. One set of tests on wet asphalt was run with the BMW, a motorcycle equipped with antilock brakes. Skid marks documented in this study are consistent with Bartlett's findings, and Dunn et al. concluded "it is difficult to determine which brakes the rider used from only the observation of braking marks."

The skid marks generated in the testing by Peck et al. [8] are also consistent with Bartlett's findings. Figure 5.3 shows rear-wheel skid marks from all six of their braking tests. These tire marks vary in length between 52.5 and 59.5 ft. At least three of them are

FIGURE 5.3 Skid marks from six rear-brake only tests (photo by Eric Deyerl).

Photo courtesy of Eric Deyerl

straight, although no front-wheel braking was employed while they were being deposited. Thus, the literature does not support using the shape of a rear-wheel skid mark to discern the level of front-wheel braking employed by the motorcycle operator.

On a somewhat related note: Hurt et al. [9] concluded that a common braking error by motorcyclists was to use the rear brake only during a crash avoidance attempt. These authors stated: "This failure to use the front brake is a critical element in collision avoidance because proper use of the front brake would have avoided many of the collisions or greatly reduced the severity."

In a more recent study, Lenkeit et al. [10] utilized a motorcycle riding simulator with conventional brakes to study the braking responses of nonexpert riders in emergency braking situations. This testing utilized sport-touring and cruiser motorcycle configurations. The 68 subjects were tested on the type of motorcycle they would typically ride. The subjects were exposed to situations requiring a range of deceleration, from normal slowing to emergency braking. The emergency braking behavior of the subjects was tested with two scenarios involving an opposing vehicle "moving rapidly into the rider-subject's lane" from the right, requiring a braking response to avoid a collision. The authors noted that "a previous 1981 NHTSA motorcycle study indicated that as many as 83% of US motorcycle riders involved in crashes did not use their front brakes prior to the crash." However, Lenkeit et al. reported that "over the range of scenarios there were some cases of a rider-subject using just the rear brake, though many more where just the front brake was used. There were a few riders who, essentially, never used the rear brake. When the focus was narrowed to the two emergency situations, there were no cases where the rider-subject used just the rear brake, though again, some riders used only the front brake."

©2022, SAE International

Gouges, Scrapes, Scuffs, and Tire Marks

When a motorcycle falls to the ground and slides or tumbles to rest, components of the motorcycle or the rider's gear will gouge, scrape, or scuff the road surface. Scuff marks can be deposited along with material from the motorcycle tires, components, or the rider's clothing and gear. An example of scrapes and tire marks from a motorcycle on a roadway is shown in the photographs of Figure 5.4. Another example of scrapes on the road surface from a motorcycle is shown in Figure 5.5. A third example, this one of both scrapes and tire marks on the roadway from a motorcycle, is depicted in Figure 5.6. During scene documentation, gouges, scrapes, tire marks, and material transfer on the road should be documented. If there is a need to understand what component of the bike or rider gear deposited specific evidence, then further documentation of the characteristics of the scrapes or gouges can be performed. These will be integral to determining the distance over which the motorcycle slid or tumbled to rest and determining what components made which marks. Also, for single-vehicle loss of control motorcycle crashes, the distance between the end of tire marks and the first gouging may reveal whether the motorcycle experienced a low-side or a high-side fall. Thus, the location of scrapes and gouges and the distance between them can have significance to the reconstruction. Also, the changing orientation of the scratches or gouges over the course of the path can help the reconstructionists determine the specific motion of the motorcycle along its path, particularly if specific motorcycle components can be related to specific scrapes or gouges.

As McNally [11] observed, some motorcycles have "a variety of protrusions that tend to interact with the roadway surface to gouge or scrape while the motorcycle is sliding. There is only limited test information available at this time for sport motorcycles, which tend to slide more easily across the roadway due to the sleek bodywork that prevents the

FIGURE 5.4 Scrapes and tire marks from a motorcycle sliding on the roadway.

CHAPTER 5 Physical Evidence from Motorcycle Crashes 105

FIGURE 5.5 Scrapes from a motorcycle sliding on a roadway.

FIGURE 5.6 Tire marks and scrapes on a roadway from a motorcycle.

deep gouging on other bikes." This implies that the motorcycle type and the extent of scraping and gouging may be useful in determining a reasonable range of decelerations for the motorcycle during the sliding and tumbling on the ground. McNally continued: "Application of any particular coefficient of friction in a real-world collision should be done with consideration for the manner in which the motorcycle traveled across the roadway surface and the degree of scraping or gouging that occurred during the slide. In general, the greater the degree of roadway gouging and scraping, the higher one would expect to be the coefficient of friction." This points to the importance of distinguishing between scrapes and gouges when documenting scene evidence.

Site Inspection Checklist

The evidence that is documented during a site inspection depends, in part, on what issues are being analyzing and on how much time has elapsed between the crash and the site inspection. The list that follows assumes a site inspection that is taking place before any evidence has significantly deteriorated. The items on this list are suggestions and are not intended to be a complete list.

- **Take overall photographs of the crash site.** Typically, this involves starting at the beginning of the physical evidence and walking along the full trajectories of the involved vehicles, taking photographs along the way. It can also be useful to walk and photograph the evidence from the end to the beginning. Aerial photography from a UAV can also be useful for overall documentation of a crash site.
- **Map the physical evidence from the crash.** Mapping will typically be accomplished with a total station, a laser scanner, or through image-based scanning or photogrammetry.
  - Tire marks
  - Gouges
  - Scrapes
  - Fluid deposits
  - Debris
- **Map the site geometry.**
  - Roadway striping
  - Terrain
  - Signage
- **Document site characteristics.**
  - Speed Limit
  - Latitude and Longitude
  - Signage
  - Note if any surveillance cameras are present.
  - Document the lighting conditions
  - Document changes to the site between the crash and the inspection

Evidence on the Motorcycle

Following a crash, the physical evidence on the motorcycle can include components that are scraped, scratched, and abraded, punctures or deformation to the fuel tank, wheel deformation, fork deformation, flat spots or abrasions on tires from skidding (Figure 5.7), broken fairings, and broken forward and signal lighting. During an inspection of the motorcycle, this evidence can be documented and preserved for later use in the reconstruction. Other information about the motorcycle can also be obtained during the inspection. For example, what gear the motorcycle is in can be documented (of course, keeping in mind that this may not be the gear the motorcycle was in immediately following the crash). The braking system components can be documented. The presence or absence of antilock brakes can be documented. The presence or absence of integration or linking between the front and rear brakes can be documented.[1]

When an upright motorcycle impacts another vehicle, there are characteristic damage patterns that will often be evident on the motorcycle. For instance, the front wheel may be deformed in the area where it first contacted the vehicle, the forks may be bent rearward and sometimes broken, the wheel may be deformed rearward into the frame or engine components behind it, and the brake rotor(s) may be deformed. This deformation can be photographically documented and measured during an inspection or with photogrammetric analysis. The photographs of Figure 5.8 show examples of front wheel and fork deformation from motorcycle collisions with other vehicles. These photographs, which were provided by Lou Peck, are from the motorcycle collisions conducted at the World Reconstruction Exposition (WREX) held in 2016.

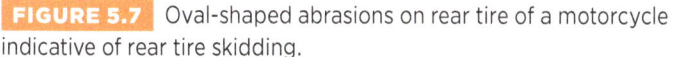

FIGURE 5.7 Oval-shaped abrasions on rear tire of a motorcycle indicative of rear tire skidding.

© SAE International

[1.] As Baxter and Robar [12] have noted: "all motorcycles that are equipped with [an integrated braking] system require the front wheel be equipped with dual disc brake calipers and rotors. The fastest way to check for this system is to follow the brake lines from each caliper. If the left and right brake hoses meet in a t-connection under the headlight area, this is a good indicator that only the front lever operates just the front brakes. Another method is to slip a piece of paper between the brake pad and face of the rotor. An assistant applies the rear brake pedal while an attempt is made to pull out the sheet of paper. If it's integrated, the front brake pad will hold the sheet in place against the rotor face."

©2022, SAE International

FIGURE 5.8 Motorcycle front wheel and fork damage and deformation, WREX 2016 collisions.

Photo courtesy of Louis Peck

One issue that often comes up is whether the motorcyclist was braking prior to the collision. Tire marks prior to the collision will, of course, provide certainty that there was braking. But under optimal braking, the motorcycle will not deposit skid marks, and so the analyst is left to identify other pieces of evidence that may indicate braking. Smith et al. [13] and others have pointed out another type of evidence that may reveal preimpact braking, noting that "because the result of a hard brake application is the transfer of weight from the rear wheel to the front wheel, this information can assist a crash investigator with understanding whether the rider was applying the brakes prior to the impact. When weight transfers to the front wheel, the front springs compress and the fork tubes slide into the shock housing. If the bend of the fork tubes occurs at the entrance to the housing, the investigator can then determine the extent to which the front wheel was loaded at the moment of impact. As an extension to this observation and based upon this test and the large number of additional motorcycle-to-car impact tests conducted by the authors, there is little or no fork compression that results from the impact itself. Therefore, evidence of fork compression observed on the motorcycle post-impact is very likely due to pre-impact weight transfer caused by braking."

When the components of a motorcycle slide along the ground scratch marks may be created on these components. Oftentimes, there are multiple families of scratches that overlap with different orientations. The orientations and order of these scratches can be documented during a postcrash motorcycle inspection. Assuming the motorcycle is not tumbling during the sliding phase of the crash, the scratches will be parallel to and opposite in direction of the ground plane velocity of the motorcycle [14]. These scrapes can be used to determine the yaw orientation of the motorcycle at various times during the sliding phase. The photographs of Figure 5.9 show examples of damage to one motorcycle from sliding on the ground. The photograph on the left shows abrasions on the fuel tank and the photograph on the right shows abrasions to the crash bar. Figure 5.10 shows another example of abrasions on the fairings of a sports motorcycle from sliding on the roadway. This motorcycle also exhibited lateral abrasions on the tires from their interaction with the roadway during the slide.

CHAPTER 5 Physical Evidence from Motorcycle Crashes 109

FIGURE 5.9 Damage and abrasions from the motorcycle sliding on the ground.

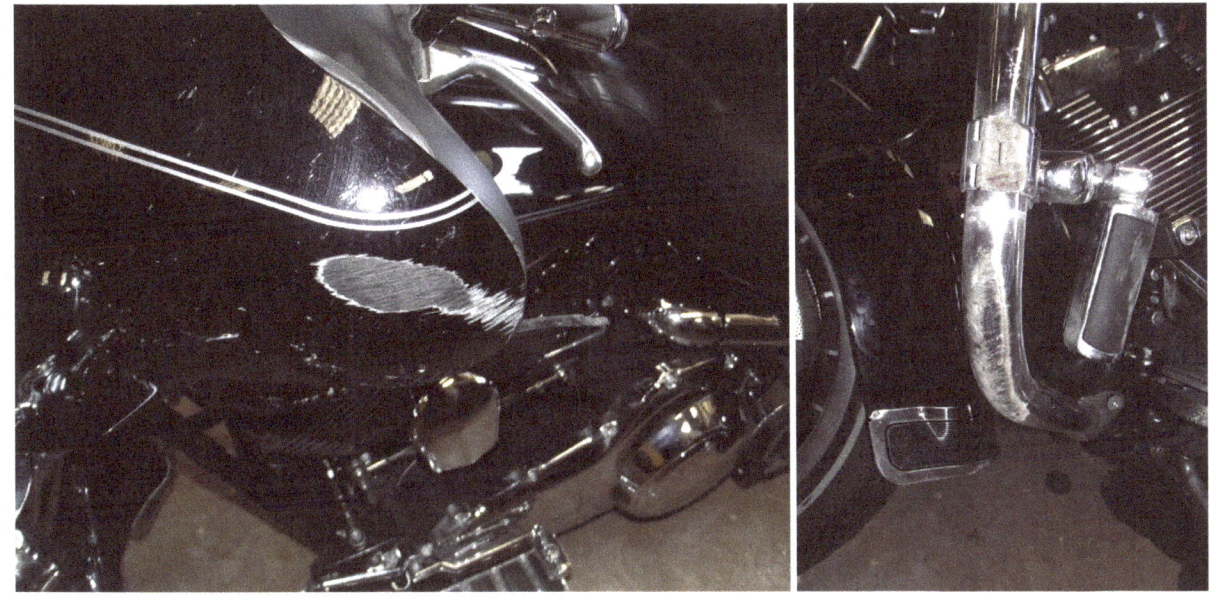

FIGURE 5.10 Damage and abrasions from the motorcycle sliding on the roadway.

It can also be useful to inspect a motorcycle rider's helmet and gear (jacket, pants, gloves, and boots). Impact damage, scuffing and abrasions, or material transfer to the helmet or gear can illuminate what side the motorcycle and rider capsized onto or how the rider was positioned at the time of an impact with a vehicle or the ground.

Motorcycle Inspection Checklist

Following is a list of items that can be included as part of a postcrash inspection of a motorcycle. This list may be more or less extensive than what is necessary for a reconstruction, depending on the issues involved. In cases requiring more extensive inspection, refer to the article by Kubly and Buse [15], which offers detailed guidelines for documenting damage to or failures of specific motorcycle components. You will encounter variability in how crash-involved motorcycles are stored and the degree to which they are damaged. Ideally, you will be able to use motorcycle front- and rear-wheel stands or a center stand or jack to position the motorcycle vertically and achieve stability. This will also position the tires of the motorcycle up off the ground for inspection. This will not always be possible, and often it will be sufficient to position the motorcycle on its stand.

- Position the motorcycle in an area with good lighting.
- Take overall photographs of the motorcycle. Typically, this can consist of eight photographs—one taken from each side of the vehicle and one from each corner of the vehicle. This could be expanded to 16 photographs by taking eight photographs around the motorcycle at two different vertical levels.
- Document the vehicle identification number (VIN) of the motorcycle. The VIN is often located on the steering head or on the frame downtube.
- If it is possible to power up the motorcycle, document the mileage, the fuel level, and the reading of the gear indicator, if applicable.
- Document the tires and wheels of the motorcycle.
  - Write down or photograph the tire sidewall information. This will include the manufacturer, type, size, and serial number.
  - Measure the tread depth of each tire.
  - Measure the tire pressure of each tire. Damage to a crash-involved motorcycle may prevent the wheels and tires from rolling, and so it can be useful to have a tire pressure gauge with a flexible hose to avoid or work around geometric interference with other components of the motorcycle.
  - Photograph damage to the wheels.
  - Document flat spots, scuffing, or abrasions from heavy braking and other damage to the tires.

Depending on how the motorcycle has been stored, you may need to clean dust or dirt off of the tire to see or feel damage to the tire.

It can help to put the motorcycle up on stands to inspect the tires.

For the rear tire, you will need to either put the motorcycle in neutral or apply the clutch lever. If you put the motorcycle in neutral, determine what gear it is in during the process of shifting in to neutral. It may not always be possible to put the motorcycle in neutral or to depress the clutch. In these instances, you can clean the tire and then use your bare hands to feel for damage around the perimeter of the tire.

Even if you can rotate the tire, it can help to run your hand along the surface of the tire while rotating the wheel. You will likely be able to feel any flat spots, scuffing, or abrasions that are present on the tire.

Depending on the lighting at the inspection location, a flashlight can also help for seeing and identifying tire damage.

- Document the throttle and braking system.
  - Is there damage to the throttle? If you twist the throttle and release it, does it spring back?
  - Identify and document components (or the lack of components) for antilock brakes.
  - Determine whether the front and rear brakes are independent or integrated.
  - Document any damage to the braking system from the crash.
  - Document the condition of the linkages, hoses, tubes, and the brake fluid level.
- Identify aftermarket components on the motorcycle.
- Document the damage to the motorcycle, including deformation, contact markings, and fractures, taking photographs with a wide range of zoom levels.
  - Identify which components of the motorcycle exhibit wear from sliding on the road surface.
  - Determine if the damaged condition was altered when the motorcycle was removed from the crash scene.
  - Document the postcrash geometry of the motorcycle. This documentation could utilize laser scanning, photogrammetry, or hand measurements. This documentation should include documentation of any wheelbase shortening, if applicable.
  - Determine the height at which the front forks of the motorcycle bent. This could enable a determination of the magnitude of front suspension compression at the time of an impact.
    - Is the instrument panel intact? If the motorcycle has an analog speedometer and tachometer, are they "frozen" in a particular position? Document this position.
- Determine which gear the motorcycle is in. Keep in mind that this may or may not be the gear that the motorcycle was in immediately after (or prior to) the crash. A rider may downshift during braking, or the shifter pedal may engage with the ground and shift while the motorcycle is sliding along the ground. This possibility could be ruled in or out by documented damage, or lack of damage, to the shifter pedal. The direction of this damage will also be relevant to whether the motorcycle could have shifted during sliding and in which direction (up or down). In addition, investigating police officers sometimes determine and report which gear the motorcycle was in when they inspect it, and they will change the gear while making this determination.
- Inspect and document the instrument panel. For motorcycles with analog speedometers or tachometers, it is possible that these will be "frozen" in a nonzero position. Document these positions.
- Document the condition of the headlamp assemblies, bulbs, and signal and brake lights.
  - What lights are present on the motorcycle?

- Do the front bulbs exhibit hot shock?
- If it is possible to power up the motorcycle, do the signals and brake lights function?
- Weigh the motorcycle.
- Determine if the motorcycle was carrying any cargo at the time of the crash. Crash scene photographs will assist with this determination.
- Document the rider's helmet and gear.
  - Any impact damage, scuffing and abrasions, or material transfer to the helmet?
  - Any tearing or scuffing of the jacket, pants, gloves, or boots?

Damage to the Struck Vehicle

Reconstructing a motorcycle collision will often involve determining the impact configuration between the motorcycle and the struck or striking vehicle. In many instances, the motorcycle will be upright at the time of impact, and there will be clear indication on the vehicle of the impact configuration in the form of damage and material transfer from motorcycle components, including the wheels or tires. For example, a motorcycle will often deposit a tire mark on the struck vehicle that will indicate the point of first contact. Several examples of such tire marks are shown in Figure 5.11. These photographs are from posttest documentation of the WREX 2016 collisions. An example from a real-world collision is shown in Figure 5.12. Hurt et al. [9] observed that "a pattern of motorcycle front tire striations on a car door might indicate use or non-use of the front brake: nearly horizontal scuffs in broadside impacts show a predominance of car motion but little tire motion, indicating that the tire had nearly stopped rolling as a result of braking at the moment of impact."

Another example from a real-world crash is included in Figure 5.13. This photograph shows damage to the driver's side rear door of an SUV from impact by the front of a large touring motorcycle. A circular imprint is evident near the center of the door from

FIGURE 5.11 Motorcycle tire marks on struck vehicles from the WREX 2016 collisions.

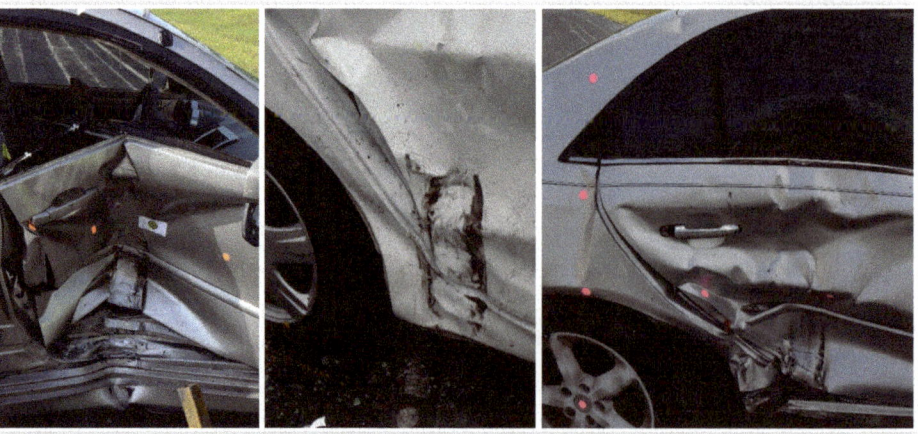

Photo courtesy of Louis Peck

FIGURE 5.12 Motorcycle tire mark on struck vehicle from real-world collision.

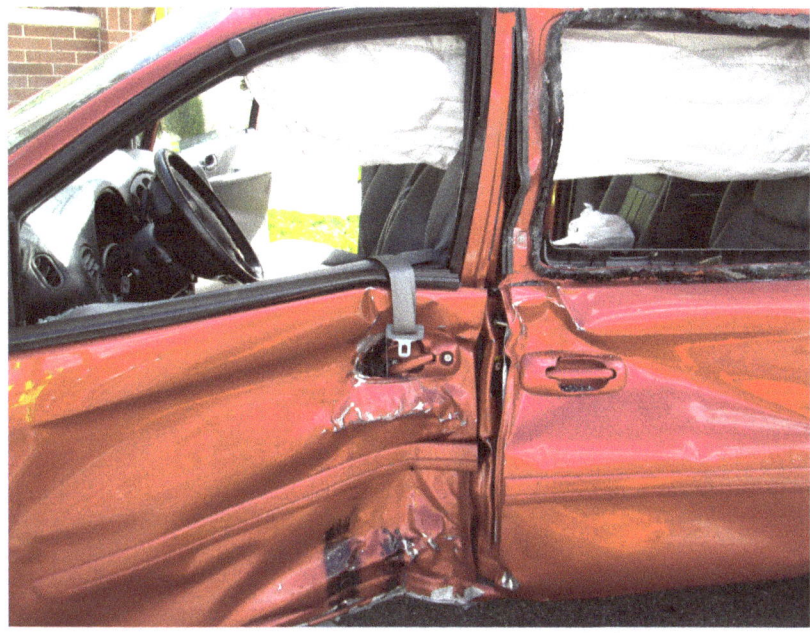

FIGURE 5.13 An imprint from the headlamp assembly, front fairing, and handlebars of a large touring motorcycle on the driver's side rear door of an SUV.

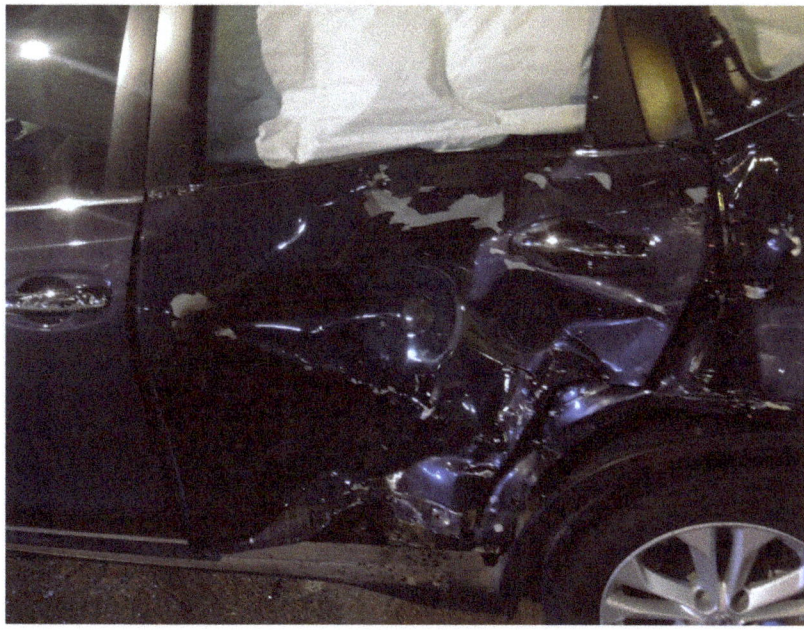

the motorcycle's headlamp assembly. Dents from the front faring of the motorcycle and handlebars fan out from this circular imprint. This evidence demonstrates that the motorcycle was still upright when it impacted the side of this SUV and provides the reconstructionist with a precise alignment between the motorcycle and the SUV at impact.

In some instances, a motorcycle will capsize prior to the collision. In these cases, determining the impact configuration can still involve identifying damage from specific motorcycle components on the struck vehicle. Computer models of the motorcycle and struck vehicle can be used to determine the alignment to produce the documented damage. In addition to specific motorcycle components leaving imprints on a struck vehicle, the rider's helmet, jacket, pants, gloves, or boots can also deposit material, scuffs, abrasions, or fabric patterns on the struck vehicle.

Evidence Documentation Methods

Mapping with LIDAR

Lidar, a term that is an acronym for light detection and ranging, is a method of *remote sensing* that can be used to map the geometry of an accident site or of a damaged or undamaged vehicle. Remote sensing refers to methods for the acquiring physical data and measurements of objects without physically touching them. These methods sense the physical characteristics of objects based on the characteristics and behavior of electromagnetic radiation interacting with these objects. "As [radiation from the sun] approaches the Earth, it passes through the atmosphere before reaching the Earth's surface. Some is reflected upward from the Earth's surface; it is this radiation that forms the basis for photographs and similar images. Other solar radiation is absorbed at the surface of the Earth and is then reradiated as thermal energy. This thermal energy can also be used to form remotely sensed images, although they differ greatly from the aerial photographs formed from reflected energy. Finally, man-made radiation, such as that generated by imaging radars, is also used for remote sensing" [16]. Lidar falls in the third category of remote sensing that utilizes man-made radiation.

Campbell describes lidar as follows:

> *Lidar … can be considered analogous to radar imagery, in the sense that both families of sensors are designed to transmit energy in a narrow range of frequencies, then receive the backscattered energy to form an image of the Earth's surface. Both families are active sensors; they provide their own sources of energy, which means they are independent of solar illumination. More important, they can compare the characteristics of the transmitted and returned energy—the timing of pulses, the wavelengths, the angles—so they can assess not only the brightness of the backscatter but also its angular position, changes in frequency, and the timing of reflected pulses. Knowledge of these characteristics means the lidar data, much like data acquired by active microwave sensors, can be analyzed to extract information describing the structure of terrain and vegetation features not conveyed by conventional optic sensors.*

It is common for crash reconstructionists to utilize a lidar (laser) scanner to map the geometry of an accident site or the postcrash condition of the motorcycle or another vehicle involved in a collision [1]. This can be particularly useful if an exemplar motorcycle or vehicle is also scanned, and the damaged and undamaged scans can be compared to determine relevant measurements, such as wheelbase shortening or crush. As an example, Figure 5.14 shows the colorized scan data of an exemplar motorcycle obtained with a FARO laser scanner.

FARO is a company that manufactures and sells scanners that are widely used by accident reconstructionists [17]. When using these scanners to document accident sites where physical evidence is still present, DiTallo et al. [18] recommended marking evidence

FIGURE 5.14 Sample scan data of an exemplar motorcycle.

© SAE International

with chalk or paint to make its location more evident in the resulting scan data. They conducted an experiment with a FARO Focus^{3D} X 330 and offered recommendations on scanner placement and settings (resolution and quality). In relationship to scanner placement, they noted that "The most effective and efficient methodology we have found is to stagger our scanner placement from one side to another, maintaining a 70-ft (21.34-m) radius between placements … Changing the vertical height from a standard tripod (6 ft+ /1.83 m+) to an extra tall tripod (9 ft+ /2.74 m+) did not gain much distance in the horizontal plane. In fact, the further the scan goes out, the more separation there is between the scan points. Your scanned data at the extremes will certainly have greater separation." In relationship to the resolution setting, they noted that "the FARO Laser Scanners provide the ability to adjust several settings which greatly affect the time required to complete a scan and the quality of the data captured. Resolution is a setting that determines the density of the scan points. Choosing a small value for resolution means there will be a larger distance between points scanned and a lower density of points in the resulting point cloud. Choosing a larger value for resolution means the distance between scanned points is smaller, so the captured points in the cloud will be dense … Choosing a higher resolution setting means each scan takes more time to complete. For most scenes, a resolution value of 1/4 or 1/5 gives good results in a reasonable amount of time." In relationship to the quality setting, they noted that "By adjusting the quality of the scan, the user is able to reduce the amount of noise (extraneous unwanted points) in the scan data. The higher the quality setting, the less noise in the scan data. You can choose from 1X to 8X depending on the resolution setting. In our experiment and demonstration for this article, we used a 3X Quality setting."

Photogrammetry

When the reconstructionist inspects a site after some or all the evidence has deteriorated or disappeared, that evidence can still be placed on an evidence diagram using photogrammetric analysis (if the evidence was documented through photographs or video). This situation is not at all unusual, considering that reconstructionists are often asked to analyze crashes long after they have occurred.

The term *photogrammetry* is defined by the American Society of Photogrammetry and Remote Sensing (ASPRS) as the art, science, and technology of obtaining reliable information about physical objects and the environment through process of recording, measuring, and interpreting photographic images and patterns of recorded radiant electromagnetic energy and other phenomena [19]. Defined in this way, the term *photogrammetry* has considerable overlap with the term *remote sensing*. Within the field of accident reconstruction, the term *photogrammetry* has typically been defined more narrowly as mathematical and graphical techniques used for making accurate measurements from photographs [20]. Baker [21] defined photogrammetry as the process of "obtaining reliable information about physical objects and the environment through processes of recording, measuring, and interpreting photographic images." Tumbas [22] defined photogrammetry as "a scientific method for determining the dimensions of an object by measuring a photographic image of the object and transforming measurements of the image to the actual surface." The term *close range photogrammetry* refers to applications of photogrammetry where the camera-to-object distances are on the order of "tens and hundreds of yards as opposed to the tens and hundreds of miles associated with aerial uses" [23].

One photogrammetric technique, referred to as *camera matching* in the more recent literature, involves reconstructing the location from which a photograph was taken and determining the field of view of the camera that took it. Once the camera location and field of view are obtained, objects (physical evidence) within the photograph can be located. This graphical technique, which is often used for diagramming scene or vehicle evidence, involves the following steps:

- The reconstructionist selects a photograph for analysis. For a photograph to be analyzed, it needs to show objects or geometry that still exist and can be physically documented. For analysis of scene evidence, such geometry would typically be roadway striping, curbs, signs, or trees.

- After selecting a photograph for analysis, the reconstructionist makes note of objects or geometry shown in the photographs that still exist at the scene. The reconstructionist then physically documents this geometry at the scene using a total station, a laser scanner, or by employing image-based scanning.

- After the scene mapping data have been processed, the reconstructionist uses a computer modeling software package to create a virtual camera and to view the survey or scan data from a perspective that is visually similar to that shown in the photograph that is being analyzed.

- Lens distortion is removed from the photograph that is to be analyzed. All camera lenses produce some level of distortion in the resulting image, and this distortion can potentially introduce errors into a photogrammetric process if it is not removed. Reference [24] examined the lens distortion caused by 35 different makes and models of camera lenses and then discussed methods and software available for removing the distortion from the resulting images. This reference noted that for a particular focal length, two lenses of the same make and model will produce the same distortion. Lens distortion occurs in two basic forms—barreling and pin cushioning. Barreling is distortion that squashes the edge of the image, whereas pin cushioning is distortion that stretches the edge of the image. In general, barreling occurs from wide focal lengths and pin cushioning occurs from zoom focal lengths, although lenses can also produce a mix of barreling and pin cushioning.

- The corrected photograph is then imported into the modeling software and is designated as a background image for the virtual camera.

- The analyst then adjusts the location, focal length, and viewing plane of the computer-modeled camera until an overlay is achieved between the survey or scan data and the scene geometry shown in the photograph. Once a match is obtained, then the analyst has reconstructed the location, focal length, and viewing plane of the camera used to take the original photograph.
- Once the camera location and characteristics are obtained, the evidence visible in the photograph can be traced or modeled such that it also overlays what is visible in the photograph.

A sample of the results of this technique is shown in Figure 5.15. The image in this figure shows a photograph of scene evidence from a motorcycle sliding on the roadway. This is the evidence depicted on Figure 5.1. In the image of Figure 5.15, a site survey has been overlaid on the roadway geometry, including the fog lines, the center lines, and the guardrail posts. The process through which this overlay was achieved resulted in a reconstructed position and characteristics for the camera that took the photograph. Once this camera position and its characteristics were obtained, the geometry of the photograph becomes known and the tire marks, gouges, scrapes, and police paint could be traced in the photographs and placed on the evidence diagram.

The camera-matching technique is a single-image photogrammetric method, in the sense that a single image is sufficient for its application. However, multiple photographs depicting the same evidence from different vantage points can be analyzed, and when the results are averaged, this may increase the accuracy of the technique. Campbell and Friedrich [20] first described the camera-matching technique in the accident reconstruction literature in 1993. Using a mock crash scene, they compared the positions of points obtained with the technique to the actual positions of these points. They reported a minimum position error of 0.1 ft and a maximum position error of 1.74 ft. In a 1999 study, Massa [25] demonstrated that the technique could be applied to analyze and critique an animation. Reference [26] described and tested the camera-matching technique for obtaining dimension from another mock crash scene. This reference compared dimensions obtained from camera matching with the same dimensions obtained with a digital survey and reported an average error of around 2%, with a range of errors between 0.48% and 8.23%. The dimensions ranged from approximately 5 ft up to approximately 140 ft.

FIGURE 5.15 Sample of the camera-matching photogrammetric technique.

Coleman et al. [27] examined the accuracy of the camera-matching technique when implemented with laser scan data rather than a digital survey. They tested the accuracy of the camera-matching technique by using it to locate evidence placed at a mock crash site, and then, comparing the obtained evidence locations to the actual locations of the evidence. They noted that "in many places the evidence location error was near zero." Their worst-case error for the camera-matching technique was about 13 cm (5.1 in.).

The camera-matching technique can also be applied to frames from a video. Chou et al. [28] reported application of the technique for tracking the motion (position, roll angle, and roll velocity) of a vehicle from video of a dolly rollover crash test. The tracked roll angle and calculated roll velocity were compared to the roll velocity measured with a sensor and the roll angle obtained from integration of the sensor data. Chou et al. reported excellent agreement between the roll angles determined from camera matching and those determined from the sensor data. There were discrepancies between the roll velocities obtained from camera matching and those obtained from the sensor, though the trend of the sensor data was generally matched with the camera-matching technique.

Rose et al. [29] reported additional application of the camera-matching technique to analysis of a dolly rollover crash test, attempting to improve on the results from Chou et al. through more extensive knowledge of the locations and characteristics of the cameras that recorded the test and the geometry of the test facility. In this study, a digital survey was conducted to locate the cameras, to document the test facility geometry, and to locate targets place on the vehicle. This reduced the number of unknowns in the camera matching. In addition to that, knowing the characteristics of the cameras made it easier for lens distortion to be removed from the video prior to the analysis.

Rose et al. reported the motion of the vehicle for a 2-sec segment of the test and compared the roll velocity results to sensor data. The resulting roll velocities from the camera matching were generally bracketed by the signals from the two roll velocity sensors on the vehicle. In other words, the discrepancy between the roll velocity from the camera matching and the roll velocity from either sensor was less than the discrepancy between the two sensors. Rose et al. also reported a comparison of two separate analysts using the camera-matching technique on the same 2-sec segment. Around 96% of the time, the vehicle position obtained by the two analysts was different by less than an inch. This propagated to an uncertainty in the calculated over the ground and vertical speeds of 1 mph with a confidence of 96%. For the vehicle roll angle, the difference between the two analysts was less than 1f, approximately 85% of the time. For an analysis time step of 40 ms, this propagated to an uncertainty in the roll velocity of approximately 18 deg/s with a confidence of 85%.

Manuel et al. [30] tested the camera-matching technique on frames of video from a moving camera (they referred to the technique as videogrammetry). They tracked the motion of the camera—and thus, the motion of the vehicle to which the camera was attached—and the motion of another vehicle captured by the camera. They conducted their analysis with varying levels of scene documentation to test the influence of the available scene data. These levels were as follows: (1) aerial photography as the only source of data about the site; (2) total station data defining the geometry of the site; and (3) lidar scan data defining the geometry of the site. For the vehicle carrying the camera, they reported that when aerial photography was used, "the reconstructed speed of the camera vehicle can be overestimated by 8%"; when total station data were used, "the reconstructed speed of the camera vehicle can be overestimated by 3%"; and when scan data were used, "the reconstructed speed of the camera vehicle can be underestimated by 2%." For the motion of the vehicle captured by the camera, the authors reported that when aerial photography was used, "the reconstructed speed for a moving vehicle captured by a moving camera can be underestimated by 12.8%"; when total station data were used, "the reconstructed speed for a moving vehicle captured by a moving camera can be overestimated by 10.4%"; and when scan data were used, "the reconstructed speed for a moving vehicle captured by a moving camera can be underestimated by 2.4%."

Camera Reverse Projection

The camera-matching technique is essentially a digital implementation of a technique which the earlier literature referred to as *camera reverse projection*. According to Tumbas [22], this term refers to the fact that once the location and characteristics of the camera are known, points from the photograph can be projected onto the scene. This technique followed a similar method to that described for camera matching, but the implementation did not use computer modeling programs or virtual cameras. Instead, the reconstructionist would overlay a transparency of the original photographic print, and trace geometry that was visible in the photograph that would still be visible at the site. A small version of this transparency would then be created and inserted into the viewfinder of what the Northwestern University *Traffic Collision Investigation* manual referred to as a "dummy" camera. While at the crash site, the reconstructionist would then change their viewing location and the camera focal length until the features traced onto the transparency overlaid on the corresponding features at the crash site. Once the viewing location, viewing plane, and focal length for the original photograph were reconstructed, the reconstructionist could then locate physical evidence that was no longer visible at the site.

One limitation in this technique is that the analyst would need to use a camera conducive to inserting a small transparency into the viewfinder. This would usually mean that the analyst would be using a different camera with different internal geometry than the camera that took the photograph. Another limitation is that the analyst would have to stand in the same place as the original photographer for an extended period. If a photograph was taken from the middle of a lane on the interstate, for example, this would not be possible without a road closure. Digital implementation of the camera-matching technique does not suffer from these limitations, because the analyst has greater control over the characteristics of the computer modeled camera and does not have to stand in the road. Another limitation of the older implementations of camera reverse projection was that there was no real way to account for or eliminate lens distortion from the photographic image. Digital implementation of the camera-matching technique can correct for or eliminate distortion from many photographic images.

Despite these limitations, many reconstructionists successfully and accurately implemented the camera reverse projection technique. Breen and Anderson [31] described an implementation of reverse camera projection to determine the postcrash shape of a damaged windshield. Woolley et al. [32] applied camera reverse projection with two photographs to quantify vehicle crush. He reported average errors of less than 1 in., noting though, that this level of accuracy "depends on the experience, patience, and care taken in all phases of the process. The more times camera reverse-projection is applied, either to a damaged vehicle or to a scene, the greater will become the skill of the analyst." This statement also applies to a digitally implemented camera matching process. However, what Woolley et al. has left unsaid is the important fact that the application of the reverse camera projection and camera-matching techniques typically results in a graphical output that allows others to evaluate the quality of the camera match.

Husher et al. [33] applied the camera reverse projection technique to a mock crash scene and reported that it "was able to locate the subject skid to within 20 cm (8 in) for the least favorable camera view, and within 5 cm (2 in) for the more favorable views. The selection of the camera view determines the accuracy of this method." They also stated that "when the results of the three [camera reverse projection] views were averaged, the determined position was almost on top of the true position for both the skid and the vehicle." This means that, though camera reverse projection is methodologically a single-image technique, applying it to multiple images and averaging the results improve the accuracy. In this study, camera reverse projection was also applied to determine vehicle crush and the authors reported that the "longitudinal crush determined was within 2.5 cm (1 inch) of the measured value."

Husher et al. also described an implementation of the camera reverse projection technique that involved "real time analysis with video equipment … Two video cameras view the original photo and the exemplar vehicle (or scene) simultaneously and a video mixer is used to overlay the two separate views to provide the alignment information …." Husher et al. applied this method to determining vehicle crush and reported that "accuracy was within 2.5 cm (1 inch) of the crush." Bruce and Knopf [34] describe an implementation of the camera reverse projection technique using physical scale models of the crash site and vehicles. Though computer implementation of the camera-matching technique began appearing in the literature as far back as 1993, the technique described by Main and Knopf is methodologically a direct precursor to the camera-matching technique.

Smith and Allsop [35] and Tumbas [22] noted that reverse camera projection can be accomplished analytically with two photographs. Tumbas stated that this would be accomplished by "surveying selected points at the site, constructing a mathematical model of the site surfaces, mathematically locating the camera, mathematically projecting the points of interest onto the model and determining the location of the intersecting rays with the model's surface." Referring to this technique as *analytical reverse projection*, Smith described the concept of this method as follows: "This method is based on the fact that once the original position of the camera lens center, the distance from the lens center to the picture-plane (photograph), and the orientation of the photograph relative to the scene are all determined, then a vector passing from that established lens center point through any point of interest in the two-dimensional picture-plane defines the ray that passes through that same point of interest in the actual three-dimensional scene. If two or more photographs, taken from different locations, are available showing the same point of interest, then the location of that point in the scene can be determined by simply calculating the point where the two rays (on from each photograph) intersect." Tumbas reported typical accuracies for locating points with reverse camera projection of less than a foot, with maximum errors around 2 ft. For analytical reverse projection he reported typical accuracies of around a half foot, with maximum errors of less than a foot. Tumbas notes that both techniques resulted in better accuracy than methods that assume the surface containing the evidence is flat. Reverse camera projection does not require such an assumption.

These studies demonstrate that analysts could achieve levels of accuracy acceptable for accident reconstruction even though they did not necessarily utilize the same make and model camera as the one used to take a photograph and they did not remove the lens distortion from the photographs. Given this, the importance of removing lens distortion from photographs should not be overestimated. In instances where such removal is not possible, acceptable results may still be possible. As with the application of most techniques in accident reconstruction, much depends on the level of accuracy required for the application. This will ultimately have to be judged on a case-by-case basis.

Removing Lens Distortion

Imperfections in the design and manufacturing processes of camera lenses cause distortion in the image captured by the sensors or by film of the camera [36]. As light is collected, focused, and transformed through the lens structure, the final collection of this light on a sensor plate or film sheet is distorted. Two common types of distortion are referred to as pin cushion and barrel distortions. Figure 5.16 illustrates these two types of distortion. The grid on the left of this figure shows the effect of barrel distortion and the grid on the right shows the effect of pin cushioning. Barrel distortion causes the image to be expanded in a barrel shape from its center [37]. This distortion type is typically associated with wide angle focal lengths. Pin cushioning causes images to be pinched at their center. Pin cushion distortion is typically associated with telephoto focal lengths. Sometimes, a combination of barrel and pin cushion distortions can be present in a photographic image. This results in a wavy distortion pattern likely to occur at the transition between wide and zoom focal lengths. Barrel and pin cushion distortions affect the positions of pixels in an image. Because photogrammetric techniques analyze the location of objects in the image, removing distortion from an image prior to analysis can improve the accuracy of the results.

For many common camera/lens combinations, there is published data that mathematically describe the distortion they create in images. This mathematical description can be applied to remove the distortion in a digital image. There are multiple programs available that use databases of distortion coefficients to automatically correct for lens distortion. For example, Epaperpress Ptlens, Hugin, and Adobe Photoshop CS5 use the information stored in a photograph's exchangeable image file format (EXIF) data to look up the distortion coefficients from a database. These distortion coefficients are published for many camera types by the manufacturer. The EXIF data of the image are information that is stored with the image when the photograph is taken, such as the make and model of the camera, the exposure setting, f-stop, and the time and date the photograph was taken (or, at least, the time reported by the clock of the camera).

Distortion correction relies on the following polynomial function that modifies the distance a pixel is from the center of the image.

FIGURE 5.16 Illustration of distortion types.

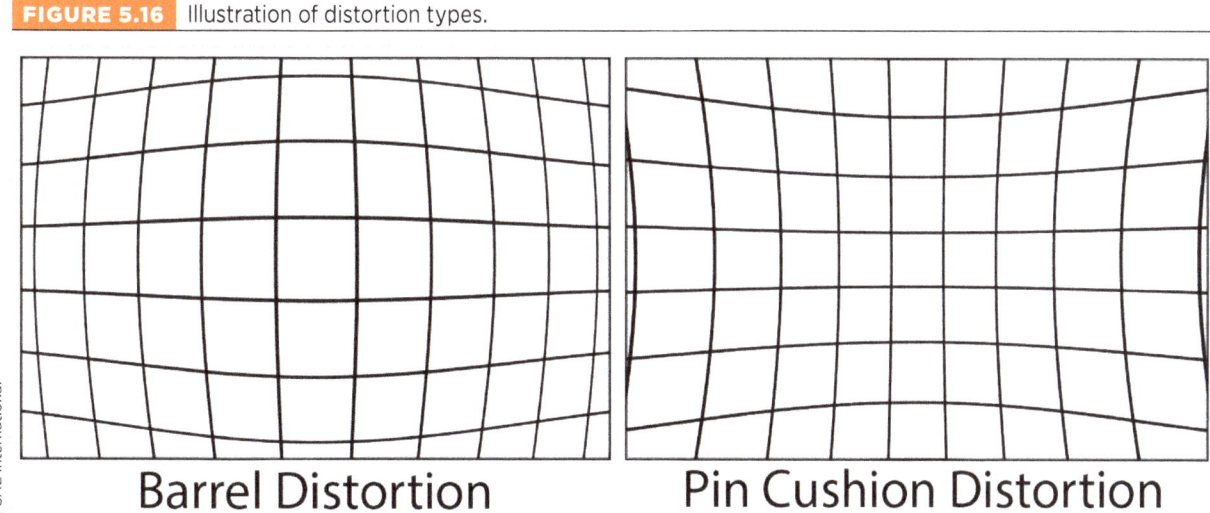

$$y = ax^4 + bx^3 + cx^2 + dx \tag{5.1}$$

In this equation, x is the distance of the distorted pixel from the center of the image and y is the distance of the undistorted pixel from the center of the image. The distances in this equation are typically normalized such that a value of 1 is equal to half the size of the shortest side of the image. The coefficients are determined empirically for any given camera/lens combination, and they control the transformation from the distorted to the undistorted pixel position. Coefficient a primarily affects the edges of the image, c affects the inside, and b affects the whole image. The d coefficient is related to the others through the following equation:

$$d = 1 - (a + b + c) \tag{5.2}$$

This equation controls the scale of the corrected image and is used to maintain the overall size of the resulting image. In addition to the published coefficients that can be obtained, image adjustment programs typically allow the user to manually enter the coefficients to perform the distortion correction.

For cameras for which coefficients are not publicly available, these coefficients can be determined with the following procedure. First, a grid is printed out and mounted on a wall as shown in Figure 5.17. Then, a photograph is taken of the grid such that the vertical and horizontal lines of the grid will be captured parallel to the digital image. The make and model of the camera and the focal length for the image can then be obtained from the EXIF

FIGURE 5.17 Setup for photographing a grid.

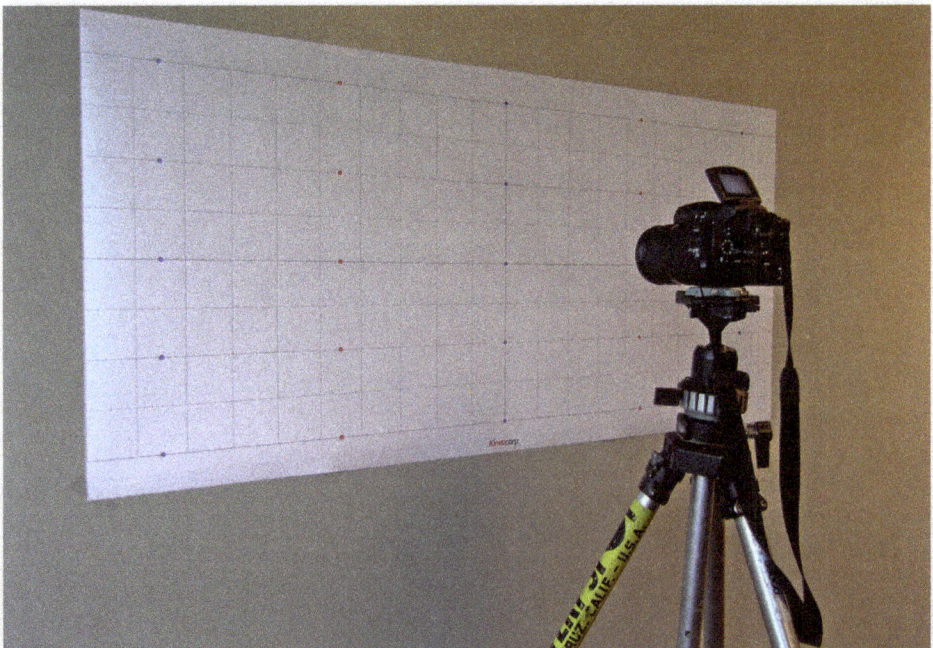

data. This information can be used in the image editing programs to complete the manual assessment of distortion. There can be differences between the focal length reported in the EXIF data and the actual focal length used when taking the photograph. After the photograph is taken and imported into the image processing program (Photoshop, for instance), distortion correction plug-ins are used that allow the user to manually input the distortion coefficients until the distortion correction is achieved (meaning, lines that are straight in the real world are also depicted as straight in the photographic image). For the case of the photograph of the grid, the coefficients would be iteratively changed until the vertical and horizontal lines match the overlaid grid lines.

Reference [37] noted a couple of trends related to lens distortion. First, the center of an image typically contains the least amount of distortion. The distortion increases moving from the center to the edges of an image. Second, the focal length at which a photograph is taken influences the resulting distortion.

Case Study: Photogrammetric Analysis of Video of a Motorcycle Accident

This section describes the reconstruction of an intersection collision involving a motorcycle and a station wagon. The crash was captured on video and camera-matching photogrammetric analysis was one of the methods employed in the analysis. The motorcycle was traveling straight westbound through the intersection and the eastbound station wagon turned left in front of the motorcycle. According to the police investigation, the motorcyclist entered the intersection on a red light and impacted the passenger side of the station wagon. The driver of the station wagon had reportedly initiated her turn near the end of the yellow light cycle. There was also a small bus that initiated a left turn in the lane next to the station wagon. The investigating officers did not take any photographs during their investigation.

The intersection where this accident occurred is depicted in the aerial photograph of Figure 5.18. This photograph, which is oriented such that north is up on the page, shows that the east-west roadway had three westbound and three eastbound through lanes. There were two turn lanes to accommodate traffic turning left from the eastbound roadway to the northbound roadway. These are the turn lanes that the station wagon and the bus were using when this crash occurred. The station wagon was turning from the leftmost turn lane and the bus was turning from the rightmost turn lane. The intersection is asphalt paved, and at the time of this accident, the roadway was dry and there were no adverse weather conditions. The speed limit for the east-west roadway was 35 mph.

A witness that was traveling westbound behind the motorcycle stated that the motorcyclist entered the intersection when the light for their direction was red. She estimated that it had been red for 5 sec when the collision occurred. She further stated that the motorcyclists "may have been traveling at an excessive speed …" Another witness, who was in a vehicle on the south side of the intersection facing north, stated that "[The station wagon] was waiting to make a left and the light for them started to turn yellow. The first left and center lane westbound started to stop. And so, [the station wagon driver] decided to make her left turn while the motorcycle was in the right lane and as soon as she got around to his lane that's when he came through trying to beat the yellow light and hit her in the front right tire."

©2022, SAE International

FIGURE 5.18 Aerial photograph of subject intersection.

The bus that was turning left next to the station wagon was equipped with a DriveCam system that captured portions of this accident on video. A DriveCam unit is an aftermarket, event-triggered video, and data recorder that is mounted to the windshield of a vehicle. DriveCam units contain accelerometers that measure longitudinal and lateral accelerations, and two cameras that record video. One of these cameras looks forward through the windshield and the other looks rearward at the vehicle occupants. If a DriveCam unit measures an acceleration that exceeds a preset threshold, it triggers an event and stores video and acceleration data. Rose et al. [38, 39] reported research related to the accuracy of these DriveCam systems and reported that they provide reliable evidence for use in accident reconstruction.

During the subject accident, an event was triggered in the DriveCam unit on the bus because the driver of the bus braked hard as the motorcycle traveled in front of him. The DriveCam unit captured video and acceleration data for 10 sec before and 10 sec after this braking and the raw DriveCam data were available for analysis. This data showed that, when the bus driver braked, the bus reached a longitudinal deceleration of 0.72 g. Figure 5.19 is one frame of the video from the forward-looking camera in the DriveCam unit. This frame shows the motorcyclist entering the intersection. This frame has a listed time of −0.25 sec. Within the DriveCam system, time is assigned in relationship to the triggering event. The triggering event is set as time zero; times before the triggering event are negative and times after the triggering event are positive. Therefore, the frame of video included below shows a point in time ¼ of a second before the DriveCam sensed the hard braking of the bus driver. This DriveCam video frame also shows smoke coming from the rear wheel of the motorcycle, indicating that this wheel locked up when the motorcyclist braked.

The intersection at which this collision occurred was inspected, documented, and mapped using a Faro Focus laser scanner. This documentation with the scanner focused on the north-east corner of the intersection because this portion of the intersection was visible in the DriveCam video. Figure 5.20 depicts some of the colorized scan data of the intersection.

Figure 5.21 is another image from the forward-facing view of the DriveCam system. In this image, the front of the bus is visible, as is a white sedan and a white van in the

FIGURE 5.19 Frame of video from forward-looking camera of DriveCam unit.

FIGURE 5.20 Colorized scan data of the north-east corner of the subject intersection.

oncoming lanes of travel. Consistent with the witness statements, it was evident from the video that these vehicles were slowing and stopping for the westbound traffic signal that was changing to yellow and then red. The scan data were used in conjunction with camera matching photogrammetry and video tracking using a software package called PFTrack. The motion of the camera attached to the bus was tracked. Once this motion was known, the motions of the white sedan and white van were also tracked. Finally, the motion of the motorcyclist was tracked. The goal of this analysis was to determine the speeds and accelerations or decelerations of these vehicles and to use this data to inform a determination of when the traffic signal for westbound traffic turned yellow and red. Unfortunately, the color of these traffic signals was not visible in the video at the significant times.

FIGURE 5.21 Another frame of video from the forward-looking camera of DriveCam unit.

The next three figures depict an example of the progression of our camera matching analysis to track the motion of the motorcycle. Figure 5.22 is the first frame from the DriveCam video in which the motorcyclist was visible (t = −0.75 sec). Lens distortion has been removed from this video frame, as it was for the rest of the video frames. Figure 5.23 is the same video image with the scan data overlaid onto the corresponding intersection geometry visible in the video image. This overlay results from reconstructing the location and characteristics of the camera and mimicking those in a computer modeling software. Figure 5.24 shows the reconstructed motorcycle position for this frame. The camera matching process was repeated for two additional frames of video in which the motorcyclist was visible (t = −0.50 sec and −0.25 sec). From the resulting reconstructed positions of the motorcycle, the speed could be determined. This led to the conclusion that the motorcyclist was traveling approximately 48 mph when he entered the intersection.

When the DriveCam video depicted the motorcyclist entering the intersection (t = −0.5 sec), the traffic light for the northbound roadway was visible in the video and it was red (see Figure 5.25). The color of the light for eastbound traffic is not visible in this frame. A half second later (t = 0.0 sec), the traffic signals for both eastbound and northbound traffic are visible and red (see Figure 5.26). At this intersection, the traffic lights for eastbound and westbound traffic had a 4-sec yellow phase and a 2-sec "all red" phase. In addition to that, the two red lights illuminated and visible in the DriveCam frame at t = 0.00 sec both remain red for longer than 2 sec, because these lights will remain red while the left turn arrow for southbound traffic cycles through to allow southbound drivers to turn left. During our site inspection, one instance was documented in which these two lights were both red for 12 sec. Thus, there is no way to determine from one frame of the DriveCam video how long these lights had been red and how long they would continue to be red.

CHAPTER 5 Physical Evidence from Motorcycle Crashes 127

FIGURE 5.22 First frame of DriveCam video in which the motorcyclist was visible.

FIGURE 5.23 Scan data overlaid onto the video image.

FIGURE 5.24 Reconstructed location of the motorcycle for this video frame.

©2022, SAE International

FIGURE 5.25 Northbound traffic signal was red when motorcyclist was entering the intersection.

FIGURE 5.26 First video frame at which eastbound traffic signal was visible.

The best indication of when the traffic light for westbound traffic turned yellow and red is the actions of westbound drivers other than the motorcyclist. The graph of Figure 5.27 shows the speed of these vehicles as they approach the intersection and stop. In this graph, the time from the DriveCam video is plotted on the horizontal axis and the speed of the approaching vehicles is plotted on the vertical axis. As this graph shows, at a time of −7.0 sec, the white van and sedan are approaching the intersection at a speed of around 15 mph, and they are accelerating. They both begin decelerating at a time of around −5.50 sec. This is an indication that, at this time, the drivers of these vehicles have seen the traffic signal turn to yellow and have begun to brake in response to that change in the

FIGURE 5.27 Results of tracking the motion of the westbound vehicles.

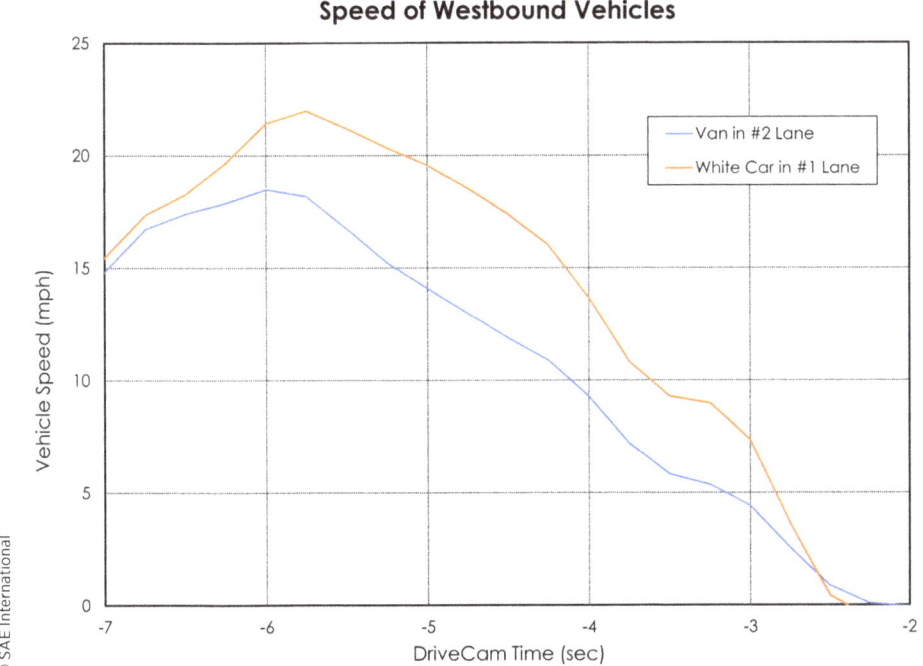

light color. It would typically take a driver around 1 to 1.5 sec to perceive the change in light color and begin applying their brakes [39, 40]. That these drivers begin braking for the light at a time of −5.5 sec indicates that the westbound traffic lights likely changed to yellow at a time between −6.5 and −7.0 sec. This means that the light turned red at a time between −2.5 and −3.0 sec. The motorcyclist was just entering the intersection at a time of −0.5 sec, and so, he did not enter the intersection until the traffic light had been red for between 2 and 2.5 sec.

Case Study: Photogrammetric Analysis of Video of Another Motorcycle Accident

This section describes the reconstruction of another collision that was captured on surveillance footage and involved a motorcycle and a blue sedan. Camera-matching photogrammetric analysis was again used to analyze frames from the surveillance video. The motorcycle was traveling westbound, and the sedan was initially sitting in a driveway facing south. There was a surveillance camera operating near the driveway. The driver of the sedan attempted a left turn in front of the motorcyclist with the intention of traveling eastbound. The motorcyclist collided with the driver's side of the sedan. The roadway was asphalt paved. At the time of the collision, the roadway was dry and there were no adverse weather conditions. The speed limit for the east-west roadway was 45 mph.

The driveway and roadway where this collision occurred was inspected, documented, and mapped using a Faro Focus laser scanner. This documentation focused on the portions of the driveway and roadway that were visible in the frames of surveillance footage. Figure 5.28 depicts some of the colorized scan data of the intersection. The next series of figures illustrates the progression of our camera matching analysis to track the motion of the motorcycle and the sedan. Figure 5.29 is a frame from the surveillance footage in which the driveway, the roadway, and the sedan were visible. Lens distortion has been removed from this video frame, as it was for the rest of the video frames. Figure 5.30 is the same video image with the scan data overlaid onto the corresponding driveway and roadway geometry visible in the video image. In this case, the camera was static, and its position was documented during our site inspection. Thus, in this case, the purpose of this overlay

FIGURE 5.28 Colorized scan data of the driveway and roadway.

FIGURE 5.29 Frame of surveillance footage.

CHAPTER 5 Physical Evidence from Motorcycle Crashes 131

FIGURE 5.30 Scan data overlaid onto the video image.

FIGURE 5.31 Reconstructed positions of the motorcycle and Chrysler for another frame.

was to establish the field of view and distortion characteristics of the camera lens. Once these were established, the positions of the sedan and the motorcycle could be tracked through a series of frames. Figure 5.31 shows a different frame of the video, in which both the sedan and the motorcycle are visible. Computer models of these vehicles have been placed relative to the virtual camera such that they overlay these vehicles in the frame of video. The result is the positions of these vehicles being located within the scan of the site. This same process was completed for a series of frames showing both vehicles, which resulted in a determination of the distances traveled by each vehicle between frames. The frame rate of the video was known (10 fps), and so, the time between frames could be determined and the speeds calculated.

©2022, SAE International

FIGURE 5.32 Reconstructed speeds from the camera-matching video analysis.

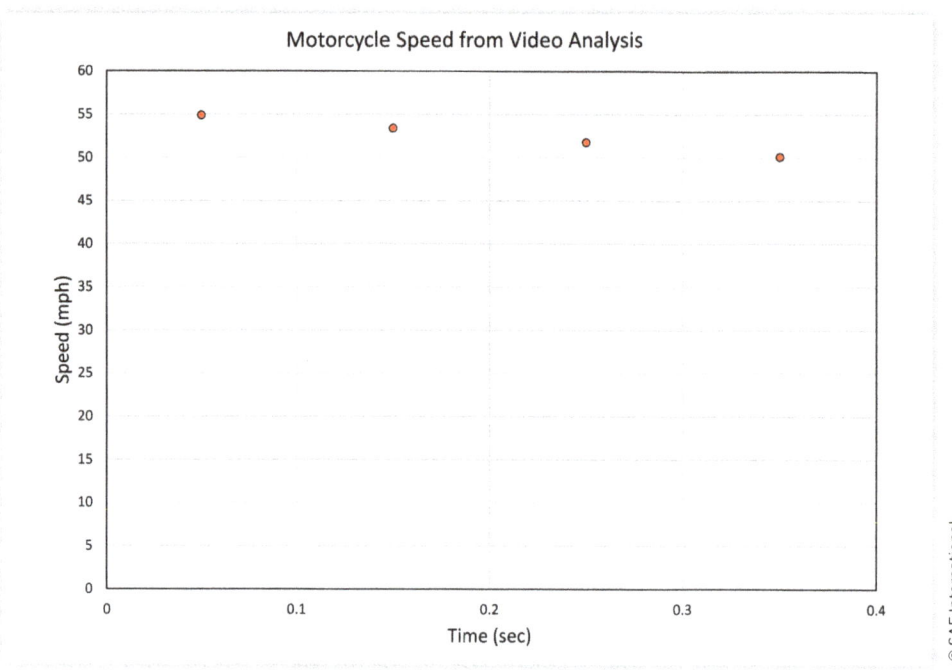

Five precollision positions were reconstructed for the motorcycle and an average speed calculated for each of the time intervals between frames. These speeds are shown in Figure 5.32. As this figure shows, the motorcycle was decelerating prior to the impact, traveling a speed of approximately 55 mph when it entered the video and approximately 50 mph during the tenth of a second preceding the collision. Smoke was visible coming off the rear tire of the motorcycle in the video. The investigating officers also documented a 73-ft long tire mark from the rear tire of the motorcycle leading up to the area of the collision, which was consistent with the smoke visible in the video and consistent with the speeds from the video analysis shown in Figure 5.32. The speeds from the video analysis implied a deceleration for the motorcycle of approximately 0.7 g prior to the collision, implying that the motorcyclist was employing both the front and rear brakes of his motorcycle.

Image-Based Scanning

Photogrammetric methods can also be used to generate point cloud data for an object like what would be generated with lidar scanning. This process is referred to as image-based scanning. As an example, aerial images captured with a small unmanned aerial vehicle (sUAV) can be used in conjunction with photogrammetric software packages such as PIX4D, VisualSFM, PhotoScan, or 123DCatch to document and model accident site geometry. This software can also be used with ground-based photographs of a damaged or exemplar vehicle to generate a point cloud of the vehicle geometry.

Vergauwen [41] presented a method to create three-dimensional (3-D) digital surface models using images without significant visual overlap. Strecha et al. [42] examined the use of image-based modeling techniques as possible replacements for lidar-based measurement systems. Dai et al. [43] compared spatial data collected with a Leica C10 laser scanner to various photogrammetry and videogrammetry methods and concluded that under certain circumstances, "image-based methods constitute a good alternative for time-of-flight-based methods.". Erickson et al. [44] explored the accuracy of using 123DCatch to create 3-D models of vehicles and concluded that "photo-based 3D scanning using 123D Catch® is a valid mode of reconstructing the geometry of a vehicle for the purposes of collision reconstruction."

Reference [45] evaluated the capabilities and accuracy of four automated photogrammetry-based software programs to accurately create 3-D point clouds of damaged and undamaged vehicles—Photomodeler Scanner, Photoscan, Pix4D, and VisualSFM. They compared the results produced by these software packages to the results to 3-D scanning. These authors reported that all four software packages produced point clouds for which an average of nearly 60% of their points were within 0.25 in. of the LiDAR point cloud data and an average of more than 80% of their points were within 0.5 in. of the LiDAR data.

Jurkofsky [46] explored the accuracy of using the software package Photomodeler Scanner in conjunction with images taken from a UAV to map accident scenes. He created mock accident scenes and captured aerial photographs of the scenes. He then used Photomodeler Scanner software to calculate the position of objects within those scenes and compared the calculated positions to positions that he measured using a total station. Jurkofsky found that this process produced accurate results, demonstrating that UAV photography could accurately reproduce the location of physical evidence, provided the physical evidence was visible in the photographs. While this is often the case for law enforcement officials, accident reconstructionists who are not involved in law enforcement may document a scene months or years after an accident occurs. By this time, the physical evidence from an accident has often deteriorated or is no longer present.

Carter et al. [47] published a study that examined the accuracy of camera-matching photogrammetry using a point cloud of a mock accident site generated with image-based scanning using aerial images from an sUAV. A mock scene was created in a parking lot with physical evidence typical of a vehicular crash. The scene was scanned with a FARO laser scanner and photographed. The evidence was then removed and video was taken of the scene from a UAV. That video footage was processed with image-based scanning software to create a point cloud, and the point cloud was used to reconstruct the positions and characteristics of the camera at the time the evidence was photographed. The evidence was then reconstructed with the camera-matching technique, and the position and size of the reconstructed evidence was compared to the position as documented by the FARO scanner. Some of the details of Carter et al.'s study are presented here as an example of applying image-based scanning to analyze the evidence at an accident scene.

The mock accident scene created by Carter et al. in a parking lot is depicted in Figure 5.33. This parking lot was chosen because it featured two vertical tiers, and this would allow testing of the image-based scanning process to capture accurate changes in elevation. Tape was placed on the pavement to represent common traffic accident investigation markings: paint indicating vehicle tire rest points, a tire mark, a series of gouge marks, and paint outlining a tire mark. After the tape was placed, the mock scene was scanned using a FARO Focus³ᴰ X 130 laser scanner, which is also depicted in Figure 5.33. The scan data, some of which is depicted in the graphic of Figure 5.34, consisted of 39 million points. Photographs of the scene were taken from many vantage points, similar to documentation of a crash scene by law enforcement officials. After the scene was scanned, the tape was removed, and video of the scene was taken from a DJI Phantom 2 sUAV equipped with a DJI Zenmuse H4-3D Gimbal and a GoPro Hero 4 camera (Figure 5.35). Video was captured at 4k resolution (3840 n 2160 pixels) at 30 fps in the GoPro "wide" setting.

FIGURE 5.33 Carter et al.'s mock accident scene [46].

FIGURE 5.34 Scan data of the accident scene.

FIGURE 5.35 Phantom 2 UAV, Gimbal, and camera used for documenting the scene.

© SAE International

The sUAV was flown in a zig-zag pattern to provide overlapping video footage and was flown at low speeds to minimize motion blur. Video was taken from three different heights, approximately 3 m, 10 m, and 20 m above ground level. The video was processed using GoPro Cinema Studio to remove the camera distortion. A total of 470 frames were extracted from the video taken with the UAV. These frames were then processed with several photo-scanning software packages: Pix4D, VisualSFM, and PhotoScan. The resulting point clouds were visually compared to determine which software package produced the most complete point cloud. Carter chose to use the point cloud created by Pix4D for the subsequent camera-matching analysis.

The camera matching process is depicted in the series of images below. Figure 5.36 is a photograph of the mock scene that was taken before the tape was removed. Figure 5.37 is a view of the point cloud created with the UAV footage and the Pix4D software. This image is taken from a perspective and location in the model similar to the perspective and location that the actual photograph was taken. Figure 5.38 depicts the image-based scanning point cloud overlaid on the photograph. Figure 5.39 then depicts the trace of the tape overlaid on an image from the FARO scan point cloud. This image visually depicts the accuracy of the evidence reconstruction.

This process was completed for four photographs of the mock scene, and a total of 32 unique points of evidence in the mock scene were reconstructed. Figure 5.40 shows the individual points that were reconstructed. The accuracy of this process was analyzed by comparing the reconstructed coordinates of the tape to the coordinates of the tape as measured by the laser scanner. Table 5.1 presents the statistical results of this analysis. In this table, the "2D Distance", "Elevation", and "3D Distance" columns represent the difference between the FARO scanner points and the reconstructed points in the x-y plane, in the z-direction, and in 3-D space, respectively. The average difference between position of the reconstructed evidence and the actual evidence, as measured by the FARO scanner, was approximately ½ in.

©2022, SAE International

FIGURE 5.36 Photograph of mock scene.

FIGURE 5.37 Point cloud of mock scene created with PIX4D.

FIGURE 5.38 Point cloud overlaid on photograph of mock scene.

CHAPTER 5 Physical Evidence from Motorcycle Crashes **137**

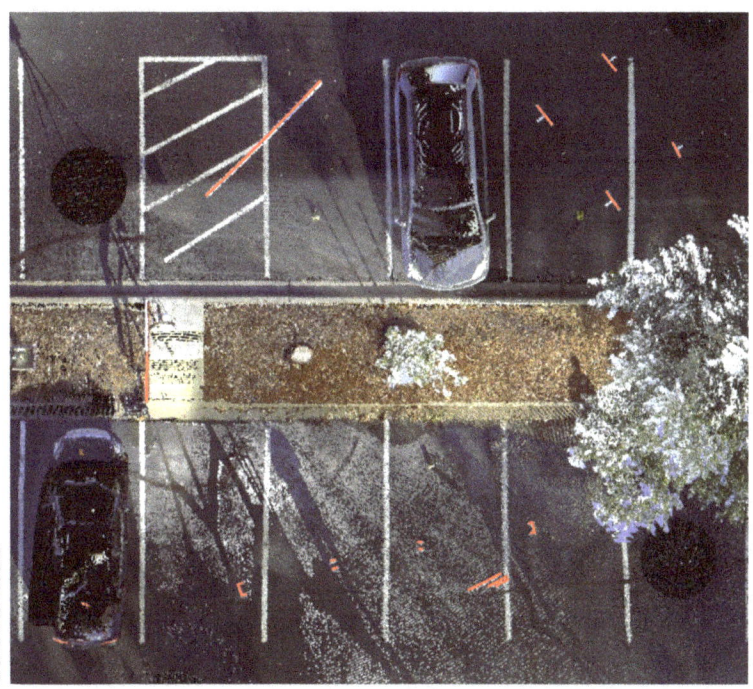

FIGURE 5.39 Reconstructed evidence overlaid on FARO scanner point cloud.

FIGURE 5.40 Points reconstructed with the camera matching analysis.

TABLE 5.1 Statistical analysis of the difference in position of reconstructed points versus their locations in the laser scan data

	2-D distance (in.)	Elevation (in.)	3-D distance (in.)
Average	0.46	0.18	0.51
Standard deviation	0.29	0.18	0.31
Maximum	1.24	0.55	1.24

© SAE International

Mapping with Small Unmanned Aerial Vehicles

The previous section discussed the use of photographs from an sUAV with methods of photogrammetry to generate a point cloud of an accident site. Prior to the wide availability of sUAVs and photogrammetry software like PIX4D, aerial photographs had been utilized by accident reconstructionists as background images for evidence diagrams, as textures for physical models, or as background images for simulations. Fay et al. [48] discussed such applications of aerial imagery taken with manned aircraft. Dilich and Goebelbecker [49] described an unmanned aerial system consisting of a stabilized 35-mm camera attached to a tethered blimp that could be used to photograph vehicular accident scenes.

More recently, Google Earth has made high-resolution aerial photography widely available for free. Wirth et al. [50] examined the accuracy of aerial images from Google Earth and found accuracy levels sufficient for accident reconstruction. Often, multiple aerial images of the same site will be available within Google Earth, each taken on a different date. Accident reconstructionists can often utilize these images to understand how a crash site has changed through time. These aerial images can also often be paired with historical images from Google Street View. These ground-level photographs can also help a reconstructionist determine what changes have occurred to a site through time. These images are also worth checking for physical evidence related to a crash. These authors have encountered instances in which some piece of physical evidence from a crash was not captured in the police photographs, but was captured in a Google Street View image.

Now that the US government has issued rules related to the commercial application of sUAVs, accident reconstructionists have begun utilizing these devices to capture custom aerial photographs and video of accident scenes. The cameras on sUAVs are often stabilized with gimbal systems to reduce camera rotation and vibration. The use of UAV imagery as an input to the image-based scanning process for accident scenes offers many advantages over ground-level imagery. The scene can be documented from different vertical perspectives, which increases the number of images of the scene that contains overlapping geometry. The imagery is less subject to sun glare issues than ground level images. Sun glare presents detection challenges for image-based scanning programs because the same object will appear differently in frames where the sun is reflecting directly off the object than in frames where the sun is not reflecting off it. Ground-level imagery may also be limited by physical structures or access issues that can be overcome using a UAV. For instance, an accident scene on a bridge may only allow for a ground-level photographer to document the bridge from one end or the other. With UAVs, documentation of the entire bridge may be possible, and the resolution of the imagery would be superior to that of photographs taken from a long distance. Aerial imagery may also avoid problems with obstacles that can visually occlude key pieces of a scene.

References

[1] Green, T., "3D Laser Scanners in Crash Testing," *Collision: The International Compendium for Crash Research*, 12(1): 80–86, 2017, ISSN 1934-8681.

[2] Fatzinger, E., Landerville, J., Bonsall, J., and Simacek, D., "An Analysis of Sport Bike Motorcycle Dynamics during Front Wheel Over-Braking," SAE Technical Paper 2019-01-0426, 2019, doi:10.4271/2019-01-0426.

[3] Bartlett, W., *Motorcycle Braking and Skidmarks*, Mechanical Forensics Engineering Services, LLC, 2000, unpublished article, https://www.mfes.com/motorcyclebraking.html, accessed on June 1, 2017.

[4] Bartlett, W., "Interpretation of Motorcycle Rear-Wheel Skidmarks for Accident Reconstruction," *Proceedings, 4th International Conference on Accident Investigation, Reconstruction, Interpretation and the Law*, Vancouver, BC, Canada, August 2001.

[5] Fricke, L.B., *Traffic Crash Reconstruction*, 2nd ed., Northwestern University Center for Public Safety, Evanston, IL, 2010, ISBN 0-912642-03-3.

[6] Baxter, A.T., *Motorcycle Crash Investigation*, Institute of Police Technology and Management, Jacksonville, FL, 2017, ISBN 978-1-934807-18-7.

[7] Dunn, A.L., Dorohoff, M., Bayan, F., Cornetto, A. et al., "Analysis of Motorcycle Braking Performance and Associated Braking Marks," SAE Technical Paper 2012-01-0610, 2012, doi:10.4271/2012-01-0610.

[8] Peck, L., Deyerl, E., and Rose, N., "The Effect of Tire Pressure on the Deceleration Rate of a Motorcycle Under Application of the Rear Brake Only," *Accident Reconstruction Journal*, 27(4): 19–21, 2017, ISSN: 1057-8153.

[9] Hurt, H.H., Ouellet, J.V., and Thom, D.R., "Motorcycle Accident Cause Factors and Identification of Countermeasures, Volume 1: Technical Report," DOT HS-5-01160, January 1981.

[10] Lenkeit, J.F., Hagoski, B.K., and Bakker, A.I., "A Study of Motorcycle Rider Braking Control Behavior," DOT HS 811 448, US Department of Transportation, National Highway Traffic Safety Administration, March 2011.

[11] McNally, B., "Summary of Motorcycle Friction Tests," *Accident Investigation Quarterly*, 76(12): 31–36, 2006, ISSN 1082-6521.

[12] Baxter, A. and Robar, N., "An Examination of the Performance of Motorcycle Brake Systems," *Accident Investigation Quarterly*, 47: 28–31, 2007, ISSN 1082-6521.

[13] Smith, J., Frank, T., Bosch, K., Fowler, G. et al., "Full-Scale Moving Motorcycle into Moving Car Crash Testing for Use in Safety Design and Accident Reconstruction," SAE Technical Paper 2012-01-0103, 2012, doi:10.4271/2012-01-0103.

[14] Bready, J., May, A., and Allsop, D., "Physical Evidence Analysis and Roll Velocity Effects in Rollover Accident Reconstruction," SAE Technical Paper 2001-01-1284, 2001, doi:10.4271/2001-01-1284.

[15] Kubly, K.D. and Buse, C.R., "Motorcycle Post-Accident Inspection Techniques," SAE Technical Paper 850064, 1985, doi:10:4271/850064.

[16] Campbell, J.B. and Wynne, R.H., *Introduction to Remote Sensing*, 5th ed., The Guilford Press, New York, 2011, ISBN 978-1-60918-176-5.

[17] Voitel, T. and Terpstra, T., "Benefits of 3D Laser Scanning in Vehicle Accident Reconstruction," Technology White Paper, FARO Technologies, Inc., 2012.

[18] DiTallo, M., Brandt, J., and Green, T.E., "Laser Scanner Basics for Public Safety – Scene Marking, Scanner Placement, and Scanner Settings," FARO White Paper, 2017, https://media.faro.com/-/media/Project/FARO/FARO/FARO/Resources/2021/01/15/22/32/Whitepaper-Laser-Scanner-basics-for-public-safety--scene-marking-placement--settings-ENG.pdf?rev=7cfca746a4fb4af0853affc3a1b1ccdc, accessed on November 6, 2021.

[19] Wolf, P., Dewitt, B., and Wilkinson, B., *Elements of Photogrammetry with Application in GIS*, 4th ed., McGraw-Hill Professional Publishing, New York, 2014, ISBN 978-0071761123.

[20] Campbell, A.T. and Friedrich, R., "Adapting Three-Dimensional Animation Software for Photogrammetry Calculations," SAE Paper Number 930904, 1993, doi:10.4271/930904.

[21] Baker, K.S., "Chapter 9: Photogrammetry for Collision Analysis," *Traffic Collision Investigation*, 10th ed., Northwestern University Center for Public Safety, Evanston, IL, 2002, ISBN 0-912642-09-2.

[22] Tumbas, N.S., "Photogrammetry and Accident Reconstruction: Experimental Results," SAE Paper Number 940925, 1994, doi:10.4271/940925.

[23] Pepe, M.D., "Accuracy of Three-Dimensional Photogrammetry as Established by Controlled Field Tests," SAE Paper Number 930662, 1993, doi:10.4271/930662.

[24] Neale, W.T., Hessel, D., and Terpstra, T., "Photogrammetric Measurement Error Associated with Lens Distortion," SAE Paper Number 2011-01-0286, 2011, doi:10.4271/2011-01-0286.

[25] Massa, D.J., "Using Computer Reverse Projection Photogrammetry to Analyze an Animation," SAE Paper Number 1999-01-0093, 1999, doi:10.4271/1999-01-0093.

[26] Neale, W., Rose, N.A., and Hughes, C., "Determining Crash Data Using Camera-Matching Photogrammetric Technique," SAE Paper Number 2001-01-3313, 2001, doi:10.4271/2001-01-3313.

[27] Coleman, C., Tandy, D., Colborn, J., and Ault, N., "Applying Camera Matching Methods to Laser Scanned Three-Dimensional Scene Data with Comparisons to Other Methods," SAE Technical Paper 2015-01-1416, 2015, doi:10.4271/2015-01-1416.

[28] Chou, C., McCoy, R., and Rose, N., "Image Analysis of Rollover Crash Test Using Photogrammetry," SAE Paper Number 2006-01-0723, 2006, doi:10.4271/2006-01-0723.

[29] Rose, N.A., McCoy, R.W., and Chou, C.C., "A Method to Quantify Vehicle Dynamics and Deformation for Vehicle Rollover Tests Using Camera-Matching Video Analysis," *SAE International Journal of Passenger Cars—Mechanical* Systems, 1(1): 301–317, 2008, doi:4271/2008-01-0350.

[30] Manuel, E.J., Mink, R., and Kruger, D., "Videogrammetry in Vehicle Crash Reconstruction with a Moving Video Camera," SAE Technical Paper 2018-01-0532, 2018, doi:10.4271/2018-01-0532.

[31] Breen, K.C. and Anderson, C.E., "The Application of Photogrammetry to Accident Reconstruction," SAE Paper Number 861422, 1986, doi:10.4271/861422.

[32] Woolley, R.L., White, K.A., Asay, A.F., and Bready, J.E., "Determination of Vehicle Crush from Two Photographs and the Use of 3D Displacement Vectors in Accident Reconstruction," SAE Technical Paper 910118, 1991, doi:10.4271/910118.

[33] Husher, S.E., Varat, M.S., and Kerkhoff, J.F., "Survey of Photogrammetric Methodologies for Accident Reconstruction," *Proceedings of the Canadian Multidisciplinary Road Safety Conference VII*, Vancouver, BC, Canada, June 1991.

[34] Bruce, W.M. and Knopf, E.A. "A New Application of Camera Reverse Projection in Reconstructing Old Accidents," SAE Paper Number 950357, 1995, doi:10.4271/950357.

[35] Smith, G.C. and Allsop, D.L., "A Case Comparison of Single-Image Photogrammetry Methods," SAE Paper Number 890737, 1989, doi:10.4271/890737.

[36] Goldberg, N., *Camera Technology – The Dark Side of the Lens*, Academic Press Inc., San Diego, CA, 1992, ISBN 0-12-287570-2.

[37] Taylor, J.T., *The Optics of Photography and Photographic Lenses*, Adamant Media Corporation, London, 2005, ISBN 1-4212-4851-4.

[38] Rose, N.A., Carter, N., Pentecost, D., and Voitel, T., "Using Data from a DriveCam Event Recorder to Reconstruct a Vehicle-to-Vehicle Impact," SAE Technical Paper 2013-01-0778, 2013, doi:10.4271/2013-01-0778.

[39] Rose, N.A. and Carter, N., "Using Data from a DriveCam Video Event Recorder to Reconstruct a Hard Braking Event," *Collision: The International Compendium for Crash Research*, 7(1), 2012, ISSN 1934-8681.

[40] Gates, T.J., "Analysis of Dilemma Zone Driver Behavior at Signalized Intersections," Paper No. 07-3351, *Transportation Research Board, 2007 Annual Meeting*, 2030(1): 29–39, 2007, doi:10.3141/2030-05.

[41] Vergauwen, M., "Wide Baseline 3D Reconstruction from Digital Stills," ISPRS Workshop on Visualization and Animation of Reality-Based 3D Models, Tarasp-Vulpera, Engadin, Switzerland, February 2003.

[42] Strecha C., von Hansen, W., Van Gool, L., and Thoennessen, U., "Multi-view Stereo and Lidar for Outdoor Scene Modelling," *International Archives of Photogrammetry, Remote Sensing and Spatial Information Science*, 2007.

[43] Dai, F., Rashidi, A., Brilakis, J., and Vela, P., "Comparison of Image-Based and Time-of-Flight-Based Technologies for 3D Reconstruction of Infrastructure," *Construction Research Congress 2012: Construction Challenges in a Flat World*, 2012, ISBN 9780784412329.

[44] Erickson, M., Bauer, J., and Hayes, W., "The Accuracy of Photo-Based Three-Dimensional Scanning for Collision Reconstruction Using 123D Catch," SAE Technical Paper 2013-01-0784, 2013, doi:10.4271/2013-01-0784.

[45] Terpstra, T., Voitel, T., and Hashemian, A., "A Survey of Multi-View Photogrammetry Software for Documenting Vehicle Crush," SAE Technical Paper 2016-01-1475, 2016, doi:10.4271/2016-01-1475.

[46] Jurkofsky, D., "Accuracy of SUAS Photogrammetry for Use in Accident Scene Diagramming," *SAE International Journal of Transportation Safety*, 3(2): 136–152, 2015, doi:10.4271/2015-01-1426.

[47] Carter, N., Hashemian, A., and Rose, N., "Evaluation of the Accuracy of Image Based Scanning as a Basis for Photogrammetric Reconstruction of Physical Evidence," SAE Technical Paper 2016-01-1467, 2016, doi:10.4271/2016-01-1467.

[48] Fay, R., Robinette, R., and Larson, V., "Engineering Models and Animations in Vehicular Accident Studies," SAE Technical Paper 880719, 1988, doi:10.4271/880719.

[49] Dilich, M. and Goebelbecker, J., "Accident Investigation and Reconstruction Mapping with Aerial Photography," SAE Technical Paper 960894, 1996, doi:10.4271/960894.

[50] Wirth, J., Bonugli, E., and Freund, M., "Assessment of the Accuracy of Google Earth Imagery for use as a Tool in Accident Reconstruction," SAE Technical Paper 2015-01-1435, 2015, doi:10.4271/2015-01-1435.

6

Sliding and Tumbling of the Motorcycle and Rider

Most motorcycle crashes involve the motorcycle capsizing during the crash sequence and sliding for some distance. Often, when performing speed calculations, the reconstructionist will need to calculate the energy or speed loss for this capsizing and sliding phase. One approach is to assume that a motorcycle decelerates at a constant rate as it slides and tumbles along the roadway. This assumption is adequate for most crash reconstruction applications, though there will be some variability in the deceleration along the slide distance depending on which motorcycle components are engaging the road surface at any point in time. If the motorcycle slides across multiple surfaces, different decelerations may need to be assigned for each surface. In practice, the reconstructionist would determine the slide distances based on the physical evidence, and then a range of decelerations would be selected from physical tests reported in the literature for motorcycles with similar characteristics sliding on similar surfaces.

Decelerations for a Sliding or Tumbling Motorcycle

Day and Smith [1] reported motorcycle sliding decelerations in 1984, analyzing the decelerating forces produced by two downed motorcycles on various surfaces—a 1967 Honda CB305 (a standard motorcycle) and a 1973 Yamaha 550 Special (another standard motorcycle). These authors towed the motorcycles using a rope (Figure 6.1) with an inline force gauge and documented the forces required to pull the motorcycles at 1 and 40 km/h (25 mph). For pavement, they reported sliding friction factors (average decelerating forces) between 0.45 and 0.58 during the tests at 40 km/h (25 mph). For gravel, the friction factor was between 0.68 and 0.79 and for grassy earth, 0.79. Day and Smith noted that "during

FIGURE 6.1 Day and Smith's setup for towing motorcycles across the pavement [1].

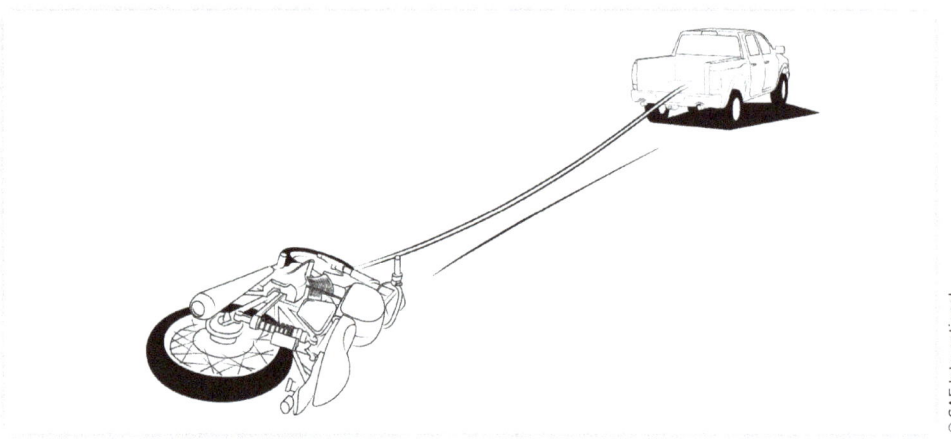

the testing, it was observed that projecting elements, such as the foot pegs and handlebars would tend to plow into the soil at low speeds, creating momentary high drag factors." In this testing, the motorcycle started in a capsized position on the ground, and so deceleration from a fall was not included in the friction factors.

Lambourn [2] explored how the average decelerations experienced by sliding motorcycles differed between tests where the vehicle was already capsized and dragged at low speeds and tests where the motorcycle was dropped at a higher speed and allowed to slide to rest. For the higher speed tests, motorcycles were dropped from a low platform (already on their side) or allowed to fall to their side from an upright position. The photographs of Figure 6.2 depict the test setup for the upright drops and one of the tests from this setup. In his literature review, Lambourn noted that some studies had reported a speed-dependence

FIGURE 6.2 Setup and sample test using the upright drop method [2].¹

---

1. Much thanks to Richard Lambourn for his efforts digging these photographs up from his archives and for giving me permission to use them.

on the sliding deceleration, with the rate decreasing with increasing speed. He examined this issue of speed dependence in his testing.

Lambourn reported that the decelerations measured "in the low-speed drag tests gave a value close to the high-speed sliding value. The friction was affected by the road surface texture, the presence of prominent side projections, and the wearing away of these projections during the slide. Some speed dependence was noted in the upright-launch tests which appears to be due to the 'digging-in' of the machine as it falls to the road, rather than an effect of the sliding friction itself." He also concluded that "the reason for there being a clear speed dependence in the results of Becke and of Ashton, but not in the tests reported here, is almost certainly due to the fact that in both their experimental methods the motorcycles were dropped a distance onto the road surface. This would subject the machines to a large decelerating impulse as they struck the ground, which would considerably increase the average deceleration in low-speed tests but be relatively unimportant in high-speed runs." Overall, Lambourn's test produced a range of decelerations of 0.43 g ± 0.11 g. For the motorcycles equipped with crash bars the range was 0.32 g ± 0.05 g.

Other authors have also suggested that differences in test methodology account for some of the variability in the decelerations for sliding motorcycles and the apparent speed dependence in the decelerations. For instance, Baxter [3] stated: "One word of caution; read the test methodology as to how the friction values were obtained. In some drop tests, the motorcycle was pre-positioned laterally a few inches/centimeters above the road surface. [Friction values obtained] using this method are generally lower than a motorcycle falling from vertical (normal position) onto its side on the road." Hague [4], Wood et al. [5], and Walsh et al. [6] also discussed the influence of the fall (capsize) on the deceleration. Hague stated: "The methodology of a motorcycle slide test can significantly affect the measured deceleration rate, apparently due to the speed lost in the initial ground impact. Those tests in which motorcycles were dropped from an increased height resulted in increased deceleration rates. During a road traffic accident, the motorcycle will also lose speed upon ground impact and testing should therefore try to mimic this process. Bearing this in mind, the most appropriate tests for the majority of collisions would be those in which an upright motorcycle was allowed to capsize from a normal height."

This is different from the approach proposed by Wood et al. [5]. They proposed splitting the speed calculations into separate phases for the capsize and the slide. Thus, within their method, the ideal test procedure for the sliding phase would not include a fall of the motorcycle. Walsh et al. [6] presented a method for incorporating the speed loss from a fall into the calculation of the deceleration from a test or the calculation of the initial speed of the motorcycle in a reconstruction. They also partitioned the test data into the following categories: motorcycles with crash bars, unfaired motorcycles, and fully or partially faired motorcycles. They found that the sliding deceleration (with the capsize phase removed) for fully faired motorcycles was 0.408 g ± 0.036 g. For unfaired motorcycles the sliding deceleration was 0.46 g ± 0.092 g. For motorcycles with crash bars the deceleration was 0.286 g ± 0.037 g.

Donohue [7] reported sliding decelerations for a 1982 Kawasaki KZ1000 Police Special. Testing with this motorcycle, which was conducted at the Los Angeles Police Department's Specialized Collision Investigation Detail, utilized a flatbed truck with a lift gate on the rear. The lift gate was positioned parallel to the roadway and approximately 6 in. above the road surface. The motorcycle, which was facing along the direction of travel, was dropped an upright position with its front tire on the lift gate and its rear tire on the roadway. The roadway was a residential, asphalt roadway next to Dodger Stadium. The initial speed of the motorcycle was measured with radar. Donohue reported five tests with sliding decelerations between 0.38 and 0.50.

In 1995, Raftery [8] slid an unknown motorcycle wearing Suzuki Katana fairings from an initial speed of 85 km/h (53 mph) and reported an average sliding deceleration of 0.26 g. Another test, apparently from a similar speed, resulted in the same 0.26 g. Raftery tested the same motorcycle with the fairings removed. The resulting deceleration was 0.33 g. The deceleration was calculated using the initial drop speed and the documented sliding distance. Raftery's methodology involved suspending the motorcycle from a boom at the rear of a tow truck, driving the tow truck up to the test speed, and releasing the motorcycle from the boom. It appears from Raftery's description of his tests that he suspended the motorcycle from the boom with the wheels in contact with the ground and when the motorcycle was released it fell to the road surface. Raftery noted that "transfer marks from the coloured fairing were readily visible on the asphalt surface beginning at the location where the motorcycle first contacted the roadway, allowing an easy identification of the location of the start of the slide."

Carter et al. [9] tested eight motorcycles (Figure 6.3) on three surfaces (asphalt, dirt, and gravel) to determine their sliding decelerations from target speeds of 48 and 97 km/h (30 and 60 mph). The tests run on off-road surfaces utilized a target speed of 48 km/h (30 mph). The tested motorcycles included the following motorcycle types: standard, cruiser, sport, and touring. Carter et al. reported 50 tests. They concluded that "some speed effects were observed, i.e., for higher speeds, the slide coefficient was lower (likely due to heat softening of structure contact points with the pavement)." Also, "for full fairing equipped motorcycles the slide coefficient was consistently lower than for non-fairing equipped motorcycles" and "deep gouges left by the motorcycle in the surfaces corresponded with higher slide coefficients." Carter et al. attempted to improve on prior studies by developing a test rig that allowed for consistent positioning and release of the motorcycles. The motorcycles were positioned front wheel forward and on their left or right sides. The motorcycles were released from a position with the lowest point on the side of the motorcycle approximately 5 cm above the ground. As Carter et al. noted, "this release height was chosen to minimize the impact forces upon release, therefore restricting (to the extent possible) the tests only to energy dissipated during sliding."

Medwell et al. [10] performed four motorcycle sliding tests using a fully faired 1992 Kawasaki ZX-7 Ninja. They stated that "the tests were designed to approximate, as closely as possible, the motion of a motorcycle falling over from an upright position. The motorcycle was positioned upright on a fabricated platform mounted on the right side of a pickup truck … The height of the platform was adjusted so that its underside was as close as possible to the roadway surface. This test setup resulted in the motorcycle tire contact surface being approximately 90 mm above the roadway. The motorcycle was held upright by an assistant riding in the bed of the pickup truck. The truck was accelerated to the test speed, then the motorcycle was released and allowed to fall over sideways onto the road surface." In two of the tests, the Kawasaki initially slid along the pavement but then traveled into a nearby area of grass, making them difficult to analyze. However, two of the tests were confined to the asphalt. Both had a release speed of approximately 80 km/h (50 mph). The motorcycles slid for 69.5 and 86.3 m (228 and 283 ft) before coming to rest. The calculated decelerations were 0.36 g and 0.29 g. The 0.36 value was obtained during the test involving the right side of the motorcycle, which was the exhaust side.

Bartlett et al. [11] reported motorcycle drop tests from Motorcycle Crash Reconstruction classes conducted at the Institute for Police Technology and Management (IPTM) from 1987 to 2006. These tests were conducted on asphalt or concrete, but the surfaces varied from class to class. The drop techniques also varied from class to class. These authors observed that "the results are a chaotic mix of sliding and tumbling, not unlike real motorcycle crashes." The dataset initially consisted of 237 drop tests using 107 different motorcycles. Twenty tests were discarded because the reported drop speed, slide distance, and deceleration were inconsistent with each other. Additional tests were excluded in which the motorcycle was dropped from a pickup bed or in which the motorcycle slid off the road surface onto the off-road terrain. The final dataset included 162 tests with 99 different motorcycles. These authors reported that the decelerations trended slightly higher with increasing speed and that the overall average deceleration for all the tests was 0.521 g ± 0.140 g. Bartlett et al. also combined their dataset with other available datasets. This resulted in 386 tests for which they reported decelerations of 0.480 g ± 0.134 g. These authors noted that in their dataset there were nine tests of sport motorcycles. They combined the results of these tests with sports bike tests from other studies and reported decelerations for these motorcycles of 0.374 g ± 0.082 g. These authors also reported that "the comments for 19 tests included the words bounce, tumble, or flip. Taken together, they ranged from 0.41g to 0.93g and averaged 0.619 ± 0.161 g. The comments for 6 tests specifically noted words to the effect of 'smooth slide.' Taken together, their average was 0.370 ± 0.093 g. Tests of motorcycles that were equipped with crash bars were noted to generate drag values of 0.26 g, 0.27 g, 0.37 g, and 0.60 g. One test noted to be 'wet' came in at 0.49 g."

FIGURE 6.3 Motorcycles tested by Carter et al. [1996].

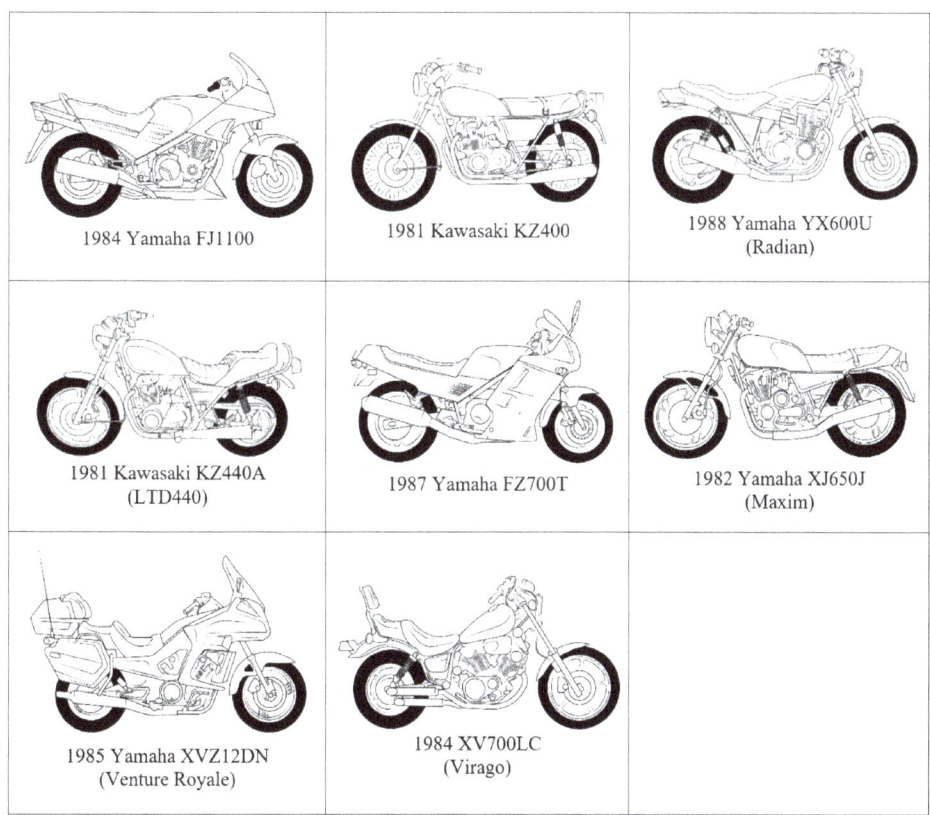

In 2003, McNally and Bartlett [12] slid a fully faired Suzuki Katana at IPTM's Special Problems and analyzed the results via frame-by-frame video and field data (known initial speed and measured slide distance). Video analysis yielded a deceleration of 0.42 g, while the sliding distance and known initial speed yielded a result of 0.39 g. Bartlett et al. [11] have also reported nine additional tests performed using fully faired motorcycles during IPTM classes over the years. The individual results were not detailed in the paper, but combined with the data from Raftery, Medwell et al., and McNally and Bartlett, the total set of 14 tests had an average coefficient of friction of 0.37 g with a standard deviation of 0.08 g.

In 2004, Hague [4] compiled data from prior studies where motorcycles capsized and then slid to rest. Hauge concluded that "the analysis shows that a more accurate estimation of deceleration rate can be made if the motorcycles are split into two different categories, based on the presence of fairing, crash bars and/or panniers." He noted that, while "one might expect partially faired motorcycles to have lower deceleration rates than unfaired machines … the two categories exhibit similar deceleration rates. Perhaps also initially surprising is that fully faired machines equipped with panniers gave similar results to the partially/unfaired motorcycles." In relation to crash bar equipped motorcycles, the only available tests were those conducted by Lambourn [2], in which the decelerations varied between 0.25 and 0.35. Hague noted that, "As expected, fully faired and crash bar equipped motorcycles decelerated at relatively low rates. The crash-bar-equipped results are probably artificially low because they were all dropped from a very low height. If they had capsized from an upright position, they would have lost additional speed on striking the ground which would increase the average deceleration rate, more so at lower speeds." Hague reported an average deceleration for partially faired and unfaired motorcycles of 0.39 and for fully faired motorcycles of 0.27.

Peck et al. [13] analyzed 15 actual crashes of sport motorcycles equipped with frame sliders and 14 controlled tests of motorcycles of various types not equipped with frame sliders. Frame sliders, usually composed of a plastic composite, are mounted to the sides of motorcycles to mitigate damages during a fall. The crashes occurred during track days or races at the New Hampshire Motor Speedway and the New Jersey Motorsports Park. The motorcycles were equipped with GPS data acquisition systems measuring the motorcycle speed at either 5 or 10 Hz. The authors reported sliding decelerations of 0.45 g (SD = 0.09) for the track crashes and 0.48 g for the controlled tests. According to the authors, their data showed that "frame sliders do not lower the drag factor of a sport bike, but actually increase it." The authors cited prior data for fully faired sports bikes without frame sliders with decelerations significantly lower than what they found for the motorcycles with frame sliders. Three of the track crashes involved sections where the motorcycle slid on hard packed dirt. These dirt slides had decelerations of 0.45 g, 0.61 g, and 1.11 g. This study also noted that there is variability in the severity of first contact of a motorcycle with the ground, and thus, variability in how the initial ground contact contributes to the overall deceleration. For example, a motorcycle would generally strike the ground with less severity during a low-side fall than during a high-side fall.

DiTallo et al. [14] examined three different test methods for determining the drag factor for motorcycles sliding on their sides. This testing utilized 26 motorcycles sliding on an asphalt roadway in North Las Vegas, Nevada. The following three methods were utilized: (1) dragging an already capsized motorcycle across the pavement; (2) releasing a motorcycle from the rear hydraulic lift of a box truck and allowing it to fall to the pavement and slide to rest; and (3) towing a motorcycle behind a moving vehicle with its front tire held in a pneumatic clamp until its release. After release, the motorcycle would fall to the ground and slide to rest. DiTallo et al. noted that many of the motorcycles had been previously utilized in impact testing at the 2016 ARC-CSI conference. These motorcycles included sport, touring, and motocross motorcycles and mopeds. One of the sport motorcycles had frame sliders.

A total of 36 pull tests (Method 1) were conducted with nine motorcycles. The motorcycles were pulled in varying orientations. This series of tests resulted in a range of drag factors between 0.36 and 0.64 and an average of 0.51. DiTallo et al. also noted that the tests with tires leading exhibited an average drag factor of 0.54 and the tests with the tires trailing exhibited an average drag factor of 0.51. Drop tests (Method 2) were conducted with 12 motorcycles at speeds ranging from 30.7 to 41.1 mph. These tests produced drag factors between 0.28 and 0.70. Clamp tests (Method 3) were conducted with five motorcycles at speeds ranging between 29 and 43.9 mph. These tests resulted in average drag factors between 0.28 and 0.43. This group contained the sport motorcycle with frame sliders. In addition to parsing their data by test methodology, DiTallo et al. also considered motorcycle type. They found that the sport motorcycles had drag factors between 0.28 and 0.61, the touring motorcycles had drag factors between 0.28 and 0.60, the mopeds had drag factors between 0.35 and 0.54, and the motocross motorcycle had a drag factor of 0.43.

Data from the studies reviewed here (along with a few additional studies) were compiled in a spreadsheet, and illustrative decelerations were calculated for different motorcycle types on asphalt and concrete road surfaces. All test types were combined, and no attempt was made to adjust for different test procedures (i.e., drop versus drag). Means and standard deviations of these decelerations are reported in Table 6.1. These decelerations exhibit a mild dependence on motorcycle type.

TABLE 6.1 Average sliding deceleration for different motorcycle types (asphalt/concrete)

Motorcycle type	Average sliding deceleration on asphalt/concrete (g)
Standard	0.45 ± 0.15
Cruiser/touring	0.50 ± 0.11
Sport	0.47 ± 0.10
Dirt/enduro	0.50 ± 0.12
Scooter/moped	0.49 ± 0.11

Table 6.1 does not address motorcycle-to-motorcycle differences or characteristics that could influence the decelerations (i.e., whether the motorcycle was fully or partially faired and what components were present to scrape and gouge the road surface or prevent scraping and gouging, for instance). These characteristics are addressed in Table 6.2. As this table shows, there is strong dependence between certain characteristics of motorcycles and the sliding deceleration. Motorcycles with fairings produce decelerations lower than those without fairings. Crash bars reduce the decelerations even further. The values listed in these tables are illustrative. As is evidence in some of the previously cited studies, motorcycle slide tests can be parsed into even more specific groups (i.e., flip and tumble versus smooth slide). Adding additional constraints on the grouping would, of course, alter the ranges listed in the tables. A reconstructionist who is applying tests from the literature to their reconstruction will have to make judgments regarding which tests are most applicable to specific crash they are analyzing.

©2022, SAE International

TABLE 6.2 Average sliding deceleration for different motorcycle characteristics (asphalt/concrete)

Motorcycle characteristics	Average sliding deceleration on asphalt/concrete (g)
No fairings or crash bars	0.54 ± 0.11
Partially faired	0.48 ± 0.11
Fully faired	0.46 ± 0.10
Crash bars	0.34 ± 0.07
Sports bikes with frame sliders	0.45 ± 0.10

© SAE International

Decelerations for a Sliding or Tumbling Rider

When a motorcycle capsizes, the rider will typically separate from the motorcycle and will slide or tumble along the ground. The rider will usually decelerate at a different and higher rate than the motorcycle, though this will not always be the case. If the point of separation between the rider and the motorcycle can be determined, along with the rest positions of both the rider and the motorcycle, the rider's slide/tumble distance can sometimes be used as further confirmation of the calculated speed of the motorcycle at the point of separation. As with the motorcycle, the assumption will typically be made that a sliding or tumbling rider decelerates at a constant rate. Again, this assumption is adequate for most crash reconstruction applications, though there will be some variability in the deceleration along the slide distance. As with the motorcycle, if the rider slides across multiple surfaces, different decelerations may need to be assigned for each different surface. The reconstructionist can determine the slide distance based on the physical evidence and then use a range of decelerations for the rider selected from physical tests reported in the literature.

Severy and Brink [15] observed that a sliding and/or tumbling person "has a higher effective drag coefficient than the automobile undergoing emergency braking …." In making this statement, Severy and Brink were including the speed loss that occurred when the person first struck the ground, along with the speed loss that occurs during the sliding following that initial impact. Referencing the CRASH2 User's Manual, Warner et al. [16] reported a range of decelerations for a "motorcycle skidding on [its] side" of between 0.55 and 0.7 g. They reported a deceleration for a "human body skidding" of 1.1 g and for a "human body tumbling" of 0.8 g. Haight and Eubanks [17] reported decelerations between 0.8 and 0.95 g for anthropomorphic test devices (ATDs) sliding on the ground after being involved in vehicle versus bicycle collisions. These decelerations also include the initial collision with the ground. In discussing motorcycle/vehicle collisions where the rider is thrown through the air, Collins [18] stated that "a rider launched from a motorcycle may fall to the ground or be slammed into the pavement, so on paved surfaces the ejected rider's friction coefficient will range between 0.8 and 1.2."

Fricke [19], on the other hand, reported the following drag factors for people sliding on various surfaces: grass, 0.45 to 0.70; asphalt, 0.45 to 0.60; and concrete, 0.40 to 0.65. Contrary to those in the previous paragraph, these drag factors exclude the speed loss the pedestrian experiences from landing on the road surface. Thus, if these drag factors were utilized for speed analysis, the additional speed loss would need to be accounted for separately.

Happer et al. [20] observed that the literature contains decelerations (drag factors) for people that are defined in several different ways: (1) a simple sliding coefficient of friction between the person and the ground; (2) a drag factor that includes the speed loss from when the person first impacts the ground; or (3) an average drag factor for the entire trajectory of the person, including the airborne phase. The first two of these are similar to issues that arise in different test procedures used to obtain decelerations of sliding or tumbling motorcycles. Happer et al. conducted an extensive literature review and found that "regardless of the ground surface (i.e. asphalt, grass, wet or dry), the drag factor for a pedestrian sliding along the ground ranges from about 0.40 to 0.72. The drag factor for a tumbling pedestrian [meaning the impact with ground is included] ranges from 0.7 to 1.22. The drag factor significantly increases for the tumbling pedestrian as the pedestrian will lose speed from impacting the ground after being vaulted through the air. The effective pedestrian drag factor for the full trajectory ranges from 0.37 to 0.79."

Wood [21] noted that the deceleration of a person sliding on the ground is decreased if the surface is wet. Happer et al. [20] noted that "if the road surface is not level and has a substantial grade, then this slope should be considered. In addition, the reviewed data did not include a significantly slippery road surface (e.g. ice); thus, a very slippery surface may also significantly affect the pedestrian drag factor." Brach and Brach [22] listed the following formula that can be used to adjust a flat-ground deceleration (f) for application to a sloped surface ($f_{adjusted}$). In this equation, θ is the angle of the slope in degrees. This angle is positive for an upgrade and negative for a downgrade.

$$f_{adjusted} = f \cdot \cos\theta + \sin\theta \tag{6.1}$$

Fugger et al. [23] reported a series of 160 pedestrian crash tests utilizing high-fronted vehicles that would generate forward projection trajectories. An Alderson Research Labs CG-95 dummy was utilized to represent the pedestrians in these tests (75.5 in tall, 169 lb). The dummy was clothed in a wetsuit covered by coveralls and standard athletic footwear. The following vans were used in this test series: a 1976 Ford Econoline 250, a 1971 Dodge B200, a 1982 Dodge B250, a 1980 Plymouth D100, a 1982 Chevrolet G20, and a 1977 Dodge Sportman. The leading edge of the hood of each of these vehicles was above the center of gravity of the dummy such that they would produce forward projection trajectories. Of the 160 tests, 56 were conducted on dry asphalt and 84 on wet asphalt. Impact speeds varied between 4 and 60 km/h, with most of the tests conducted at speeds below 32 km/h. Fugger et al. reported average decelerations of the sliding pedestrians of 0.43 for the dry asphalt and between 0.31 and 0.41 for the wet asphalt, depending on the water depth. These decelerations excluded the initial impact with the ground.

Baxter [3] states that "the occupants of a motorcycle thrown to a highway surface will decelerate quite rapidly, tumbling or rolling at first, then sliding to a stop. Various test by Hart and Baird [24], using cadavers or specialized instrumented dummies, show a sliding friction range of 0.90 g to 1.2 g. Clothing worn does have an effect on the value. Leather clothing and a full coverage helmet reduce the friction value to a level of 0.60 g to 0.70 g … Values for other clothing, such a polyester (0.70 g) and cotton or wool (0.70 g to 0.85 g), were also established." In reviewing the Hurt and Baird study, Baxter appears to be citing incorrect values. That study actually states that "when the rider or passenger falls to the pavement, the deceleration to the point of rest will usually be rapid. When the rider is wearing soft cloth apparel, the deceleration will be on the order of 0.9 to 1.2 g, with the mode including violent tumbling and rolling. Such violent motion and high abrasion may tear the soft clothing from the downed rider. Only when the rider is clad with full leathers, boots, and protective headgear is the mode of deceleration likely to change. Then the lower friction of the leather clad body will reduce the tumbling and rolling tendency and also reduce the deceleration to the range of 0.7 to 0.9 g." Thus, the deceleration for a rider clothed in leather clothing and a full coverage helmet was 0.7 to 0.9 g, not 0.6 to 0.7 g.

Determining the Initial Speed for a Sliding Motorcycle or Rider

The slide distance of a motorcycle would be measured from the initiation of sliding evidence on the roadway to the rest position of the motorcycle or rider and should include gaps in the visible marks [4]. Once the slide distance has been determined and a range of decelerations assigned, the speed of the motorcycle or rider at the beginning of the slide (v_{slide}) can be determined with the following equation.

$$v_{slide} = \sqrt{2gf_{slide}d_{slide}} \tag{6.2}$$

In this equation, d_{slide} is the slide distance, f_{slide} is the deceleration during the slide, and g is the gravitational constant. Equation (6.2) can be applied with decelerations that already include the speed loss from capsizing. However, if the reconstructionist wants to account for this speed loss separately, the following equation could be used [6, 25, 26]:

$$v_{land} = f_{slide}v_z + \sqrt{2gf_{slide}d_{slide}} \tag{6.3}$$

In Equation (6.3), v_z is the vertical impact velocity of the center of gravity of the motorcycle. If this equation was being utilized, then the selected deceleration rates should not already include the speed loss from capsizing. Walsh et al. [6] presented the following equation for estimating the downward velocity of motorcycle that capsizes inline (slip angle = 0) under the influence of gravity without a rider:

$$v_z = \sqrt{\frac{2gh_{cg} - gf_w}{1 + \left(\dfrac{k_r}{h_{cg}}\right)^2}} \tag{6.4}$$

In this equation, g is the gravitational constant, h_{cg} is the center of gravity height of the motorcycle, f_w is the frame width, and k_r is the roll radius of gyration of the motorcycle. This equation could be applied to calculate the downward velocity of the motorcycle in motorcycle drop tests, but it may not apply to many real-world crashes, where the motorcycle motion will also be influenced by tire forces and interaction with the rider. Equation (6.4) could be adapted to include a rider. The center of gravity height would become the combined center of gravity height for the motorcycle and the rider and the roll radius of gyration would be for the motorcycle and the rider about the combined center of gravity height.

How to Calculate Speed Loss for a Sliding Motorcycle

- Measure the distance from the first evidence of motorcycle capsizing (usually a scrape or a gouge) to the postcrash rest position of the motorcycle.

- Identify the road surface type for the subject crash.
    a. Pavement (asphalt or concrete)
    b. Wet pavement
    c. Dirt
    d. Gravel
    e. Grass
- Identify relevant features of the subject motorcycle.
    a. Fairings
    b. Crash Bars
    c. Panniers
    d. Frame sliders
- Identify tests from the literature that share a similar road and motorcycle features to those in the subject crash. Determine the range of decelerations from these tests. What is the mean? Standard deviation?
- A conservative approach would be to use the entire deceleration range or to apply the mean and standard deviation. On the other hand, the reconstructionist could ask: Are there any features of the subject crash that could push the deceleration to the lower or higher end of the range? For example, if the motorcycle flips from one side to the other during the sliding phase, this will introduce an additional motorcycle impact with the ground, which would increase the overall deceleration.
- Consider the test procedure in the tests you are using. Do they include or exclude the impact between the motorcycle and the ground? Often, there will be a mix. Choose a set of tests that fits with how you are modeling the sliding phase. Does your model include the speed loss from the first impact with the ground in the sliding phase or does it treat it as a separate phase?
- Use the distance for the sliding phase, along with a reasonable range of decelerations, to calculate the energy or speed loss during the slide phase. This could utilize Equation (6.2).

References

[1] Day, T. and Smith, J., "Friction Factors for Motorcycles Sliding on Various Surfaces," SAE Technical Paper 840250, 1984, doi:10.4271/840250.

[2] Lambourn, R., "The Calculation of Motorcycle Speeds from Sliding Distances," SAE Technical Paper 910125, 1991, doi:10.4271/910125.

[3] Baxter, A.T., *Motorcycle Crash Investigation*, Institute of Police Technology and Management, Jacksonville, FL, 2017, ISBN 978-1-934807-18-7.

[4] Hague, D., "Calculation of Speed from Motorcycle Slide Marks," *Impact: The Journal of the Institute of Traffic Accident Investigators*, 13(1): 10–16, 2004, ISSN 0959-4302.

[5] Wood, D.P., Alliot, R., Glynn, C., Simms, C.K. et al., "Confidence Limits for Motorcycle Speed from Slide Distance," *Proceedings of the IMechE Vol. 222, Part D: J. Automobile Engineering*, 1349–1360, 2008, doi: 10.1243/09544070JAUTO731.

[6] Walsh, D.G., Wood, D.P., Alliot, R., Glynn, C. et al., "Motorcycle Capsize Mechanisms and Confidence Limits for Motorcycle Capsize Speeds from Slide/Bounce Distance," *18th EVU Conference*, Hinckley, UK, 2009.

[7] Donohue, M.D., "Motorcycle Skidding and Sideways Sliding Tests," *Accident Reconstruction Journal*, 3(4): 43, 1991, ISSN 1057-8153.

[8] Raftery, B., "Determination of the Drag Factor of a Fairing Equipped Motorcycle," SAE Technical Paper 950197, 1995, doi:10.4271/950197.

[9] Carter, T., Enderle, B., Gambardella, C., and Trester, R., "Measurement of Motorcycle Slide Coefficients," SAE Technical Paper 961017, 1996, doi:10.4271/961017.

[10] Medwell, C., McCarthy, J., and Shanahan, M., "Motorcycle Slide to Stop Tests," SAE Technical Paper 970963, 1997, doi:10.4271/970963.

[11] Bartlett, W., Baxter, A., and Robar, N., "Motorcycle Slide-to-Stop Tests: IPTM Data through 2006," *Accident Investigation Quarterly*, 46: 18–23, 2007, ISSN 1082-6521.

[12] McNally, B. and Bartlett, W., "Motorcycle Sliding Coefficient of Friction Tests," *Presentation at Institute of Police Technology and Management's (IPTM) Special Problems in Accident Reconstruction*, Jacksonville, FL, 2003

[13] Peck, L., Focha, W., and Gloekler, T., "Motorcycle Sliding Friction for Accident Investigation," *Proceedings of the 10th International Motorcycle Conference*, Institute for Motorcycle Safety, Essen, Germany, 62–67, 2014.

[14] DiTallo, M., Munyon, B., Green, T., Paul, E. et al., "3 Different Methodologies for Determining the Drag Factor for Motorcycles Sliding on Their Sides," *Collision: The International Compendium for Crash Research*, 12(1): 8–22, 2017, ISSN 1934-8681.

[15] Severy, D. and Brink, H., "Auto-Pedestrian Collision Experiments," SAE Technical Paper 660080, 1966, doi:10.4271/660080.

[16] Warner, C., Smith, G., James, M., and Germane, G., "Friction Applications in Accident Reconstruction," SAE Technical Paper 830612, 1983, doi:10.4271/830612.

[17] Haight, W. and Eubanks, J., "Trajectory Analysis for Collisions Involving Bicycles and Automobiles," SAE Technical Paper 900368, 1990, doi:10.4271/900368.

[18] Collins, J.C., *Accident Reconstruction*, Charles C. Thomas Publisher, Springfield, IL, 1979, ISBN 0-398-03907-0.

[19] Fricke, L.B., "Vehicle-Pedestrian Accident Reconstruction," *Accident Reconstruction Journal*, 1992, ISSN 1057-8153.

[20] Happer, A., Araszewski, M., Toor, A., Overgaard, R. et al., "Comprehensive Analysis Method for Vehicle/Pedestrian Collisions," SAE Technical Paper 2000-01-0846, 2000, doi:10.4271/2000-01-0846.

[21] Wood, D., "Application of a Pedestrian Impact Model to the Determination of Impact Speed," SAE Technical Paper 910814, 1991, doi:10.4271/910814.

[22] Brach, R.M. and Brach, R.M., *Vehicle Accident Analysis and Reconstruction Methods*, SAE International, 2005, ISBN 0768007763.

[23] Fugger, T., Randles, B., Wobrock, J., and Eubanks, J., "Pedestrian Throw Kinematics in Forward Projection Collisions," SAE Technical Paper 2002-01-0019, 2002, doi:10.4271/2002-01-0019.

[24] Hart, H.H. and Baird, J.D., "Accident Investigation Methodology Peculiar to Motorcycles and Minibikes," *2nd International Congress on Automotive Safety*, Paper No. 73051, 1973.

[25] Searle, J.A., "The Trajectories of Pedestrians, Motorcycles, Motorcyclists, etc., Following a Road Accident," SAE Technical Paper 831622, 1983, doi:10.4271/831622.

[26] Searle, J., "The Physics of Throw Distance in Accident Reconstruction," SAE Technical Paper 930659, 1993, doi:10.4271/930659.

7

Motorcycle Falls

The previous chapter provided data and methods for analyzing the sliding and tumbling phase of a motorcycle crash. This phase is always preceded by a fall of the motorcycle and the rider to the ground, either because of a collision or a loss of control. This chapter discusses the dynamic processes that lead to the motorcycle and the rider falling to the ground. The previous chapter discussed a simple capsize, a fall that would be most likely to occur in a parking lot or in a laboratory setting. This chapter analyzes falls that are more frequent in real-world crashes—low-side falls, high-side falls, falls because of front-wheel over-braking, impact-induced capsizing, and falls because of interaction with deteriorated portions of the roadway.

Low-Side Falls

Low-side falls typically occur during acceleration or braking, typically while the motorcycle is traversing a curve. The sequence begins when the longitudinal force at the rear tire, from either acceleration or braking, consumes enough of the available traction to leave insufficient traction available for maintaining lateral stability. This can occur during cornering, when the motorcycle needs lateral traction to maintain stability around the curve. It can also sometimes occur during straight-line braking if the rider brakes with sufficient force to lock the rear wheel. In either instance, the rear wheel begins sliding laterally, the motorcycle develops a yaw rotation, and the motorcycle and the rider roll to the inside of the yaw and fall to the ground. In this type of fall, the motorcycle will lead in the slide and, because the rider typically decelerates at a higher rate during sliding than the motorcycle, the motorcycle will usually come to rest further down the road than the rider. Figure 7.1 illustrates the motorcycle and rider motion for a low-side fall that results from the rider locking the rear wheel of the motorcycle.

FIGURE 7.1 Graphic of low-side fall sequence from locking the rear tire.

Steady-state riding on a straight road.

The rider applies the brakes, locking the rear tire.

The rear tire kicks out to the right. The motorcycle and rider lean to the left.

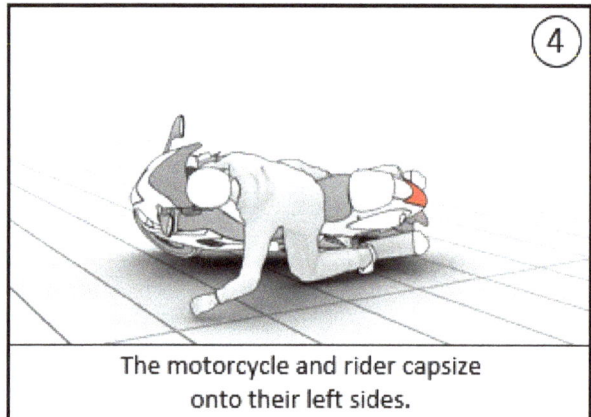
The motorcycle and rider capsize onto their left sides.

Case Study: A Low-Side Fall

To further explore the dynamics of low-side falls, consider a motorcycle crash that occurred in the Santa Monica Mountains of California on a curve of the Mulholland Highway called Edwards Corner. This crash was captured on video by Ken Snyder, a videographer who captures video at Edwards Corner and posts footage of crashes on his YouTube channel under the name RNickeymouse. To aid in this analysis, we mapped Edwards Corner using both a Sokkia total station and a Faro laser scanner. This enabled determination of the initial speed of the motorcycle, identification of rider actions, quantification of the motion of the motorcycle and rider, and characterization of the roadway radius and superelevation throughout the curve.

The following images depict the geometry and characteristics of Edwards Corner. Figure 7.2 is an aerial image showing the curve. This photograph is oriented such that north is up. Figure 7.3 is a photograph that shows the geometry of the curve from the vantage point of a nearby hillside. Riders traveling westbound through the curve would be traveling toward the viewer of this image. Figure 7.4 is another photograph of the ascent into the curve from the westbound direction. Motorcyclists traveling westbound around Edwards Corner would

FIGURE 7.2 Aerial photograph of Edwards Corner.

Map Data: Google and its Data Providers

be traveling into this photograph and traversing a leftward curve. Motorcyclists traveling eastbound would be coming toward the viewer of this photograph and would be traversing a rightward curve.

Survey and scan data were used to quantify the radius, slope, and cross-slope of this curve. Riders traveling westbound through Edwards Corner encounter a left-hand curve with an upslope. During the first half of the curve, the upslope is approximately 3.2° and during the second half it is 4.9°. There is a total elevation gain of 27.6 ft over a 390.8-ft travel distance. At the westbound entry to the curve, the cross-slope is negligible. It increases to a maximum cross-slope of 5.7° (with the inside of the curve being lower than the outside). The cross-slope then begins to decrease again, reaching a value of 1.9° at the exit of the curve. The radius of the curve in the center of the westbound lane is approximately 82 ft. Given the radius and cross-slope, a rider traveling through this curve at a speed in the range of 35 to 45 mph would need a leftward lean angle of approximately 40° to 55° relative to the roadway while traversing the curve [1, 2].

©2022, SAE International

FIGURE 7.3 Photograph of Edwards Corner.

FIGURE 7.4 Photograph of the ascent into Edwards Corner.

Figure 7.5 contains a series of video frames from the video of this crash. The crash involved a Kawasaki Ninja ZX-10R traveling westbound through Edwards Corner. The frames of Figure 7.5 begin after the motorcyclist has entered the curve and developed maximum lean. The frames end after both the rider and the motorcycle have begun sliding on the ground. The motorcycle and the rider are still in motion at the end of these video frames.

FIGURE 7.5 Video sequence (low-side fall).

Photo courtesy of Ken Snyder

The development of a right-side leading sideslip at the rear tire becomes evident by the second frame of video in Figure 7.5. The motorcycle begins yawing in a counterclockwise direction. During this yaw, the rear tire of the motorcycle generates a tire mark. Eventually, the rider and the motorcycle go down onto their left sides and begin sliding along the roadway. Oftentimes, motion like this will result in the motorcycle depositing at least one tire mark on the roadway prior to capsizing and scrapes or gouges on the roadway during the slide. For a low-side fall, there will typically be little distance between the end of the tire mark(s) and the beginning of scrapes and gouges on the roadway. Also, for a low-side fall, the motorcycle will typically lead in the slide, with the rider following behind.

High-Side Falls

Like low-side falls, high-side falls typically occur during acceleration or braking. The sequence begins similar to a low-side fall. The longitudinal force at the rear tire, from either acceleration or braking, consumes enough of the available traction to leave insufficient traction available for maintaining lateral stability. This typically occurs during cornering, where the motorcycle needs lateral traction to maintain stability. It can also sometimes occur during straight-line braking if the rider brakes with sufficient force to lock the rear wheel. In either instance, the rear wheel begins sliding laterally, the motorcycle develops a yaw rotation, and the motorcycle and the rider begin to lean to the inside of the yaw. If the rider does not release the throttle or brake, the motorcycle and the rider will continue to fall to the inside of the yaw until they strike the ground, resulting in a low-side. High-sides, on the other hand, occur if the rider releases the throttle or rear brake. This allows the rear tire to regain traction and a sudden increase in the available lateral traction results. This spike in lateral force at the rear tire can lead to the motorcycle and rider rolling to the opposite direction of the initial lean. The rider is then often projected upward and thrown ahead of the motorcycle [3, 4].

Figure 7.6 illustrates the motorcycle and rider motion that would be typical of a high-side fall. For a high-side fall, there will be a gap between a tire mark deposited by the rear wheel of the motorcycle and the scraping or gouging from the motorcycle impacting and sliding on the ground [5]. In addition to that, for a high-side the rider will initially be thrown out ahead of the motorcycle and will initially lead in the slide. That said, because the rider will typically decelerate at a higher rate than the motorcycle, the motorcycle may ultimately pass the rider and come to rest downstream of the rider.

FIGURE 7.6 High-side fall from locking the rear wheel.

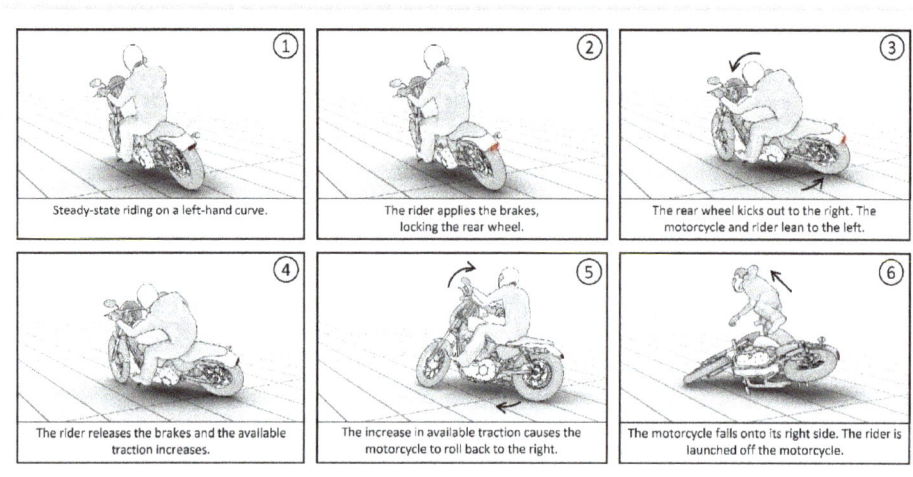

Case Study: A High-Side Fall

This section examines another motorcycle crash that Ken Snyder captured on video at Edwards Corner. This crash involved a Honda CBR1000RR (MY 2006 or 2007) traveling westbound through Edwards Corner. Figure 7.7 contains a series of frames from the video of this crash showing the motorcycle and rider motion. Snyder provided the video for this crash, which was captured at 59.94 fps. The time intervals between the frames shown in Figure 7.8 are not constant, but instead were chosen so that the sequence of images would give the reader a sense for the motion of the motorcycle and the rider during the loss of control and fall. The frames of this figure begin after the motorcyclist has entered the curve and developed maximum lean, which developed more than approximately 1 sec. The sequence of frames in this figure concludes with the rider and the motorcycle having struck the ground but before coming to final rest.

Tar marks and dark asphalt patches on the roadway, visible in Figure 7.8, were used as reference points to determine travel distances of the motorcycle and the rider across a series of video frames. Calculations based on the distances traveled by the motorcycle between reference points and the known frame rate established that the motorcycle was traveling at a speed of 43 mph as he approached the exit of the curve. Prior to this, the rider had been applying throttle to increase his speed, as was indicated by the audio of the video. Beyond this point, the speed began to decrease, as expected.

The development of a right-side leading sideslip at the rear tire becomes evident in the video, and the motorcycle begins yawing in a counterclockwise fashion. During this yaw, the rear tire of the motorcycle generates a tire mark on the roadway. As the yaw progresses,

FIGURE 7.7 Video sequence (high-side fall).

the rider steered the front wheel to the right and decreased the throttle input. When the rider rolled off the throttle, the longitudinal force demand on the rear tire decreased, and the available lateral traction increased. This caused the rear tire to seek alignment with the front tire, and the leftward lean of the motorcycle began to decrease. Eventually, the counterclockwise yaw ceased, and the motorcycle yawed clockwise as the motorcycle righted itself. The rear tire mark terminated as the motorcycle came back to vertical.

As the heading of the motorcycle came into alignment with its velocity direction and the motorcycle righted itself, the rider steered the front tire to the left. The momentum of the motorcycle initially continued to carry it toward a leftward lean/fall with a clockwise yaw rotation. A left leading sideslip developed. As this occurred, the rider's buttocks separated from the motorcycle. Eventually, the leftward steer input reversed the yaw direction, and the motorcycle again developed a right side leading sideslip. The rear tire of the motorcycle generated a tire mark on the roadway, and the lateral tire forces threw the motorcycle into a rightward lean. As the motorcycle transitioned from a leftward lean to a rightward lean, the rider's groin and left thigh impacted the fuel tank of the motorcycle. The tires lost contact with the roadway and the rider was projected upward, out ahead of, and to the right of the motorcycle.

The motorcycle landed on its right side, and then the rider landed in a seated posture facing rearward, opposite his initial direction of travel. The rider was launched in the air approximately 54 ft. The motorcycle stayed on its right side throughout its entire slide on the ground and did not tumble. It decelerated at an average rate of approximately 0.4 g and then came to rest on its right side facing opposite its initial direction of travel. As the rider traveled to rest, his torso and head rotated toward the ground and ultimately struck it. He then tumbled to rest. The rider decelerated at a rate of approximately 1.07 g, a rate significantly higher than the rate at which the motorcycle decelerated. This includes the speed loss from the rider impacting the ground. Because of this high deceleration of the rider, the motorcycle came to rest further down the road than the rider.

The video footage included documentation of the motorcycle after it had been removed from the roadway. This footage revealed abrasions to the following components of the motorcycle because of interaction with the roadway: the end of the right handlebar, the front brake lever, the right-side cowling, fuel tank cover, right side engine crash protector, and the right-side passenger seat cowling. The video footage did not show the entire motorcycle, so other components may also have been damaged. The rear tire also showed scuffing consistent with high slip angles. The rider's helmet was also documented and showed striated abrasions to its rear.

Rose et al. [6] reported the analysis of this crash and three additional high-side crash videos captured by Snyder. These four crashes had several common features, including the following: (1) The loss of control in each case began with a loss of lateral stability at the rear tire. This loss of stability was caused either by a throttle application while cornering or by a braking force at the rear wheel of sufficient severity. (2) A yaw developed, the motorcycles and the riders leaned to the trailing side of the yaw, and the riders steered to counteract the yaw. (3) As this yaw developed, the riders released the throttle or the brake. (4) Because of the throttle or brake release, the available lateral traction quickly increased. (5) At this point the rear tires were operating with a large sideslip angle, and thus, the increase in traction caused an overturning moment to be generated that was not adequately countered by the lean of the motorcycles and the riders. This induced a roll of the motorcycle toward the leading side of the yaw. There are differences between the four crashes as to what happens next, differences that are discussed in the next paragraph. (6) Ultimately, though, the riders were projected upward and out ahead of the motorcycle. The riders were also projected to the side of the motorcycle paths, toward the outside of the final yaw. (7) The motorcycles landed on their leading sides and slid to rest, remaining on that side without losing contact with the ground. They did not tumble. (8) The riders landed and tumbled to rest.

There were distinct differences between the motorcycle motions in crashes where the loss of lateral stability was a result of accelerating excessively in a curve versus a crash where the loss of lateral stability was because of locking of the rear wheel. In all the cases, the rider countersteered to the right during the initial yawing of the motorcycle. However, in the acceleration cases, the increase in lateral traction that accompanied the throttle release was less significant than in the locked rear wheel case. This is because, in the locked rear wheel case, all the available traction was consumed. Thus, there was no available lateral traction before the brake was released. In the acceleration cases, on the other hand, the throttle application consumed some, but not all, of the available traction. Therefore, the difference in lateral force when the rear brake was released was greater than the force difference in the acceleration cases. Because of this, the riders' countersteer inputs were more effective in the acceleration cases, and the motorcycles yawed side-to-side several times before falling on their sides. In the locked rear wheel case, the motorcycle fell directly onto its side after the initial side slip developed. In the acceleration cases, between 0.87 sec and 1.13 sec elapsed from the time when the motorcycles were at maximum sideslip to the time that they impacted the roadway. For the braking case, this time was about 0.33 sec.

Case Study: A High-Side Fall

This section describes the reconstruction of a motorcycle crash involving a high-side fall. This accident occurred on a rural, two-lane highway, and the motorcyclist was driving in a westbound direction when his motorcycle capsized. According to a witness, a westbound pickup that was three cars ahead of the motorcyclist had slowed down and made a U-turn. This required an SUV traveling behind the pickup to "slam on his brakes." The witness, who was traveling behind the SUV in a sedan then applied her brakes "pretty hard" but did not have to use "emergency braking or drive off the road to avoid a collision." The motorcyclist was behind this sedan, and he applied his brakes. The witness "heard brakes squealing from behind her and saw the motorcycle crash on the road." The witness stated that, prior to the accident, she was traveling 60 mph and that when she braked she slowed to a speed of 5 mph. There was no contact that occurred between the witness's vehicle and the motorcycle. When the accident occurred, the weather was clear, and the asphalt roadway was dry. The posted speed limit was 60 mph.

Prior to capsizing, the motorcycle deposited a 106-ft long skid mark on the roadway. The motorcyclist came to rest 184 ft west of the beginning of this skid mark and the motorcycle came to rest on its right side approximately 204 ft west of the beginning of the skid mark. Given the length of the skid mark, it was deposited by the rear tire of the motorcycle. There was a 31-ft gap between the end of the tire mark and the beginning of these scrapes. This gap is an indication that the motorcycle and the rider experienced a high-side fall during this crash. The scrapes from the motorcycle were 67-ft long and led up to the point of rest. The motorcycle exhibited scrape marks on the left- and right-side crash bars, the left side of the gas tank, and the right side of the windscreen. The motorcycle came to rest on its right side. The diagram of Figure 7.8 depicts the evidence from this crash, along with relevant dimensions of that evidence.

Equation (6.2) was used to evaluate the speed the motorcycle was traveling when it began sliding on the roadway. The inputs into this analysis were the distance traveled by the motorcycle from the time it began sliding on the road surface until the time it came to rest, the slope of the roadway, and the deceleration rate of the motorcycle while it was sliding. The motorcycle slid on the road for approximately 67 ft, and the section of roadway along which this sliding occurred was essentially flat. The deceleration of the motorcycle while sliding was estimated using the data summarized in the previous chapter. In evaluating these studies, tests involving motorcycles and road surfaces like those involved in this accident were utilized. The motorcycles in these tests decelerated at an average rate of approximately 0.47 g, with a standard deviation of approximately 0.10 g.

Utilizing this range of decelerations, Equation (6.2) yielded a speed range for the motorcycle at the beginning of sliding of between 27.2 and 33.8 mph. These calculations are illustrated below for both the low and high end of this range.

$$v_{slide} = \sqrt{2gf_{slide}d_{slide}} = \sqrt{2 \cdot 32.2 \cdot \begin{bmatrix} 0.37 \\ 0.57 \end{bmatrix} \cdot 67} = \begin{bmatrix} 40.0 \\ 49.6 \end{bmatrix} \text{fps} = \begin{bmatrix} 27.2 \\ 33.8 \end{bmatrix} \text{mph} \quad (7.1)$$

FIGURE 7.8 Physical evidence diagram.

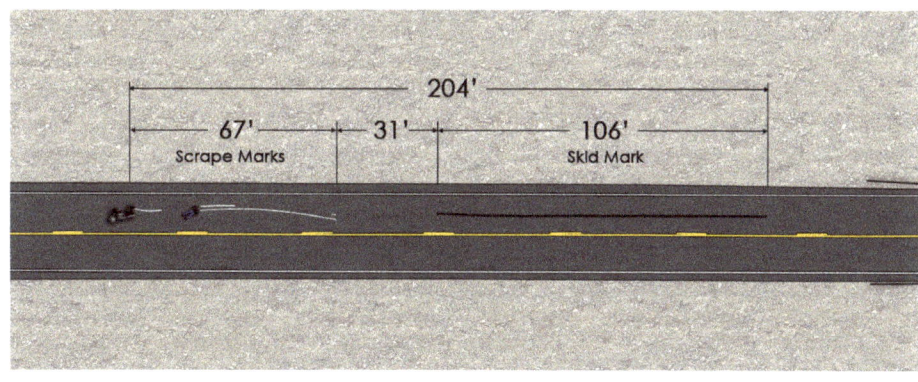

If the motorcyclist was initially traveling the same speed as the witness (60 mph), which she indicated he was, the motorcyclist applied the brakes of his motorcycle at a sufficient rate to decelerate from 60 mph down to a speed in the range of 27.2 to 33.8 mph before his motorcycle began sliding on the ground. This implies that when his motorcycle was upright, and he was braking (depositing a rear-wheel skid), the rider was braking with sufficient force to decelerate his motorcycle at a rate between 0.77 and 0.9 g. This level of deceleration implies that the motorcyclist employed both the front and rear brakes. This implied deceleration rate is on the high end of what a typical motorcyclist would be able to achieve.

Falls Because of Front-Wheel Over-Braking and Lock-Up

Sometimes a motorcycle capsizes from the operator over-braking the front brake. This can result in locking of the front wheel and in the motorcycle capsizing without first experiencing significant yaw. In these falls, the rider and the motorcycle will fall to the ground and slide side by side. A short tire mark is often deposited by the front tire, which may exhibit a hook at the end from lateral sliding of the front tire during capsizing. Another possibility from front-wheel over-braking is that the motorcycle can pitch-over. Frank et al. [7] conducted testing of pitch-overs induced by braking (using a 1985 Kawasaki Ninja) and found that decelerations exceeding 1.0 g produced pitch-overs.

Fatzinger et al. [8] reported a study of front-wheel over-braking with sport motorcycles. Testing was conducted with a 2002 Kawasaki ZRX1200R, a 2006 Yamaha YZF-R6, and a 2013 Ninja EX300. Thirteen tests were completed, with initial speeds ranging from 50 to 60 mph. All three motorcycles had independently actuated front and rear brakes without antilock brakes. Testing was conducted on a flat asphalt surface. Brake pressure was applied to the front brake lever or the rear brake pedal with elastic straps. Electronically controlled valves installed in each brake line prevented this pressure from being applied to the calipers until the motorcycle was up to speed. Of the 13 tests, three were performed with a 6 ft, 1 in., and 175 lb dummy on the motorcycle. In some of the tests, rear-wheel braking was applied in addition to the front braking, and in some of the tests, no rear braking was applied. Front-wheel lockup was achieved immediately after brake application in eight of the tests. Based on these tests, Fatzinger et al. reported that the motorcycles in these eight tests deposited front-wheel tire marks that were between 26 and 57-ft long. The fall times varied between 0.34 and 1.19 sec and "the shorter duration fall times corresponded to the tests where the motorcycle lean angles were the highest. And similarly, the tests with the highest lean angles at brake application produced the lowest average deceleration rate throughout the braking event."

Fatzinger et al. reported that the deceleration achieved by the motorcycle with front-wheel lockup depended on the lean angle of the motorcycle at the beginning of the braking. They reported that the average deceleration when the front wheel locked up and the initial lean was approximately 2° or less was in the range of 0.69 and 0.8 g. The average deceleration when the initial lean was around 3° or 4° was between 0.51 and 0.67 g. The average deceleration when the initial lean angle was between 8° and 9° was 0.32 to 0.39 g. A test where the front-wheel braking resulted in a pitch-up had a deceleration of approximately 0.8 g. A test that resulted in pitch-over had a deceleration of 0.98 g, consistent with what reference [7] reported. Rear brake application did not significantly increase the deceleration of the motorcycles when front-wheel lock had been achieved. Also, there were "no significant differences noted in the peak and average decelerations between the tests" with and without the dummies.

Impact-Induced Capsize

Capsizing of the motorcycle can occur because of the motorcycle and/or the rider being struck by another vehicle or object. In these instances, the motorcycle and the rider will typically end up falling and separating. The specific motion that the motorcycle and the rider experience will depend on many factors, including the degree to which they interact with the struck vehicle or object and the direction of travel between them. As an example, if a motorcycle strikes the front fender of a passenger car, the rider may be thrown forward over the hood of the vehicle with minimal interaction with the vehicle. On the other hand, if a motorcycle strikes the front door of a sport utility vehicle, the rider may be thrown forward and experience full engagement with the vehicle. These two different interactions would result in significantly different postimpact motion for the riders and significantly different rest positions. In addition, these two different interactions would call for different ways of handling the rider's weight in mathematical analysis of the collision.

Motorcycle Interactions with Potholes, Roadway Deterioration, and Debris

A motorcycle fall is sometimes attributed to interaction with a pothole, pavement deterioration, a pavement edge, or roadway debris. Some of these can cause a motorcyclist to lose control and capsize. However, many are traversable or avoidable without causing control or stability issues. In evaluating these crashes, the reconstructionist may need to evaluate if a pothole, area of deteriorated road, or debris on the road was an adequate mechanism to cause a loss of control and if these were traversable or avoidable by the motorcyclist. This evaluation will need to consider the characteristics of the pothole, deterioration, or debris, along with the characteristics of the motorcycle, the rider, and the maneuver required to traverse or avoid. The visibility of the pothole, deterioration, or debris and the time and distance it would take for the motorcyclist to maneuver around it may also need to be considered. As with any other issue of causation, this is a determination that will have to be made on a case-by-case basis. This section offers principles and data that might help in evaluating the role of a particular roadway irregularity in causing a motorcyclist to capsize.

A Motorcycle Encountering a Longitudinal Pavement Edge

One example of a roadway feature that could contribute to a fall by a motorcyclist is a longitudinal pavement edge (aligned with the roadway travel direction). A longitudinal edge could be encountered as one side of a long pothole or area of road deterioration. Another example occurs on a roadway that is being repaved, particularly when one lane has already been repaved and the adjacent lane has not. In such instances, the repaved lane may sit higher than the adjacent lane. A motorcyclist that maneuvers from the lower lane to the higher lane could encounter a longitudinal pavement edge of sufficient height to contribute to a loss of control and a capsize.

Passenger vehicle drivers sometimes encounter longitudinal pavement edges as well. A classic scenario is when a driver allows their passenger side tires to exit the paved portion of the roadway and their vehicle encounters the pavement edge when the driver attempts to steer back onto the roadway. While there are clear differences between a motorcycle and a car encountering a longitudinal pavement edge, significant research has been conducted related to the loss of control that can occur for passenger vehicle drivers in these situations [9–17]. There are aspects of this research that can inform an evaluation of the interaction of a motorcycle with a longitudinal pavement edge.

These passenger-vehicle-related studies have revealed the following scenario that can lead to a loss of control:

- The vehicle exits the roadway at a shallow angle, and the passenger side tires drift onto the right shoulder.
- The driver steers to the left to reenter the roadway with a shallow attack angle relative to the pavement edge.
- The passenger's side front tire fails to reenter the roadway and begins scrubbing on the pavement edge.
- Realizing the front tire is not climbing, the driver may increase their steering input until the tire climbs the pavement edge.
- The steering input that was just enough to accomplish the climbing of the pavement edge now quickly becomes more than what is necessary to return to the driver's initial lane of travel. This steering input may send the vehicle back to the left beyond the boundary of the lane.
- The driver may respond with a countersteer, but may still lose control or travel beyond their lane boundary where they can collide with other traffic or fixed objects on the far side of the roadway.

A key feature of this scenario is the scrubbing that occurs between the car tire(s) and the pavement edge. Several studies have shown that it is when such scrubbing occurs on an edge of sufficient height and severity that control problems can result. When scrubbing does not occur, the pavement edge typically will not cause control problems for the driver, though a driver can still lose control because of an overcorrection that is independent of the pavement edge. The following factors will determine if a tire scrubs along the pavement edge: (1) the height and shape of the pavement edge; (2) the tread and sidewall heights of the tire; (3) the tire pressure (which could affect the shape of the tire); (4) how worn the tire is; (5) the angle at which the tire approaches the pavement edge (the attack angle); and (6) the speed of the vehicle. An edge with a 90° square profile will be more severe and likely to produce scrubbing than one with a rounded or beveled transition from the shoulder to the pavement. The greater the height of the pavement edge, the more likely the tire is to scrub along the edge. The smaller the tread and sidewall heights of the tire, the more likely the tire is to scrub. A worn tire climbs sooner than a new tire. This may be partly because a worn tire has reduced tread height, bringing more of the tire sidewall into contact with the edge. Klein and Johnson [10] also attribute it to a larger cornering force generated by a tire when it is worn. The lower the angle of approach between the tire and the pavement edge, the more likely scrubbing is to occur. And finally, the higher the vehicle speed, the more likely the driver is to experience difficulty controlling the vehicle after the scrubbing tire finally mounts the pavement edge.

Some of these findings are conceptually applicable to motorcycles. Consider a motorcyclist traveling on a straight section of roadway and entering a long pothole or area of roadway deterioration. Assume the following:

- The pothole is longitudinally traversable without instability, as long as the tires of the motorcycle do not interact with one of the longitudinal edges (on the left or the right).
- The pothole is of sufficient width that the motorcycle tires will not simultaneously interact with both the left and right edges.
- The depth of the pothole is such that, if the tire interacts with either longitudinal edge, the tire will be engaged on its sidewall.

Now, assume that the front tire of the motorcycle does interact with the left edge of the pothole and begins scrubbing along that edge. This would apply a lateral reaction force to the side of the tire and a longitudinal scrubbing force that would tend to steer the tire to the left. A leftward steer would induce a rightward lean that could destabilize the motorcycle and cause it to capsize. On the other hand, the rider could respond to the leftward steer by inputting a rightward steer. If the scrubbing is of short duration, and the rider's clockwise steering response is of the necessary magnitude, a fall could be prevented. The rider could also respond more severely than what is needed and the rightward steer could induce a lean back to the left and the motorcycle could capsize onto its left. If the pothole is of sufficient length, this process could repeat itself, with the rider eventually becoming unable to control the motorcycle and capsizing could occur.[1] It is also possible in this scenario that the longitudinal edge itself could prevent the rider from countersteering sufficiently to counteract the scrubbing-induced steer.[2]

This example also illustrates that, for a motorcycle, the instability from interaction with a longitudinal pavement edge would likely occur in roll rather than yaw (as it would for a car). For a car, the roll moment applied to the vehicle by the lateral force from the pavement edge will typically be offset by the opposing roll moment applied about the contact point by the weight of the vehicle, because the center of gravity (CG) of the vehicle is laterally distant from the contact area. For a motorcycle, the steering induced by scrubbing can induce roll instability.

The ultimate outcome of an interaction between the front tire of the motorcycle and the left longitudinal edge of the pothole will depend on the length and depth of the edge, the shape of the tire, the speed of the motorcycle, and the rider's skill level. For a reconstructionist analyzing a situation like this, the following questions might be worth asking and exploring via the physical evidence:

- Does the pothole exhibit a longitudinal edge?
- What is the shape and depth of the longitudinal edge?
- What is the shape of the tire and where is the tire likely to interact with the edge?
- What forces would this apply to the tire, and would this induce steering?

---

[1] Broker and Hottman give a good description of this scenario for bicycles, referring to it as a front-wheel diversion. See: *Bicycle Accidents, Crashes, and Collisions: Biomechanical, Engineering, and Legal Aspects*, 2nd ed., by Jeffrey Broker and Megan Hottman, Lawyers and Judges Publishing Company, Inc., 2017, ISBN 9781936360581.

[2] Hough has observed: "Two-wheelers are particularly vulnerable to pavement edges or grooves. Remember, a two-wheeler is balanced mostly by steering the front wheel … For example, if a motorcycle starts to fall over to the left, you can steer the front wheel more to the left to rebalance. The term for this balancing act is *countersteering* … Countersteering explains why edge traps are so hazardous to two-wheelers while only a jarring, wheel-bending inconvenience to other vehicles. A car or a side-car rig can slide sideways without losing balance, but if the rider of a two-wheeler loses steering for more than a couple of seconds, it becomes very difficult to maintain balance. Easing up to a curb, you can maintain balance right up to the point where the front wheel contacts the edge. After that, with the tire scrubbing along the edge of the curb, you can't countersteer to maintain balance … the trick is to cross it aggressively at a maximum angle rather than attempting to ease over, and to use a little power to bounce the front wheel up." See: Hough, David L., *Proficient Motorcycling: The Ultimate Guide to Riding Well*, 2nd ed., Fox Chapel Publishers, 2008.

- What speed was the motorcycle traveling when it interacted with the edge?
- How long (in terms of time) would the interaction have persisted?
- Is there evidence on the front or rear tire of the motorcycle of scrubbing on the longitudinal edge?
- Where is that scrubbing? Does it make sense dimensionally with the pavement edge?
- Is there material transfer from the tire onto the pavement edge?

These questions will not necessarily be answerable. Oftentimes, by the time a reconstructionist is involved, the pothole or roadway deterioration has been repaired and the precise shape and depth cannot be determined from photographs. If the pothole was not photographed, historical aerial and Streetview photographs on Google can sometimes help for quantifying the shape and dimensions of the pothole. State, city, and county departments of transportation also sometimes document their roadways photographically at regular intervals, and these photographs can help for quantifying the shape and dimensions of roadway deterioration.

Now consider the scenario in which a motorcyclist encounters the need to change lanes on a road that is being repaved. If one lane has been repaved and the adjacent lane has not, the repaved lane may sit higher than the adjacent lane. This scenario could also generate steering forces on the front tire of the motorcycle, but here the lateral force is also likely to come into play and can potentially cause instability. To prevent capsizing, the motorcyclist will need to approach the pavement edge with a sufficient attack angle (and lean) to minimize both the longitudinal and lateral forces from the pavement edge. But what attack angle is necessary? The answer to this will relate to features of the edge and the speed of the motorcycle. The greater the height discrepancy between the low lane and the high lane, the greater the attack angle that will be necessary. The more aggressive the transition (a 90° square edge versus a gradual or beveled transition), the greater the attack angle that will be necessary. And, the greater the speed of the motorcycle, the greater the attack angle that will be necessary. Braking or longitudinal acceleration present during the traversal of the pavement edge could also influence the outcome because these will consume some of the available traction.

At low speeds, achieving significant attack angles to a pavement edge is not a problem. At higher speeds, though, this will be more difficult. Consider, for example, a scenario in which a motorcyclist is attempting a lane change at a highway speed of 60 mph from a lane that sits lower than the adjacent lane. Based on the lane change modeling discussed in Chapter 4, at a speed of 60 mph, a motorcyclist making a 12-ft lateral movement with a peak lateral acceleration of 0.1 g would achieve a maximum heading angle of approximately 3° halfway through the lane change with a lean angle of zero. This would correspond to a maximum lean angle of approximately 6° when the motorcycle is ¼ and ¾ of the way through the lane change. With a peak lateral acceleration of 0.2 g, the maximum heading angle would be approximately 4° to 5°. This corresponds to a maximum lean angle of approximately 11°. With a peak lateral acceleration of 0.3 g, the maximum heading angle would be approximately 5° to 6°. This corresponds to a maximum lean angle of approximately 17°.

Metz [18] used analysis and low-speed experiments to examine the behavior of a motorcycle that encounters a pavement edge parallel to its direction of travel during a lane change from a lower to a higher lane. Consistent with the prior paragraph, Metz points out that at highway speeds the attack angle of the motorcycle to the pavement edge is likely to be small. However, in his analysis, Metz assumed a very low lateral acceleration of 0.058 g for the motorcyclist's lane change, which at 55 mph yields an attach angle of 1.3°. This leads him to the conclusion that the motorcycle would have negligible lean during the maneuver, and thus, that the lean angle of the motorcycle need not be considered in the modeling. While a motorcyclist could, of course, make a casual lane change across a pavement edge, this seems to represent a rider who was unaware of the presence of the edge. A rider who recognizes the edge would more likely make some attempt to increase their attack angle by utilizing a higher lateral acceleration and greater lean. If the rider was able to time their lane change to encounter the edge at the middle point of the maneuver, they could traverse the edge with negligible lean but would likely have a higher attack angle than what Metz assumes. On the other hand, such an ideal lane change might not be achieved by the motorcyclist, and they could encounter the edge during a lower attack angle—but higher lean angle—portion of the lane change.

Based on his analysis and experiments, Metz concluded that "a motorcyclist encountering a step of any significant height at a small angle of attack will be highly unlikely to successfully negotiate the maneuver. If speeds are low, step height is low and/or angle of attack is significant (say, > ~ 5-10 deg), success is much more likely." Unfortunately, the experimental speeds in Metz's study were low. He acknowledged that "additional experiments at higher speeds would be useful" but he hypothesized that these experiments "would be dangerous to the test rider to conduct." The accuracy of Metz's conclusions depends on the height and profile of the pavement edge, and more research is needed to establish more precisely what would constitute "a step of any significant height."

As with a motorcycle encountering a longitudinal edge while driving straight, the ultimate outcome of an interaction between the front tire of the motorcycle and a longitudinal edge during a lane change will depend on the length, profile, and height of the edge, the shape of the tire, the speed of the motorcycle, and the rider's skill level. A similar list of questions can be proposed for a reconstructionist analyzing such a situation:

- What is the shape and depth of the longitudinal edge?
- What is the shape of the tire and how is the tire likely to interface with the edge?
- What forces would this apply to the tire, and would this induce steering?
- What speed was the motorcycle traveling when it interacted with the edge?
- How long (in terms of time) would the interaction have persisted?
- Is there evidence on the front or rear tire of the motorcycle of scrubbing on the edge?
- Where is that scrubbing? Does it make sense dimensionally with the pavement edge?
- Is there material transfer from the tire onto the pavement edge?

A Motorcycle Encountering a Pothole

In addition to a longitudinal pavement edge, a motorcyclist could lose control and capsize because of interaction with the far face or edge of a pothole. The loss of control mechanism in this scenario could be a significant front-wheel diversion or significant slowing of the front wheel that causes pitching. To cause either, the contact or collision between the tire and the far edge of the pothole would have to be of sufficient severity to cause such a diversion or slowing. Figure 7.9, which shows a pothole in a parking lot, graphically defines the "near edge" and "far edge" of a pothole. The pothole depicted in this figure would likely be traversable at most any speed. When evaluating a specific pothole and the potential for it to contribute to a motorcyclist losing control, it can help to parse the pothole into regions, because different features of a pothole can have different proclivities to cause a motorcyclist to lose control. As an example, consider Figure 7.10 where the pothole depicted in Figure 7.9 is parsed into regions. In this example, the regions are bare dirt and alligator cracking. Other possibilities could be areas of loose material or asphalt chunks within the pothole, or a vertical edge at one of the extents of the pothole.

FIGURE 7.9 Graphically defining "far edge" and "near edge."

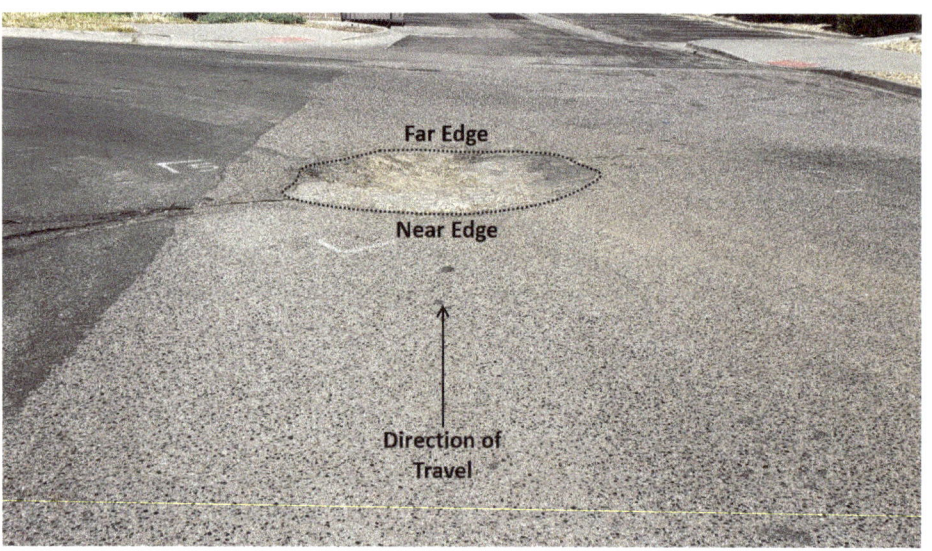

FIGURE 7.10 Parsing a pothole into regions.

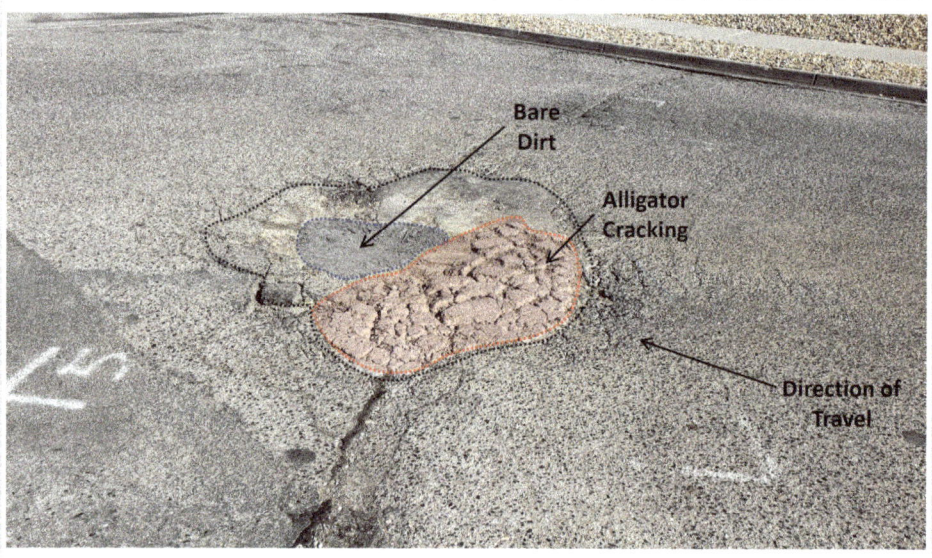

If a pothole contains a vertical edge perpendicular to the direction of travel, the following factors will determine the severity of the interaction between the tire and the edge: (1) the shape and depth of the edge in the area of contact; (2) the length of the pothole; (3) the dimensions of the tire and what portion of the tire first contacts the edge; (4) the inflation pressure and stiffness of the tire; and (5) the speed of the motorcycle. The influence of the edge characteristics on the possibility of a loss of control is illustrated in Figure 7.11. This figure illustrates a motorcycle tire encountering three different edges at the far side of a pothole. The edge depicted in "A" illustrates an extreme case where the edge

FIGURE 7.11 Pothole far edges of varying severity.

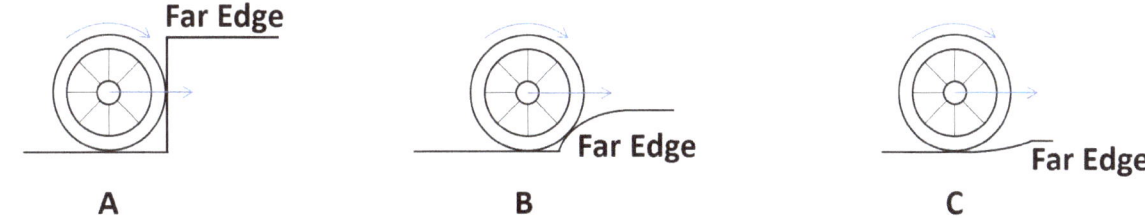

is completely vertical, and the depth of the edge is as large as the tire. A vertical edge of this magnitude could clearly cause significant slowing of the wheel, pitching of the motorcycle, and a loss of control. Few pothole edges will be so severe. The edge depicted in "B" is less severe both in terms of its shape and its depth. The potential for this edge to cause control problems is more ambiguous and depends on the dimensions of the edge relative to the tire and wheel. The edge depicted in "C" would be easily traversable and shows the other end of the spectrum from "A."

Wahba et al. [19] examined motorcycle accelerations occurring during successful, perpendicular traversals of several roadway depressions and a 3.5-in. speed bump. They tested speeds up to 40 mph (64.4 km/h) with a single motorcycle—a 2003 Yamaha R6—and a single rider. The depressions had generally smooth and gradual transitions to their base, and could not be accurately characterized as potholes. The speed bump was smooth and free of debris. These authors concluded that the higher the speed of approach, the shorter the duration of the interaction of the motorcycle with the depression or bump and the larger the magnitude of the accelerations experienced by the motorcycle. They also observed that "video analysis revealed that the rider was the most displaced from his riding position while traversing the speed bumps. As the rear tire mounted the leading incline of the speed bump, the rider was functionally catapulted several inches off the seat toward the front handlebars."

Analyzing the Motion of Projected Riders

Determining the speed at which a rider was thrown from a motorcycle can sometimes help in a reconstruction. Searle [20, 21] derived the following formula for determining a pedestrian's projection velocity, based on the conceptual model depicted in Figure 7.12. This model is sometimes used to calculate the velocity of a projected motorcycle rider. In this equation, μ is the coefficient of friction between the pedestrian and the ground, g is the acceleration because of gravity, s is the throw distance, including the airborne and sliding/tumbling phases, θ is the projection angle, and h is the change in the person's CG height off the ground from before the collision to their point of rest. A drop in CG position is negative. Searle reported a typical coefficient of friction for a person sliding on asphalt of 0.66. This model improved on the Collins model by incorporating a launch angle and incorporating the ground plane speed loss because of the person landing on the ground after vaulting through the air.

$$v_{proj} = \frac{\sqrt{2\mu g(s + \mu h)}}{\cos\theta + \mu\sin\theta} \qquad (7.2)$$

Searle observed that the projection angle is often not known, and so, he derived a version of this equation that would yield the lowest possible value and one that would yield the highest possible value. These equations are as follows:

$$v_{impact,min} = \sqrt{\frac{2\mu g(s+\mu h)}{1+\mu^2}} \tag{7.3}$$

$$v_{impact,max} = \sqrt{2\mu g s} \tag{7.4}$$

FIGURE 7.12 Searle's model.

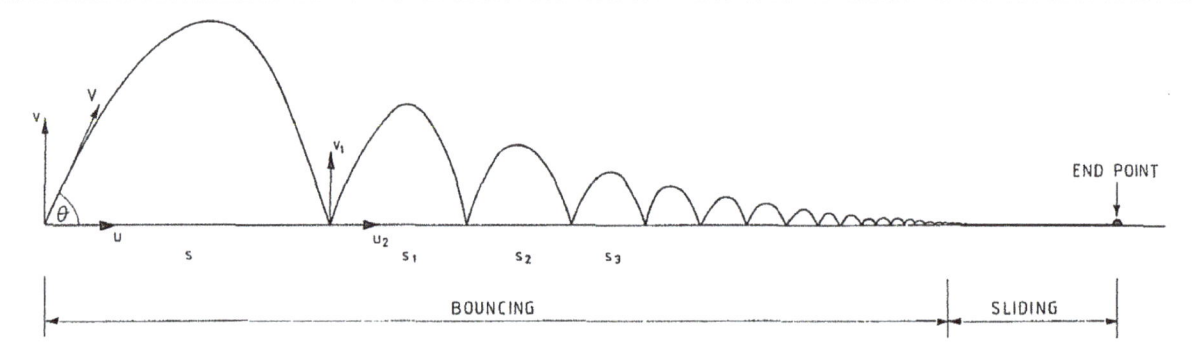

Reprinted with permission from Ref. [20]. © SAE International

Application of Equation (7.4) requires the analyst to ensure that the maximum projection angle could not have been greater than a critical projection angle calculated with the following equation.

$$\theta_{crit} = 180° - 2\arctan\frac{1}{\mu} \tag{7.5}$$

If projectile analysis is being applied to analyze the motion of a motorcycle rider that was thrown from their motorcycle because of an impact with another vehicle, then it is important to keep in mind that these formulas (like the Collins model) would yield the speed of the rider upon separation with that other vehicle, not the initial impact speed of the rider. The analyst would need to consider the degree to which the rider interacted with the struck or striking vehicle and how that would have altered the rider's speed before separation.

References

[1] Rose, N., Carter, N., and Pentecost, D., "Analysis of Motorcycle and Rider Limits on a Curve," *Collision: The International Compendium for Crash Research*, 9(1): 28-36, 2014, ISSN 1934-8681.

[2] Carter, N., Rose, N., and Pentecost, D., "Validation of Equations for Motorcycle and Rider Lean on a Curve," *SAE International Journal of Transportation Safety*, 3(2): 126-135, 2015, doi:10.4271/2015-01-1422.

[3] Cossalter, V., Bellati, A., and Cafaggi, V., "Exploratory Study of the Dynamic Behaviour of Motorcycle-Rider During Incipient Fall Events," *Proceedings of the 19th International Technical Conference on the Enhanced Safety of Vehicles*, Paper Number 05-0266, June 2005.

[4] Cossalter, V., *Motorcycle Dynamics*, 2nd ed., self-published, 2006, ISBN 978-1-4303-0861-4.

[5] Baxter, A.T., *Motorcycle Crash Investigation*, Institute of Police Technology and Management, Jacksonville, FL, 2017 ISBN 978-1-934807-18-7.

[6] Rose, N., Carter, N., Pentecost, D., and Hashemian, A., "Video Analysis of Motorcycle and Rider Dynamics During High-Side Falls," SAE Technical Paper 2017-01-1413, 2017, doi:10.4271/2017-01-1413.

[7] Frank, T., Smith, J., Hansen, D., and Werner, S., "Motorcycle Rider Trajectory in Pitch-Over Brake Applications and Impacts," *International Journal of Passenger Cars—Mechanical Systems*, 1(1): 31–42, 2009, doi:10.4271/2008-01-0164.

[8] Fatzinger, E., Landerville, J., Bonsall, J., and Simacek, D., "An Analysis of Sport Bike Motorcycle Dynamics during Front Wheel Over-Braking," SAE Technical Paper 2019-01-0426, 2019, doi:10.4271/2019-01-0426.

[9] Nordlin, E.F., Parks, D.M., Stoughton, R.L., and Stoker, J.R., "The Effect of Longitudinal Edge of Pave Surface Drop-Off On Vehicle Stability," CA-DOT-TL-6783-1-76-22, California Department of Transportation, CA, March 1976.

[10] Klein, R.H. and Johnson, W.A., "Vehicle Controllability in a Pavement/Shoulder Edge Climb Maneuver," SAE Technical Paper 780620, 1978, doi:10.4271/780620.

[11] Zimmer, R.A. and Ivey, D.L., "Pavement Edges and Vehicle Stability – A Basis for Maintenance Guidelines," Research Report 328-1, Texas Transportation Institute, TX, September 1982.

[12] Ivey, D.L. and Sicking, D.L., "Influence of Pavement Edge and Shoulder Characteristics on Vehicle Handling and Stability," Transportation Research Record 1084, 1986.

[13] Olson, P.L., Zimmer, R., and Pezoldt, V., *Pavement Edge Drop*, The University of Michigan Transportation Research Institute, UMTRI-86-33, July 1986.

[14] Glennon, J.C., "Effect of Pavement/Shoulder Drop-Offs on Highway Safety," State-of-the-Art Report 6, Transportation Research Board, 1987, ISSN: 0892-6891.

[15] Rudny, D.F. and Sallmann, D.W., "Analysis of Accidents Involving Alleged Road Surface Defects (i.e., Shoulder Drop-offs, Loose Gravel, Bumps and Potholes)," SAE Technical Paper 960654, 1996, doi:10.4271/960654.

[16] Deyerl, E. and Cheng, L., "Computer Simulation of Pavement Edge Traversal," SAE Technical Paper 2009-01-0464, 2009, doi:10.4271/2009-01-0464.

[17] Ivey, D., Zimmer, R.A., Julian, F., Sicking, D.L. et al., "Pavement Edges," Article contained in *Influence of Roadway Surface Discontinuities on Safety – State of the Art Report*, Transportation Research Circular Number E-C134, Transportation Research Board, May 2009.

[18] Metz, L.D., "Behavior of a Motorcycle after an Encounter with a Road Irregularity Parallel to its Direction of Travel," SAE Technical Paper 2006-01-1561, 2006, doi:10.4271/2006-01-1561.

[19] Wahba, R., Nelson, J., Timbario, T., Jordan D. et al., "Motorcycle Accelerations while Successfully Traversing Roadway Irregularities and Traffic Calming Devices (speed bumps) at Small Lean Angles," SAE Technical Paper 2019-01-0434, 2019, doi:10.4271/2019-01-0434.

[20] Searle, J.A., "The Trajectories of Pedestrians, Motorcycles, Motorcyclists, etc., Following a Road Accident," SAE Technical Paper 831622, 1983, doi:10.4271/831622.

[21] Searle, J., "The Physics of Throw Distance in Accident Reconstruction," SAE Technical Paper 930659, 1993, doi:10.4271/930659.

8

Motorcycle Collisions with Vehicles and Roadside Barriers

Analysis Based on Motorcycle Wheelbase Reduction

In 1970, Severy et al. [1] reported seven staged collisions involving motorcycles and passenger cars. The motorcycles were carrying rider dummies, and they struck a front door or fender of a stationary 1964 Plymouth Fury (Figure 8.1). Three different Honda motorcycle models were tested (CL-90, CB-350, and CB-750) with impact speeds of 20, 30, and 40 mph. Table 8.1 summarizes these tests, including the motorcycle used in each test, the motorcycle impact speed, the total motorcycle deformation (wheel deformation plus wheelbase reduction), the motorcycle wheelbase reduction, the car deformation, and the total deformation (wheelbase reduction + maximum car deformation).

These authors made observations that continue to be developed and tested by crash reconstructionists. For example, they observed that "the permanent shortening of the motorcycle wheelbase as a result of collision varied linearly with the speed of collision and did not appear to be significantly affected by variations in size of the motorcycle or in location of impact." This concept has been developed, disputed, and refined over the past five decades. Later research has confirmed that the wheelbase shortening of the motorcycle can be used for motorcycle speed analysis. However, later studies have found that the location of impact on the struck vehicle (door, wheel, or pillar, for example) does make a difference to the postcollision deformation exhibited by the motorcycle and that the calculations can be improved by including the car deformation.

FIGURE 8.1 Honda 350 impacting 1964 Plymouth at 30 mph, experiment 128 (Figure 10 from reference [1]).

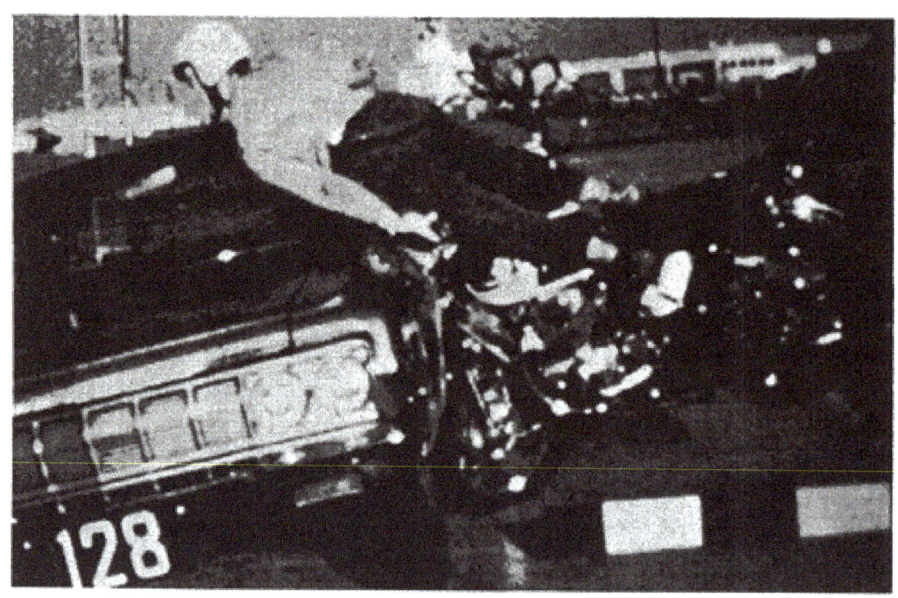

TABLE 8.1 Summary of the motorcycle-to-vehicle tests reported by Severy et al. [1]

Test #	Cycle	Cycle weight (lb)	Cycle speed (mph)	Impact location	Total cycle deformation (in.)	Wheelbase reduction (in.)	Car door deformation (in.)	Car sill deformation (in.)	Car fender deformation (in.)	Wheelbase reduction + car deformation (in.)
127	CL-90	200	30	Driver's door	17.0	9.2	5.0	–	–	14.2
128	CB-350	350	30	Driver's door	15.0	8.5	6.5	–	–	15.0
129	CB-350	350	20	Passenger's side front door	4.7	4.7	3.4	–	–	8.1
130	CB-350	350	30	Passenger's side front door	16.7	9.5	5.5	2.7	–	15.0
131	CB-350	350	30	Right front fender	9.5	9.5	–	–	6.3	15.6
132	CB-750	480	30	Passenger's side front door	16.7	9.3	4.2	2.0	–	13.5
133	CB-350	350	40	Driver's door	21.0	13.0	11.5	4.7	–	24.5

Figure 8.2 is a series of photographs included reference [1]. These photographs show the postcrash condition of each motorcycle. The authors reported that the test at 20 mph produced 4.7 in. of wheelbase reduction because of fork deformation and 8.1 in. of total deformation. The five tests at 30 mph, on the other hand, exhibited varying levels of wheelbase reduction (9.5 to 17.0 in.) and total deformation (15.8 to 22.2 in.). These different levels of wheelbase reduction and total deformation with the same impact speed can be attributed, at least in part, to the different stiffnesses of the vehicle structures impacted by these motorcycles. Stiffer passenger vehicle structures will produce greater

FIGURE 8.2 Wheel and fork deformation from the Severy collisions [1].

wheelbase reduction and total deformation to the motorcycle than softer ones, given the same impact speed.

Severy et al.'s testing also enabled observations related to the motion of the riders during a collision. First, the rider and the motorcycle are not coupled in the same way that a passenger car and its belted occupants would be. During an impact, the rider may interact with the motorcycle, but is likely to also move forward and separate from the motorcycle entirely. The rider may continue forward and collide with the struck car, depending on what part of the car is impacted. Second, there is typically a much smaller weight disparity between a motorcycle and its rider than between a passenger car and its occupants. This becomes relevant when analyzing collisions between motorcycles and passenger cars because the analyst will typically have to decide how much of the weight of the rider participated in the collision, and this will be consequential in the calculations. Table 8.2 is a timeline of the collisions from reference [1]. As this table demonstrates, the collision between the motorcycle and the car is essentially complete prior to the occurrence of the collision between the rider and the car. "In general, motorcycle collapse is complete 40–60 ms following contact, with maximum penetration into the side of the struck car by 80 ms." The collision between the riders' bodies and the cars occurred between at times between 80 and 195 ms following first contact between the motorcycle and the car.

TABLE 8.2 Timeline for the tests reported by Severy et al. [1]

Experiment number	First contact (ms)	Motorcycle forward movement stopped (ms)	First contact of rider with vehicle (ms)	Rider's body impacts vehicle (ms)
127	0	55	75	130
128	0	55	75	120
129	0	53	185	195
130	0	UNK	70	195
131	0	80	90	160
132	0	55	120	125
133	0	63	55	80

© SAE International

In 2002, Adamson et al. [2] reported 17 staged collisions that involved 1989 to 1993 model year Kawasaki 1000 police motorcycles. These tests were conducted at the World Reconstruction Exposition in September of 2000 (WREX 2000). In the first seven tests, the motorcycles impacted a concrete block at speeds varying between 10 and 42 mph. The concrete block had a steel bottom surface and weighed 11,080 lb. The reported static and dynamic coefficients of friction between the runway and the block were 0.46 and 0.40, respectively. The block was nondeformable but was light enough that it moved during most of the tests. In the remaining ten crash tests, the motorcycles impacted one of two stationary 1989 Ford Thunderbirds at various locations on the vehicle. These vehicles weighed approximately 3600 lb. Motorcycle impact speeds for these tests varied between 25 and 49 mph and were measured with a hand-held Stalker radar gun. All tests were run on an abandoned concrete runway with a reported coefficient of friction for vehicle tires of 0.72. Table 8.3 summarizes the results from these tests.

Reference [2] used linear regression to obtain a best-fit line describing the relationship between the motorcycle impact speed and the wheelbase reduction. This reference did not include the car crush in the regression analysis. Separate linear fits were obtained for the motorcycle-to-concrete block impacts and the motorcycle-to-vehicle impacts. For the motorcycle versus concrete block impacts, a reasonably strong linear fit ($R^2 = 0.87$) was obtained. However, for the motorcycle versus vehicle impacts, the fit was weak ($R^2 = 0.47$). The weak correlation between the wheelbase reduction and the impact speed for the motorcycle-to-car impacts is to be expected because the magnitude of the wheelbase reduction depends not only on the impact speed of the motorcycle, but also on the stiffness of the impacted object (among other factors). The correlation could potentially be improved if the tests were grouped according to the portion of the vehicle they struck. For example, motorcycles striking a door would need to be categorized separate from impacts where the motorcycle strikes a wheel.

Baxter [3] noted this issue and observed: "Depending on the area of the automobile that is struck, the wheel and fork will begin to deform. If a stiff portion of the car is struck (a wheel for example), the motorcycle will collapse more severely than if contact had been made with a softer sheet-metal area." For impacts involving one of the front wheels of the passenger car, even the location of the impact on the wheel can influence the wheelbase reduction of the motorcycle. This is because an impact to the wheel forward or rearward of the wheel centerline can cause the wheel to steer to the left or to the right. This effectively makes the wheel less stiff, because up to a point, it gives way to the motorcycle. This would not occur if the centerline of the wheel was struck.

TABLE 8.3 Summary of the staged collisions reported by Adamson et al. [2]

Motorcycle #	Target	Impact location	Cycle speed (mph)	Wheelbase reduction (in.)	Max car deformation (in.)	Wheelbase reduction + car deformation (in.)
1	Block	Vertical face	42	11.75	–	11.75
2	Block	Vertical face (Motorcycle leaning 30° left at impact)	10	1.13	–	1.13
3	Block	Vertical face	31	10.25	–	10.25
4	Block	Vertical face	20	5.25	–	5.25
5	Block	Vertical face	24	8.25	–	8.25
6	Block	Vertical face	21	8.0	–	8.0
7	Block	Vertical face	35	13.0	–	13.0
8	Car	Body between B-post and LR wheel well	46	10.75	13.75	24.5
9	Car	Body LR, between wheel well and bumper	39	7.63	10.5	18.13
10	Car	Rear bumper, 17 in. left of right end	34	8.25	9	17.25
11	Car	Right side, between front wheel well and door	25	5.63	No crush data	
12	Car	Right front wheel	30	3.25	3.5	6.75
13	Car	Right door, center	42	6.81	12	18.81
14	Car	Front bumper, 6 in. right of center	30	5.75	3.75	9.5
15	Car	No target impact	–	–	–	–
16	Car	Right front fender between wheel well and bumper	41	7.5	5.75	13.25
17	Car	No target impact	–	–	–	–
18	Car	Front bumper, right of center	45	8.81	15	23.81
19	Car	Body left rear fender, between wheel well and bumper	49	7.25	13.5	20.75

In 2009, Bartlett [4] published an analysis that considered the variability in stiffness of the impacted structures. He began by gathering the staged collision data that were available at the time of his article. When the data were analyzed without consideration of what portion of the vehicle was struck, he found a weak correlation between wheelbase reduction and impact speed. Bartlett made the following observation: "This [weak correlation] is certainly influenced by the variety of vehicles tested, and the variety of impact locations on the car sides: body panels such as fenders and doors do not have the same crush characteristics as axles and wheels. Another limitation of the analysis is its relative insensitivity at higher speeds: after the front wheel and fork have collapsed to the frame and engine block there is little additional motorcycle deformation possible."

Bartlett parsed the available data into two categories: (1) motorcycles striking a door or fender and (2) motorcycles striking at a pillar or within a foot of an axle. He noted that "the pillar/axle tests produced 3 to 5 in. less total crush [motorcycle wheelbase reduction plus maximum car crush], supporting the notion that pillar/axle locations were stiffer." Bartlett presented the following two equations for when the motorcycle strikes the *door or fender*:

$$S = 2L + 1.8C + 2 \tag{8.1}$$

$$S = 1.4534(L + C) + 10.124 \tag{8.2}$$

In these equations, L is the motorcycle wheelbase reduction in inches, C is the maximum crush to the struck vehicle, and S is the impact speed of the motorcycle in miles per hour. Bartlett noted that the functional form of Equation (8.1) was originally derived based on testing conducted by the California Association of Accident Reconstruction Specialists (CA²RS) in 2004. He updated the coefficients to the equation by adding additional data. The functional form of Equation (8.2) was originally presented by Eubanks in 1991 [5]. Bartlett used the same form but again updated the coefficients by adding additional data. Bartlett presented the following two equations for instances in which the motorcycle struck a *pillar or axle* (within 1 ft of the wheel centerline):

$$S = 2.5L + 1.9C + 4.5 \tag{8.3}$$

$$S = 1.5875(L + C) + 14.72 \tag{8.4}$$

Equation (8.3) follows the form of the original CA²RS equation, but uses updated coefficients obtained by adding additional data. Equation (8.4) follows the form of the Eubanks equation, and again, Bartlett updated the coefficients with additional data. When compared to the available data, Bartlett reported that the modified Eubanks equations yielded "slightly more accurate" predictions than the modified CA²RS equations. He reported that, for the modified Eubanks equations, "the door/fender collision speeds were found to be predicted with a 95% confidence range of plus or minus 20% of the nominally calculated value, while the pillar/axle speeds were predicted with a range of plus or minus 28%."

Two other observations by Bartlett are worth mentioning. First, "evaluating the crush energy or wheelbase reduction of just the motorcycle does show a slight variation based on the type of wheel, with the cast-wheeled motorcycles typically experiencing 5 to 2 inches less crush at a given impact speed. The variation goes down as the speeds increase. However, when the total crush experienced by both the motorcycle and the car is evaluated this differentiation disappears, and the wheel type becomes less relevant … It appears that while the stiffer cast wheels deform less, they produce additional crush in the cars they are striking." Second, "collisions involving cars moving at high enough speeds to affect the motorcycle's trajectory appear to reduce the total measured crush, such that using the relationships developed here will underpredict speeds, typically by 15%. The additional complication of a moving target vehicle also increases the data spread significantly."

In 2013, Bartlett et al. [6] reported data from a series of 25 motorcycle to stationary car collisions which were conducted in 2009 by the CA²RS. These tests are summarized in Table 8.4. The tests were conducted on an asphaltic concrete runway, which was smooth, dry, flat, weathered, and traffic-polished. The authors reported a coefficient of friction for the surface of 0.64, a value that was obtained by averaging the results of three skid tests with a 2001 Ford Crown Victoria Police Interceptor. The struck vehicles were passenger cars and the motorcycles included standard, cruiser, touring, and sport motorcycles.

TABLE 8.4 Summary of the staged collisions reported by reference [6]

Motorcycle	Target	Impact location on target	Motorcycle (MC) speed (mph)	Wheelbase reduction (in.)	Max car deformation (in.)	Wheelbase reduction + car deformation (in.)	Car rotation (deg)
1976 Suzuki GS550E	1986 Honda Accord	82 in. behind front axle	45.5	9.5	10.5	20.0	19
1983 Yamaha XJ750M	1986 Honda Accord	17 in. forward of front axle	40.1	8.0	3.75	11.75	61
1983 Honda VT750C	1986 Honda Accord	54 in. forward of rear axle (near CG)	35.9	9.5	6.5	16.0	5
1981 Kawasaki KZ440D	1986 Honda Accord	Just ahead of RR axle	37.9	5.0	1.5	6.5	22
1983 Honda CB550 Nighthawk	1989 VW Golf	Just rear of front axle	45.5	8.25	4.5	12.75	26
1982 Honda CM450C	1989 VW Golf	B pillar	47.0	8.0	12.5	20.5	38
1981 Kawasaki KZ440D	1989 VW Golf	LR axle	41.3	8.75	6.5	15.3	68
1980 Kawasaki KZ440B	1989 Nissan Maxima	17 in. rearward of LR axle	54.0	9.0	15.0	24.0	53
1986 Suzuki LS650 Savage	1989 Nissan Maxima	34 in. rearward of RF axle	55.0	14.5	10.25	24.75	10
2004 Kawasaki EX250-F	1989 Nissan Maxima	RR axle	54.6	12.0	6.0	18.0	38
1988 Honda CBR600F	1989 Nissan Maxima	38 in. forward of RR axle	55.9	7.5	14.5	22.0	5
1979 Suzuki GS550	1989 Honda Civic	23 in. forward of RR axle	56.3	11.25	13.5	24.75	40
1981 Yamaha Seca 750	1989 Honda Civic	Left A pillar, 23 in. rearward of LF axle	57.3	10.0	16.0	26.0	52
1982 Yamaha XV920 Virago	1989 Honda Civic	LR axle	56.4	11.5	5.5	17.0	89
1990 Kawasaki EX500A	1992 Mercury Tracer	36 in. rear of RF axle	60.9	11.25	17.0	28.25	1
1982 Suzuki GS750	1992 Mercury Tracer	LR axle	59.8	12.0	10.0	22.0	125
1980 Suzuki GS750	1992 Mercury Tracer	19 in. rear of RR axle	59.3	9.0	11.5	20.5	125
1986 Suzuki VS700GL Intruder	1986 Mercedes 300E	RF axle	44.3	14.5	7.0	21.5	47
1985 Honda CB650 Nighthawk	1986 Mercedes 300E	39 in. rear of LF axle	48.4	10.0	13.5	23.5	11
1987 Suzuki GSXR750	1986 Mercedes 300E	12 in. rear of RR axle	47.7	9.5	7.0	16.5	57
1983 Suzuki GS1100ES	1986 Mercedes 300E	28 in. rear of LR axle	49.1	9.0	21.0	30.0	75
1978 Honda CB750 H-matic	1988 Honda Prelude	47 in. rear of LF axle	29.8	6,75	7.0	13.75	9
1990 Honda Pac. Cst. PC800	1988 Honda Prelude	RR axle	32.9	5.5	2.0	7.5	61
1984 Honda Goldwing 1200	1988 Honda Prelude	35 in. rear of front axle, near LF pillar	58.8	9.5	20.0	29.5	22
1978 Suzuki GS1000	1988 Honda Prelude	RR	60.3	13.0	14.5	27.5	107

These tests included impacts to the struck vehicle doors, fenders, pillars, wheels, and axles. The authors reported the motorcycle wheelbase reduction and maximum car crush for the struck vehicle in each test. They also reported the yaw rotation of the struck vehicle that occurred as a result of each collision (Table 8.4). The authors plotted the total crush for each test (wheelbase reduction plus maximum car crush) against the impact speed and the data exhibited a generally linear relationship.

Another set of motorcycle-to-car collisions was conducted at the World Reconstruction Exposition in 2016 (WREX 2016). Table 8.5 summarizes the tests in this series that involved Harley-Davidson motorcycles. This table includes the motorcycle and the passenger vehicle utilized in each test, the impact configuration, the motorcycle impact speed, the resulting wheelbase reduction for the motorcycle, and the maximum crush to the vehicle. The passenger vehicles were stationary prior to the collision in these tests. Motorcycle impact speeds varied between 30.3 and 46.3 mph. These tests were conducted on an asphalt roadway near Orlando International Airport with a reported coefficient of friction of 0.7.

Peck et al. [7] analyzed the WREX 2016 collisions with several approaches that had been previously published. They concluded that the modified Eubanks equations [5] yielded the greatest accuracy, with an average error of 0.4 mph and a standard deviation of 4.8 mph. After using these previously published approaches, these authors added these tests into the existing dataset, along with four additional staged collisions conducted at ARC-CSI 2016. These tests also involved Harley-Davidson motorcycles. In three of the tests, the motorcycles struck passenger cars and in the fourth the motorcycle struck a concrete barrier. The involved vehicles and resulting deformations for these tests are summarized in Table 8.6. With this fuller dataset, these authors parsed the data into four regions (axle, bumper/pillar, door, and fender) and calculated updated coefficients for the Eubanks-form equations for each of these regions.

TABLE 8.5 Motorcycle-to-vehicle collisions performed at WREX 2016

Test	Motorcycle	Motorcycle weight (lb)	Struck vehicle	Struck vehicle weight (lb)	Impact location	Motorcycle impact speed (mph)	Motorcycle wheelbase reduction (in.)	Max crush to car (in.)
3	2013 H-D Softail Breakout FXSB	667	2006 Nissan Maxima	3449	Passenger's side front door	43.0	11.3	18.7
5	2013 H-D Dyna Street Bob FXDBA	633	2006 Nissan Maxima	3449	Driver's side rear door	36.9	12.8	8.6
8	2012 H-D Dyna Fat Bob FXDF	678	2005 Dodge Durango	4740	Driver's side rear quarter panel	46.3	3.3	9.1
11	2012 H-D Dyna Low Rider FXDL	649	2005 Dodge Durango	4740	Driver's side front door	42.9	10.7	10.3
22	1997 H-D Sportster 883	498	2006 Hyundai Sonata	3547	Passenger's side rear quarter panel	30.3	4.3	2.1
23	2002 H-D Sportster 883	500	2006 Hyundai Sonata	3547	Passenger's side rear door	42.7	8.8	7.1
24	2003 H-D Sportster 883	483	2006 Hyundai Sonata	3547	Driver's side rear door	35.5	8.9	4.2

© SAE International

©2022, SAE International

These equations are as follows:

Axle: $$S = 2.16(L+C) + 17.33 \qquad (8.5)$$

Bumper/Pillar: $$S = 1.36(L+C) + 19.50 \qquad (8.6)$$

Door: $$S = 1.50(L+C) + 9.27 \qquad (8.7)$$

Fender: $$S = 1.26(L+C) + 22.95 \qquad (8.8)$$

Most of the tests used for the development of these equations involved stationary vehicles. For a case in which the struck vehicle was moving, these equations would yield the relative speed between the motorcycle and the struck vehicle, along the direction the motorcycle was initially traveling. As the speed of the struck vehicle increases, the motorcycle wheel and forks are more likely to twist than to deform rearward.

Another issue that arises in the application of these equations is that some collisions could plausibly fit into two of the categories (axle, bumper/pillar, door, or fender). As reference [7] noted: "It can often be difficult to determine how the struck portion of the automobile should be qualified, as the motorcycle will often engage surrounding portions of the vehicle. For instance, in [WREX 2016] test 23 the motorcycle struck near the central portion of the right rear door of Hyundai … However, there was substantial damage to the adjacent C-pillar, and the rocker panel was deformed. So, while the analyst might first qualify this impact as a door strike, there seems to be an argument to qualify this as a stiffer pillar strike. Using … the door model predicts an impact speed of 33.1 mph while the pillar model predicts 41.2 mph. The actual impact speed was 42.7 mph." The damage to the Hyundai from Test #23 is shown in Figure 8.3.

Reference [7] continued: "Difficulty in qualifying the impact area, in combination with typical automotive construction, may explain the scatter associated with the … fender equations. While the average error for the … fender equation is 0.0 mph, meaning the equation is just as likely to overestimate impact speed as underestimate impact speed, the standard deviation is 7.0. The trailing portion of a typical fender will terminate at the junction with the vehicle's stiff A-pillar, near the middle of the fender will be a wheel assembly and underlying the forward portion of the fender is the end of the bumper reinforcement (stiff) or often nothing (very soft). Considering this diverse structure, the struck portion of the fender could be either very stiff or very soft."

TABLE 8.6 Summary of motorcycle collisions from ARC-CSI 2016

Test	Motorcycle	Motorcycle weight (lb)	Struck vehicle	Struck vehicle weight (lb)	Impact location	Motorcycle impact speed (mph)	Motorcycle wheelbase reduction (in.)	Max crush to car (in.)
1	2006 H-D Sportster XL1200C	579	Barrier			25.6	7.2	
3	2011 H-D Road King FLHRC	794	2012 Volkswagen Passat	3360	Wheel	31.9	5.0	6.2
5	2011 H-D Electra Glide FLHTCU	881	2012 Volkswagen Passat	3360	Pillar	29.0	6.5	2.7
9	2011 H-D Softail FLSTC	649	2011 Ford Crown Vic	4057	Fender	32.8	3.8	10.3

©2022, SAE International

FIGURE 8.3 Damage to right rear door and surrounding structure of struck vehicle (WREX 2016, Test #23).

Equations (8.5) through (8.8) are plotted in the following figures, along with the data from which they were derived. Figure 8.4 contains a point for each of the motorcycle-to-axle collisions. The horizontal axis is the motorcycle wheelbase reduction plus the maximum car crush in inches. The vertical axis is the motorcycle collision speed in miles per hour. The dark black line represents Equation (8.5). The dashed blue lines represent the range associated with a 68% confidence level (−3.3 mph, +2.9 mph). The dashed red lines represent the range associated with a 95% confidence level (−6.4 mph, +6.0 mph). Figure 8.5 contains a point for each of the motorcycle-to-bumper (squares) and motorcycle-to-pillar collisions (triangles). The dark black line represents Equation (8.6), and again, the dashed blue lines represent the range associated with a 68% confidence level (−5.5 mph, +5.3 mph) and the dashed red lines represent the range associated with a 95% confidence level (−10.9 mph, +10.7 mph). Figure 8.6 contains a point for each of the motorcycle-to-door collisions. The dark black line represents Equation (8.7), and again, the dashed blue lines represent the range associated with a 68% confidence level (−6.5 mph, +6.1 mph) and the dashed red lines represent the range associated with a 95% confidence level (−12.8 mph, +12.4 mph). Finally, Figure 8.7 contains a point for each of the motorcycle-to-fender collisions. The dark black line represents Equation (8.8), and again, the dashed blue lines represent the range associated with a 68% confidence level (−7.0 mph, +7.0 mph) and the dashed red lines represent the range associated with a 95% confidence level (−14.0 mph, +14.0 mph).

It is worth mentioning another observation that reference [7] makes: "the vast majority of data feeding current equations" are from motorcycles with conventional forks, rather than upside down (USD) forks. The authors note that "only four documented tests involving motorcycles equipped with USD forks are available at this time ... three of those are included in the source data used to create the [above] equations ... For these three USD fork tests, the data fit well with that of the traditional fork, with an average error of −2.5 mph. However, additional testing and analysis is desired to determine how confident an analyst can be when establishing impact speed for motorcycles equipped with USD forks, which are so stiff that they commonly fracture in a manner that traditional forks do not."

CHAPTER 8 Motorcycle Collisions with Vehicles and Roadside Barriers   **187**

FIGURE 8.4 Equation (8.5) plotted with the crash test points.

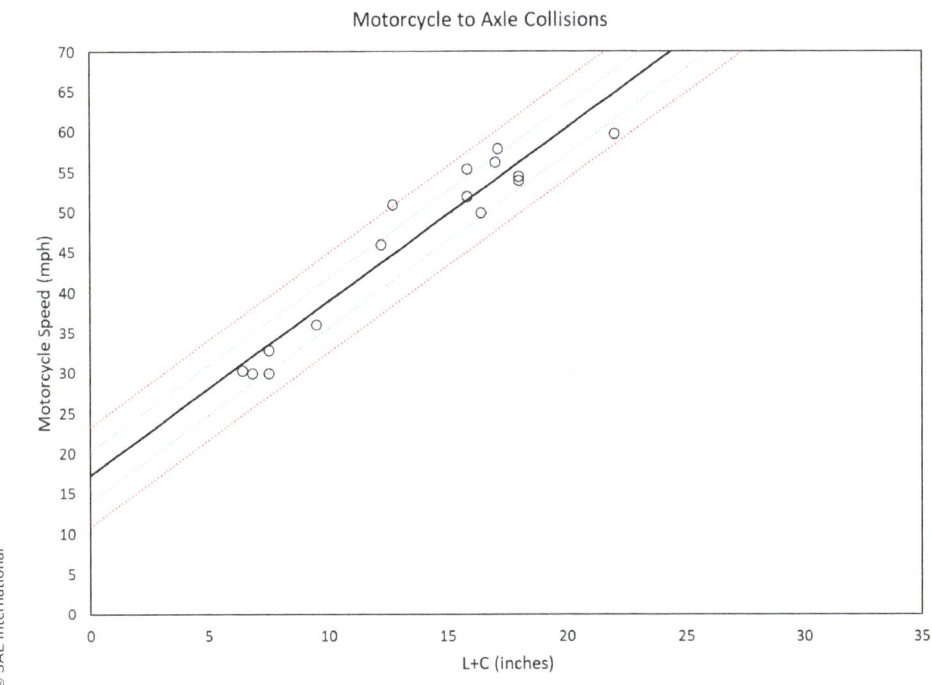

FIGURE 8.5 Equation (8.6) plotted with the crash test points.

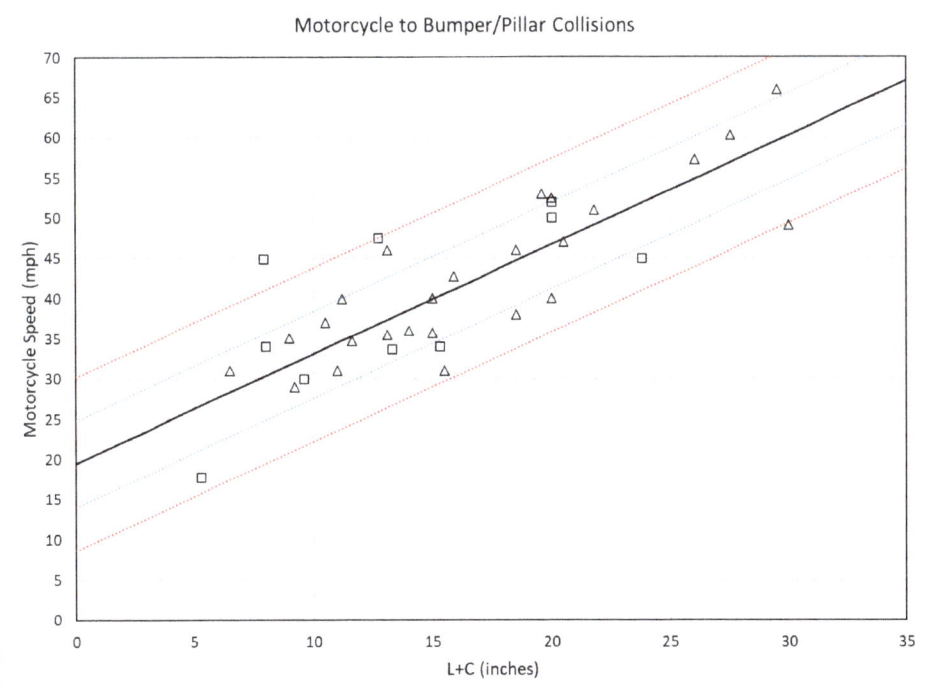

FIGURE 8.6 Equation (8.7) plotted with the crash test points.

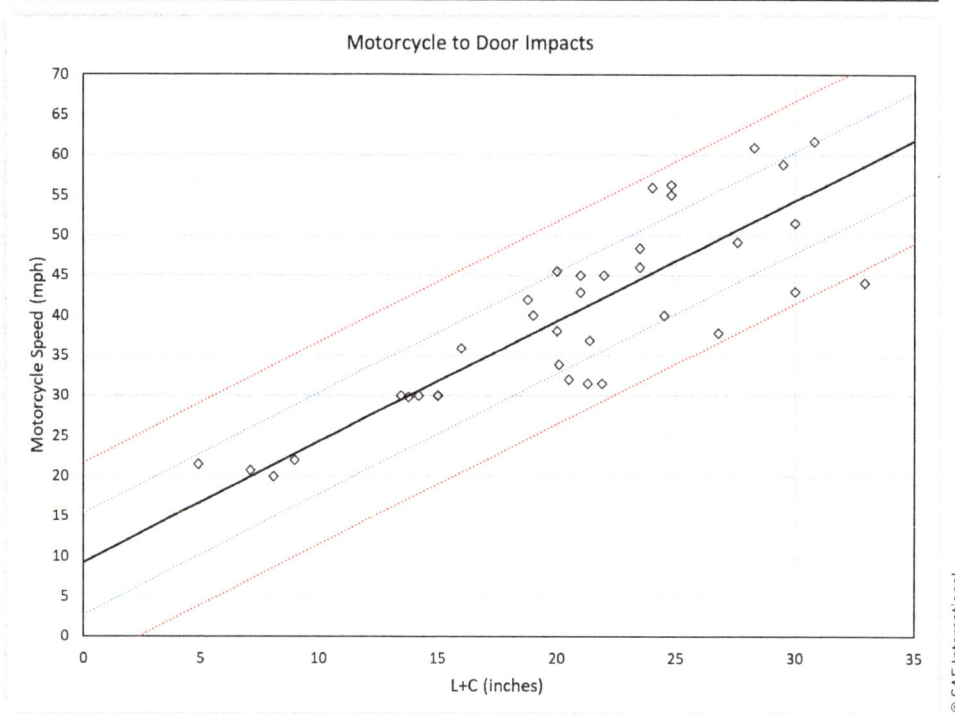

FIGURE 8.7 Equation (8.8) plotted with the crash test points.

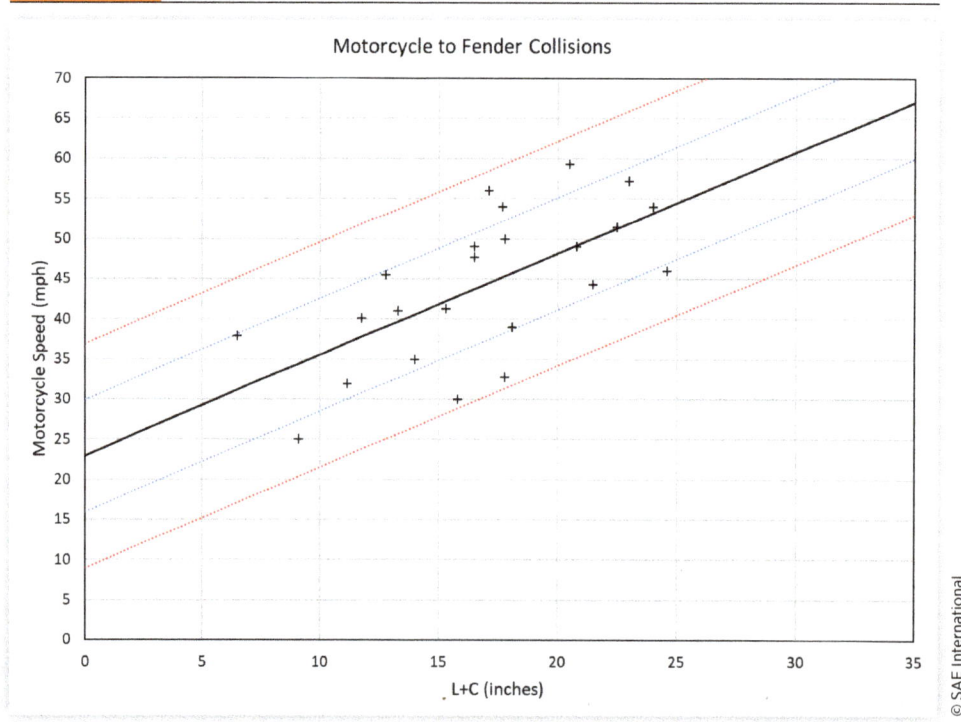

The equations covered so far have not been formulated in the way that crush analysis has traditionally been formulated for car-to-car collisions. For car-to-car collisions, the CRASH algorithm and its derivatives [8–17] have been a standard way to formulate the relationships between deformation, stiffness, impact speed, and severity. In theory, crush analysis for motorcycle collisions could also be formulated in terms of the CRASH crush analysis equations [18]. However, in practice, application of these equations relies on having a database of vehicle-to-rigid barrier collisions, where the stiffness of the structure of each vehicle (the motorcycle and whatever part of the car is struck) can be isolated and calculated. Such a database does not exist for motorcycle collisions.

Though they used different nomenclature than that used in CRASH, Wood et al. [19–21] reformulated the wheelbase reduction method along the lines of the CRASH method, utilizing relationships between deformation energy and impact speed, rather than between wheelbase reduction and speed. As a part of this, they included a means of imposing force balance (the collision force applied to the motorcycle is equal in magnitude to the collision force applied to the passenger car). These authors analyzed data from 43 motorcycle and scooter impacts with barriers and proposed the following empirical equation for calculating the energy absorbed by a motorcycle ($E_{A,mc}$) based on its wheelbase reduction. In this equation, the average wheelbase reduction (WBR) is entered in m, the mass (m_{mc}) is entered in kilograms, and the resulting energy is in units of N-m. Wood et al. observed that "the specific energy (E/M) absorption characteristics of motorcycles and scooters in frontal impacts are similar where the primary load path is through the front wheel and fork assembly." He reported an r^2 value for Equation (8.9) of 0.845.

$$E_{A,mc} = 641.7 m_{mc}(WBR + 0.1)^{1.89} \qquad (8.9)$$

Wood et al. noted that Equation (8.9) should only be used for wheelbase shortening less than 0.45 m (18 in.) and when the front forks and front wheel remain intact. "Barrier test data show that, when the front wheel rim, etc., breaks away from the wheel hub, the resulting wheelbase shortening underestimates the specific energy." Also, during motorcycle-to-car collisions, "the motorcycle-to-car impact takes place before the rider(s) collide against the car and that the motorcycle-to-car and the rider-to-car collisions can be considered as two discrete separate events. Further, the rider-to-car impact occurs at a higher level on the car. Consequently, the crush damage to the motorcycle or scooter and associated deformation to the car can be considered as being due to the motorcycle or scooter to car collision alone." Among other implications of this statement, it implies that the mass of the motorcycle used for Equation (8.9) would not include the mass of the rider(s).

Equation (8.10) is an English-unit version of Equation (8.9). In this equation, the motorcycle weight would be entered in lb and the wheelbase reduction in inches.

$$E_{A,mc} = 0.224809 \cdot 3.28084 \cdot \frac{641.7}{2.20462 \cdot} \cdot W_{mc}(0.0254 \cdot WBR + 0.1)^{1.89} \qquad (8.10)$$

Wood et al. also noted that the energy absorbed by the car ($E_{A,car}$) could be estimated with the following equation, which they derived by imposing force balance. In this equation, d_{car} is the maximum crush to the side of the car, measured in the same units used to measure the wheelbase reduction.

$$E_{A,car} = E_{A,mc} \frac{d_{car}}{WBR} \tag{8.11}$$

Wood et al. examined 31 staged tests where motorcycles impacted stationary cars at 90° and derived the following empirical equation for estimating the energy absorbed by the side of the struck vehicle. Equation (8.12) is included here in its English-unit form, such that the maximum crush to the car would be entered in inches and the resulting absorbed energy would be in ft-lb. Wood et al. reported an r^2 value of 0.845 for this equation.

$$E_{A,car} = 0.224809 \cdot 3.28084 \cdot 65,305 \cdot (0.0254 \cdot d_{car} - 0.0576) \tag{8.12}$$

In the collisions reported by Adamson et al. [2], the motorcycle struck the car at an angle of either 0°, 90°, or 180°. Clearly, this will not always be the case in the real world. Wood et al. [21] noted that, when the collision is angled, the damage energy for the car should be adjusted with the following equation:

$$E_{A,car} = \frac{E_{car,90}}{\cos\alpha} \tag{8.13}$$

In this equation, $E_{car,90}$ is the calculated energy absorption for the car assuming a 90° impact angle to the side of the vehicle and α is the angle of the collision to the side of the vehicle. The same concept could be applied for front and rear collisions to the vehicle. To apply this calculation, the deformation would be measured perpendicular to the car.

Once the absorbed energies have been obtained, they can be summed and used with the following equation to calculate the collision speed of the motorcycle.

$$S_{mc,impact} = \sqrt{\frac{\gamma_{mc} m_{mc} + \gamma_{car} m_{car}}{\gamma_{mc} m_{mc} \cdot \gamma_{car} m_{car}} \cdot 2E_{A,total}} \tag{8.14}$$

In Equation (8.14), $E_{A,total}$ is the total absorbed energy for the motorcycle and the car combined, m_{car} is the mass of the car, and γ_{mc} and γ_{car} are the effective mass multipliers. The effective mass multipliers are defined by the following equation:

$$\gamma_i = \frac{k_i^2}{k_i^2 + h_i^2} \tag{8.15}$$

In this equation, k_i is the radius of gyration for either the motorcycle or car and h_i is the moment arm of the collision force about the center-of-mass of that vehicle. In many motorcycle-to-vehicle collisions, the moment arm of the collision force for the motorcycle will be zero, and so, the effective mass multiplier for the motorcycles will be equal to 1. When the car involved in a collision is not stationary and some component of the velocity of the car is directed into the motorcycle (in other words, when the car is moving and the impact angle is something other than 0°, 90°, or 180°), then Equation (8.14) will yield the closing speed between the motorcycle and the car rather than the impact speed of the motorcycle. To apply Equation (8.15), the reconstructionist will need to calculate the radius of gyration of the struck vehicle about its yaw axis. This value is obtained by taking the square root of the yaw moment of inertia divided by the mass. The yaw moment of inertia characterizes the resistance of the vehicle to rotation about its yaw axes. MacInnis et al. [22] and Allen et al. [23] examined methods for estimating the whole vehicle (as opposed to sprung mass) moments of inertia for a passenger vehicle. The crush energy calculations proposed by Wood and his colleagues do not appear to recognize the difference in stiffness of different regions of the struck vehicle.

Determining Motorcycle Speed from the Struck Vehicle Postimpact Translation and Rotation

When a motorcycle collides with a passenger vehicle, the impact can cause a change in the translational and rotational velocities of the passenger vehicle. If these velocity changes or the magnitude of the translation and rotation of the struck vehicle can be quantified, then these can potentially be used to calculate the impact speed of the motorcycle. There are several methods that could be used for this analysis. The most general and comprehensive solution will be to use one of the widely accepted accident reconstruction software programs—PC-Crash, HVE (the EDSMAC4 or SIMON modules), Virtual CRASH, or VCRware [24–29]. These software packages incorporate equations for calculating the velocity changes that occur when the vehicles collide and equations for calculating the postimpact motion of the vehicles. These models enable the analyst to specify and incorporate the level of steering and braking at each tire (because of driver input or damage) and most of them utilize a graphical interface that allows the analyst to import a diagram of the physical evidence (tire marks, scrapes and gouges, fluid on the roadway, and vehicle rest positions, for instance) and to compare the simulated motion to the evidence on the diagram.

It would also be useful to have simple formulas for obtaining a reasonable estimate of the motorcycle impact speed based on the observed postimpact translation and rotation of the struck vehicle. All the more useful if a method could be based on the struck vehicle rotation alone. And, in fact, several such methods exist in the literature. However, these equations necessarily invoke simplifying assumptions.

Planar Impact Mechanics

The equations of planar impact mechanics (PIM) developed by Brach [30] and Brach and Brach [31–33] provide a useful starting place for modeling the impact portion of the crash. These equations utilize the principle of impulse and momentum (conservation of linear and angular momentum), and they allow for both translation and rotation of the vehicles to be considered. The full set of PIM equations are cumbersome for hand-calculations. However, there may be a series of simplifying assumptions that can be invoked to simplify them. In their complete form, the PIM equations invoke the following assumptions: (1) the collision is contained within a two-dimensional (2-D) plane (no vertical motion or pitch or roll motion is considered); (2) the collision is instantaneous (this equates to no change in position or orientation of the vehicles during the collision); (3) the deforming region of both vehicles is assumed to be small relative to the size of the vehicles (this equates to no change in inertial properties because of the collision and collision force can be assumed to be transferred at a single point); (4) forces other than collision forces are assumed negligible or the collision force is assumed very large relative to other forces (tire forces are neglected).

©2022, SAE International

The accident reconstruction text by Brach and Brach [33] contains the following diagram (Figure 8.8) defining the variables that are used in the PIM equations. First, an arbitrarily oriented x-y coordinate system is established. In practice, the orientation of this coordinate system can be oriented to simplify the resulting equations. Second, the normal (n) and tangential (t) directions are established. These directions are established, first, by using physical inspection, photographs, or photogrammetry to document and diagram the residual damage to the vehicles. Third, the documented damage is used to establish an impact configuration. Once this impact configuration is established, the tangential direction will lie along the surface of contact between the two vehicles. The normal direction is then at a 90° angle to the tangential direction. The angle Γ is the angle between the x-axis and the n-axis and between the y-axis and the t-axis. The masses of the vehicles are designated with the variables m_1 and m_2, and the yaw moments of inertia about the center of gravity (CG) are designated with the variables I_1 and I_2.

FIGURE 8.8 Definitions of variable for the PIM equations, Figure 6.5 from Brach et al. [29].

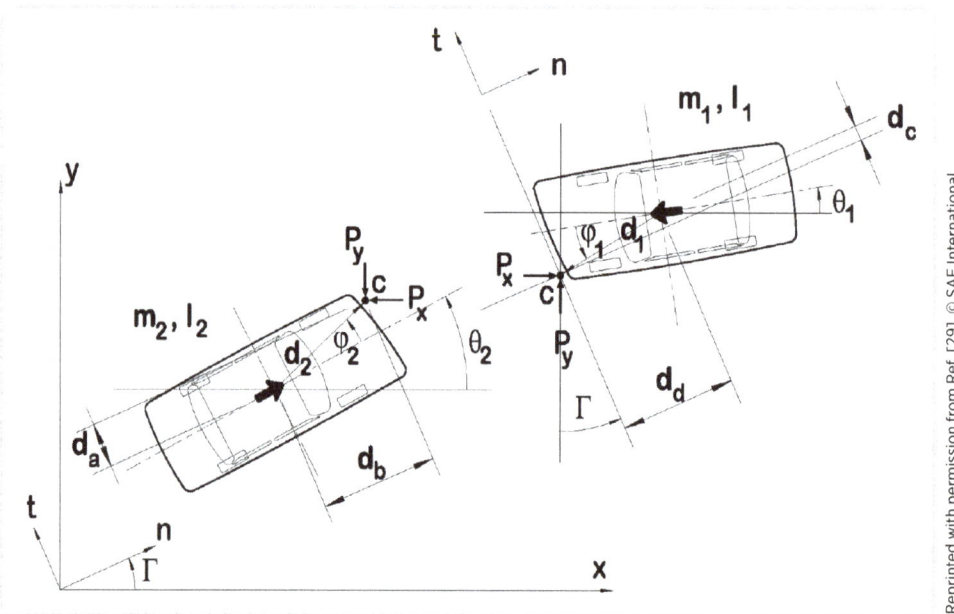

After the coordinate directions are established, the Point C needs to be established. This is the point at which the resultant collision force is applied to each vehicle. This is a single point along the intervehicle contact surface. This point will lie along the contact surface, but its precise location along that surface might not be known. Often, the location of this point can be varied and iterated to produce the best fit with the known physical evidence. Once Point C has been established, d_1 is the distance from the CG of Vehicle 1 to the Point C and d_2 is the distance from the CG of Vehicle 2 to the Point C. The heading angle of Vehicle 1, measured off the x-axis with counterclockwise being positive, is specified with the angle θ_1 and the heading angle of Vehicle 2 is specified with θ_2. These angles are measured as depicted in Figure 8.8. The angles φ_1 and φ_2 are the angles between the vehicle headings and the line that connects the CG of each vehicle to the Point C. Given these definitions, the variables d_a, d_b, d_c, and d_d are defined as follows:

$$d_a = d_2 \sin(\theta_2 + \varphi_2 - \Gamma) \quad (8.16)$$

$$d_b = d_2 \cos(\theta_2 + \varphi_2 - \Gamma) \quad (8.17)$$

$$d_c = d_1 \sin(\theta_1 + \varphi_1 - \Gamma) \quad (8.18)$$

$$d_d = d_1 \cos(\theta_1 + \varphi_1 - \Gamma) \quad (8.19)$$

Finally, the velocities will be designated as follows:

$v_{1n,i}$ = the initial (preimpact) velocity of Vehicle 1 along the normal direction
$v_{1t,i}$ = the initial (preimpact) velocity of Vehicle 1 along the tangential direction
$v_{2n,i}$ = the initial (preimpact) velocity of Vehicle 2 along the normal direction
$v_{2t,i}$ = the initial (preimpact) velocity of Vehicle 2 along the tangential direction
$\omega_{1,i}$ = the initial (preimpact) yaw angular velocity of Vehicle 1
$\omega_{2,i}$ = the initial (preimpact) yaw angular velocity of Vehicle 2
$v_{1n,f}$ = the final (postimpact) velocity of Vehicle 1 along the normal direction
$v_{1t,f}$ = the final (postimpact) velocity of Vehicle 1 along the tangential direction
$v_{2n,f}$ = the final (postimpact) velocity of Vehicle 2 along the normal direction
$v_{2t,f}$ = the final (postimpact) velocity of Vehicle 2 along the tangential direction
$\omega_{1,f}$ = the final (postimpact) yaw angular velocity of Vehicle 1
$\omega_{2,f}$ = the final (postimpact) yaw angular velocity of Vehicle 2

Given these variables, the impulse-momentum equations for the collision are given as follows:

$$v_{1n,f} = v_{1n,i} + \frac{\overline{m}}{m_1}(1+e)v_{rn}q \quad (8.20)$$

$$v_{1t,f} = v_{1t,i} + \mu \frac{\overline{m}}{m_1}(1+e)v_{rn}q \quad (8.21)$$

$$v_{2n,f} = v_{2n,i} - \frac{\overline{m}}{m_2}(1+e)v_{rn}q \quad (8.22)$$

$$v_{2t,f} = v_{2t,i} - \mu \frac{\overline{m}}{m_2}(1+e)v_{rn}q \quad (8.23)$$

$$\omega_{1,f} = \omega_{1,i} + \frac{\bar{m}}{m_1 k_1^2}(1+e)(d_c - \mu d_d)v_{rn} q \qquad (8.24)$$

$$\omega_{2,f} = \omega_{2,i} + \frac{\bar{m}}{m_2 k_2^2}(1+e)(d_a - \mu d_b)v_{rn} q \qquad (8.25)$$

In these equations:

$$\bar{m} = \frac{m_1 m_2}{m_1 + m_2} \qquad (8.26)$$

$$k_1 = \sqrt{\frac{I_1}{m_1}} \qquad (8.27)$$

$$k_2 = \sqrt{\frac{I_2}{m_2}} \qquad (8.28)$$

$$v_{rn} = \left(v_{2n,i} - d_a \omega_{2,i}\right) - \left(v_{1n,i} - d_c \omega_{1,i}\right) \qquad (8.29)$$

$$\frac{1}{q} = 1 + \frac{\bar{m} d_a^2}{m_2 k_2^2} + \frac{\bar{m} d_c^2}{m_1 k_1^2} - \mu \left(\frac{\bar{m} d_c d_d}{m_1 k_1^2} + \frac{\bar{m} d_a d_b}{m_2 k_2^2} \right) \qquad (8.30)$$

These equations also include the coefficient of restitution (e) and the impulse ratio (μ). The coefficient of restitution is defined at Point C (the impact center) and it will typically fall between 0 and 1. A range for the coefficient of restitution for a given collision can be determined based on experimental data. The impulse ratio (μ) is the ratio of the tangential to normal impulses generated by this collision, as follows:

$$\mu = \frac{P_t}{P_n} \qquad (8.31)$$

In some instances, the impulse ratio can be interpreted as a friction coefficient between the colliding vehicles, but it is not limited to this interpretation [30–34]. In addition to the effects of friction, the impulse ratio can also include the effects of forces generated by snagging between the vehicles. The "available friction" can be set at a value that reflects such snagging when it occurs. Brach [30] presented the following expression for the critical impulse ratio, which is the impulse ratio necessary to cause sliding to cease during the collision. The impulse ratio should not be set at a value that exceeds the critical impulse ratio.

$$\mu_c = \frac{rA + (1+e)B}{(1+e)(1+C) + rB} \qquad (8.32)$$

In this equation:

$$A = 1 + \frac{\overline{m}d_c^2}{m_1 k_1^2} + \frac{\overline{m}d_a^2}{m_2 k_2^2} \quad (8.33)$$

$$B = \frac{\overline{m}d_c d_d}{m_1 k_1^2} + \frac{\overline{m}d_a d_b}{m_2 k_2^2} \quad (8.34)$$

$$C = \frac{\overline{m}d_d^2}{m_1 k_1^2} + \frac{\overline{m}d_b^2}{m_2 k_2^2} \quad (8.35)$$

The relative magnitude of the available friction coefficient to the critical impulse ratio will have physical significance for an impact. The available friction coefficient represents the magnitude of friction force that *can* be recruited during the impact. The critical impulse ratio represents the magnitude of friction force that *must* be recruited for relative motion to cease along the contact surface. When the critical impulse ratio is greater than the available friction coefficient, all the available friction will be recruited during the impact, but that friction will be insufficient to cause sliding to cease in the contact region. When the available friction coefficient exceeds the critical impulse ratio, only a portion of the available friction will be recruited, and sliding will cease in the contact region. In such cases, the value of the impulse ratio for the impact model should be set at the value of the critical impulse ratio, not the available friction coefficient. This is because recruitment of the available friction depends on relative velocity being present along the contact surface. Once this relative motion ceases, no additional friction can be recruited.

To simplify the PIM equations for application to specific special cases, consider first a 90° collision of a moving motorcycle with stationary struck vehicle, as depicted in Figure 8.9. The x-direction has been aligned to the initial travel direction of the motorcycle, and because the passenger vehicle is initially stationary, it is also reasonable to assume that the normal direction is aligned to the initial travel direction of the motorcycle. This means that the angle $\Gamma = 0$.

FIGURE 8.9 90° collision between moving motorcycle (Vehicle 2) and a stationary passenger vehicle (Vehicle 1).

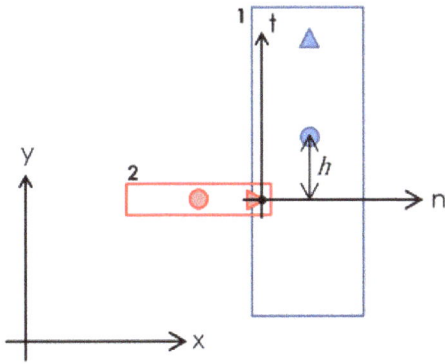

This scenario will be used to examine the relationship between the collision speed of the motorcycle and the postimpact velocity conditions of the struck vehicle. Thus, out of Equations (8.20) through (8.25), only Equations (8.20), (8.21), and (8.24) need to be considered. The discussion can be simplified even further by noting that because this collision has no initial velocity in the tangential direction, there will not be a change in velocity along that direction, and Equation (8.21) can also be dropped. The impulse ratio μ can also be set to zero because there is not tangential velocity. Equations (8.20) and (8.24) remain. These two equations can be simplified by noting that the initial translational and rotational velocities of Vehicle 1 are zero. Also, the initial rotational velocity of the motorcycle (Vehicle 2) is zero. Thus, the variable v_{rn} reduces to $v_{2n,i}$, which is the impact speed of the motorcycle. In addition, θ_2 and φ_2 are both zero, thus $d_a = 0$ and $d_b = d_2$. With the configuration of the collision and coordinate systems shown in Figure 8.9, $\theta_1 = -90°$. Considering the geometry of the collision, this means that d_c becomes the longitudinal distance from CG of Vehicle 1 to the point of collision with the motorcycle (h). With these assumptions, Equations (8.20) and (8.24) can be reduced to the following equations:

$$v_{1n,f} = (1+e)v_{2n,i}\frac{\overline{m}k_1^2}{I_1 + \overline{m}h^2} \tag{8.36}$$

$$\omega_{1,f} = (1+e)v_{2n,i}\frac{\overline{m}h}{I_1 + \overline{m}h^2} \tag{8.37}$$

These equations can be solved for the impact speed of the motorcycle, yielding the following:

$$v_{2n,i} = v_{1n,f}\frac{1}{1+e}\frac{I_1 + \overline{m}h^2}{\overline{m}k_1^2} \tag{8.38}$$

$$v_{2n,i} = \omega_{1,f}\frac{1}{1+e}\frac{I_1 + \overline{m}h^2}{\overline{m}h} \tag{8.39}$$

These equations relate the postimpact translational and rotational velocities of the struck vehicle to the impact speed of the motorcycle. Either of these equations will yield the motorcycle impact speed. Thus, if either the postimpact translational velocity or postimpact rotational velocity can be determined, then the motorcycle impact speed can be determined (assuming, of course, that simplifying assumptions are satisfied).

Relating the Rotational Displacement to the Angular Velocity

To apply Equation (8.39), a method is needed to relate the total rotational displacement experienced by the struck vehicle to the yaw velocity the vehicle possessed immediately following the collision. Factors that could influence the magnitude of postimpact angular rotation experienced by the struck vehicle for a given postimpact yaw rotation rate include:

- Postimpact translational speed of the struck vehicle
- Yaw moment of inertia of the struck vehicle
- Coefficient of friction between the roadway and the tires
- Struck vehicle wheelbase
- Wheel damage
- Impact induced steering

The following equation has been proposed for relating the angular velocity to the angular displacement [35–38]. In this equation, τ_{rot} is the average torque applied to the struck vehicle during the postimpact rotation about the point of rotation, θ is the angular displacement, and I_{rot} is the yaw moment of inertia of the struck vehicle about the point of rotation.

$$\omega = \sqrt{\frac{2\tau_{rot}\theta}{I_{rot}}} \quad (8.40)$$

This equation assumes that the angular kinetic energy can be partitioned from the translational kinetic energy and equated to the angular work done by the torque, which has been partitioned from the translational work done by the friction force. This assumption is not correct. However, given the presence of this equation in the literature, its application will be tested and illustrated in this chapter. Per references [35–38], the torque about the point of rotation is given by the following equation:

$$\tau_{rot} = fW_{rot}d_3 \quad (8.41)$$

In this equation, f is the coefficient of friction, W_{rot} is the weight on the rotating axle, and d_3 is the distance from the point of rotation to the rotating axle. So, if the point of rotation is where the center of the front axle, for example, this distance would be the wheelbase of the vehicle. The yaw moment of inertia about the point of rotation is calculated with the following equation:

$$I_{rot} = I + md_1^2 \quad (8.42)$$

In this equation, I is the yaw moment of inertia about the CG, d_1 is the distance from the CG to the point of rotation, and m is the mass of the vehicle. Once the yaw velocity of the struck vehicle is obtained, the change in velocity of the motorcycle can be calculated with the following equation:

$$\Delta v_{mc} = \frac{I_{rot}g\omega}{W_{mc}d_2} \quad (8.43)$$

In Equation (8.43), d_2 is the collision force moment arm about the center of rotation. Once the motorcycle change in velocity is obtained, the motorcycle impact speed can then be calculated with Equation (8.44).

©2022, SAE International

$$v_{mc} = \frac{d_2 \omega + \Delta v_{mc}}{1+e} \qquad (8.44)$$

To illustrate the relationship between the initial postimpact yaw velocity and the angular displacement of the struck vehicle, PC-Crash 12.0 was used to run impact scenarios between a motorcycle and a passenger car. Simulations were set up with a friction coefficient of 0.76. The TM-Easy tire model was utilized with default parameters for both the motorcycle and the struck vehicle. The struck vehicle was assumed to have a weight of 4385 lb, a yaw moment of inertia of 2408 lb-ft-s^2, a wheelbase of 108 in., a length of 185.7 in., and a width of 75.8 in. The distance from the CG to the front axle of the struck vehicle was 46.9 in. Thus, the static load on the front axle was 2481 lb and the static load on the rear axle was 1904 lb. The motorcycle was assumed to have a weight of 850 lb, thus the weight ratio between the motorcycle and the struck vehicle was 5.16. The CG height of both vehicles was set to zero (2-D simulations).

A set of simulations was run with the speed of the struck vehicle set at zero and the impact speed of the motorcycle at 44.7 mph. The moment arm of the collision force was varied in ½-ft increments between 0 and 6 ft behind the CG. This set of simulations assumed no wheel damage or impact induced steering occurred to the struck vehicle. The brake factors on all four wheels of the struck vehicle were set at 5%. The coefficient of restitution for the collisions was held constant at 0.1 and the impulse ratio was held constant at 0.6.

For the simulation with a moment arm of 6 ft, the postimpact rotation of the struck vehicle occurred approximately about the center of front axle. The wheelbase of the struck vehicle was 9 ft and the moment arm of the collision force about the point of rotation was 9.91 ft. The struck vehicle experienced a yaw displacement of 56.6° (0.99 rad). With these values, Equation (8.41) yields a torque of 13,023.4 lb-ft, as follows:

$$\tau_{rot} = 0.76 \cdot 1,904 \cdot 9.0 = 13,023.4 \; lb-ft$$

Equation (8.42) yields a moment of inertia about the point of rotation of 4490.2 lb-ft-s^2, as follows:

$$I_{rot} = 2,408 + 136.2 \cdot 3.91 \cdot 3.91 = 4,490.2 \; lb \cdot ft \cdot s^2$$

Equation (8.40) then yields a postimpact yaw velocity of 137 deg/s (2.4 rad/s), as follows:

$$\omega = \sqrt{\frac{2 \cdot 13,023.4 \cdot 0.99}{4,490.2}} = 2.4 \frac{rad}{s} = 137 \frac{deg}{s}$$

The simulated postimpact yaw velocity was 168 deg/s (2.9 rad/s). Using the calculated yaw velocity, Equation (8.43) yields a velocity change for the motorcycle of 28.1 mph (41.2 ft/s), as follows:

$$\Delta v_{mc} = \frac{4,490.2 \cdot 32.2 \cdot 2.4}{850 \cdot 9.91} = 41.2 \frac{ft}{s} = 28.1 \; mph$$

Using the yaw velocity from the simulation, Equation (8.43) yields a change in velocity of 33.9 mph (49.8 ft/s), as follows:

$$\Delta v_{mc} = \frac{4,490.2 \cdot 32.2 \cdot 2.9}{850 \cdot 9.91} = 49.8 \frac{ft}{s} = 33.9 \ mph$$

The simulated change of velocity was 31.1 mph (45.6 ft/s). Using the calculated yaw velocity, Equation (8.44) yields an impact speed for the motorcycle of 40.3 mph (4.4 mph lower than the simulated value), as follows:

$$v_{mc} = \frac{9.91 \cdot 2.4 + 41.2}{1 + 0.1} = 59.1 \frac{ft}{s} = 40.3 \ mph$$

Using the simulated yaw velocity, Equation (8.44) yields an impact speed for the motorcycle of 48.7 mph (4.0 mph higher than simulated value), as follows:

$$v_{mc} = \frac{d_2 \omega + \Delta v_{mc}}{1 + e} = \frac{9.91 \cdot 2.9 + 49.8}{1 + 0.1} = 71.4 \frac{ft}{s} = 48.7 \ mph$$

Another set of simulations was run with the speed of the struck vehicle set at 10 mph and the impact speed of the motorcycle at 44.7 mph. The moment arm of the collision force was varied in ½-ft increments between 0 and 6 ft behind the CG. This set of simulations also assumed no wheel damage or impact induced steering occurred to the struck vehicle. The brake factors on all four wheels of the struck vehicle were set at 5%. The coefficient of restitution for the collisions was held constant at 0.1 and the impulse ratio was held constant at 0.6.

For the simulation with a moment arm of 6 ft, the postimpact rotation of the struck vehicle occurred approximately about the center of the front axle. The wheelbase of the struck vehicle was 9 ft and the moment arm of the collision force about the point of rotation was 9.91 ft. The struck vehicle experienced a yaw displacement of 117.7° (2.05 rad). Equation (8.41) still yields a torque of 13,023.4 lb-ft and Equation (8.42) still yields a moment of inertia about the point of rotation of 4490.2 lb-ft-s².

Equation (8.40) yields a postimpact yaw velocity of 197.6 deg/s (3.45 rad/s), as follows:

$$\omega = \sqrt{\frac{2 \cdot 13,023.4 \cdot 2.05}{4,490.2}} = 3.45 \frac{rad}{s} = 197.6 \frac{deg}{s}$$

The simulated postimpact yaw velocity was 172.4 deg/s (3.0 rad/s). It is useful to observe that there was little difference between the postimpact rotational velocities of the struck vehicle in the two simulations with the moment arm of 6 ft. However, there were significant differences in the total angular displacement—56.6° versus 117.7°. This demonstrates that the initial translational velocity of the struck vehicle significantly affects the magnitude of the postimpact angular displacement of the struck vehicle. This effect is not captured by Equation (8.40). Instead, this equation attributes increased angular displacement of the struck vehicle to an increase in the rotational velocity, which is incorrect. This equation predicts a difference in the angular velocity between the two simulations of approximately 60 deg/s.

An additional set of simulations was run with the speed of the struck vehicle set at 20 mph and the impact speed of the motorcycle at 44.8 mph. The moment arm of the collision force was varied in ½-ft increments between 0 and 6 ft behind the CG. This set of simulations also assumed no wheel damage or impact induced steering occurred to the struck vehicle. The brake factors on all four wheels of the struck vehicle were set at 5%. The coefficient of restitution for the collisions was held constant at 0.1 and the impulse ratio was held constant at 0.6.

For the simulation with a moment arm of 6 ft, it was again reasonable to assume that the postimpact rotation of the struck vehicle occurred about the center of front axle. The wheelbase of the struck vehicle was 9 ft and the moment arm of the collision force about the point of rotation was 9.91 ft. The struck vehicle experienced a yaw displacement of 166.56° (2.91 rad). Equation (8.41) still yields a torque of 13,023.4 lb-ft and Equation (8.42) still yields a moment of inertia about the point of rotation of 4490.2 lb-ft-s².

Equation (8.40) yields a postimpact yaw velocity of 235 deg/s (4.11 rad/s), as follows:

$$\omega = \sqrt{\frac{2 \cdot 13,023.4 \cdot 2.91}{4,490.2}} = 4.11 \frac{rad}{s} = 235 \frac{deg}{s}$$

The simulated postimpact yaw velocity was 176.5 deg/s (3.1 rad/s), again not that different from the simulations with an initial struck vehicle speed of 0 and 10 mph. This further confirms that Equation (8.40) is attributing too much of the increased rotational displacement to an increase in yaw velocity and neglecting the influence of the translational speed. This also illustrates that the translational and rotational energies cannot be partitioned in the way that Equation (8.40) assumes.

Figure 8.10 is a graph showing results of applying Equation (8.40) to calculate the postimpact yaw velocity for all the PC-Crash simulations. The simulated angular velocity is plotted on the horizontal axis and the angular velocity predicted with Equation (8.40) is plotted on the vertical axis. In instances where Equation (8.40) is accurate, the resulting point would lie along the black line. As this graph shows, there are few instances where Equation (8.40) yields an accurate angular velocity. It generally overestimates the angular velocity. This tendency becomes more pronounced as the initial speed of the struck vehicle increases.

Figure 8.11 is a graph showing the total angular displacement of the struck vehicle plotted against the postimpact yaw velocity for all the PC-Crash simulations. Separate curves are depicted for struck vehicle speeds of 0, 10, and 20 mph. The data labels next to the points on this graph refer to the collision force moment arm about the CG for the simulation each point represents. This graph demonstrates that the initial speed of the struck vehicle has a substantial influence on the total angular displacement experienced by the struck vehicle. This occurs without a significant increase in the postimpact angular velocity of the struck vehicle, as demonstrated by Figure 8.12. The points in this figure are for the same set of simulations contained in Figure 8.11. In Figure 8.12, the postimpact rotational velocity of the struck vehicle is plotted versus the collision force moment arm about the CG. As the initial speed of the struck vehicle increases, only a small increase in the rotational velocity occurs.

Thus, increased yaw displacement of the struck vehicle cannot be primarily attributed to an increase in the postimpact angular velocity, as is assumed in Equation (8.40). This means that the underlying assumptions of Equation (8.40) become increasingly inaccurate as the struck vehicle speed increases.

FIGURE 8.10 Results using Equation (8.40).

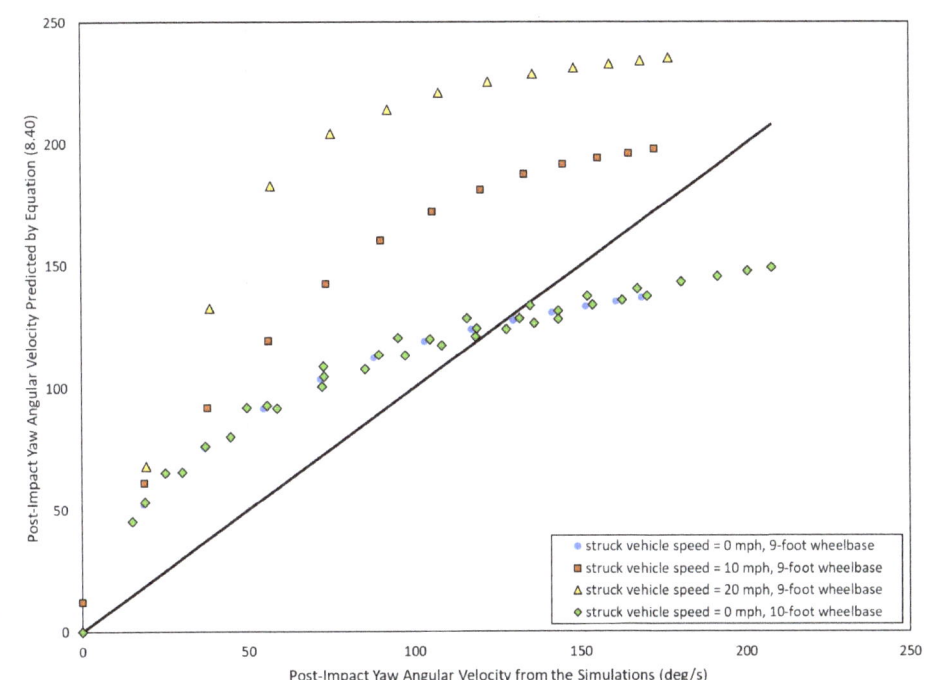

FIGURE 8.11 Angular displacement versus postimpact yaw velocity.

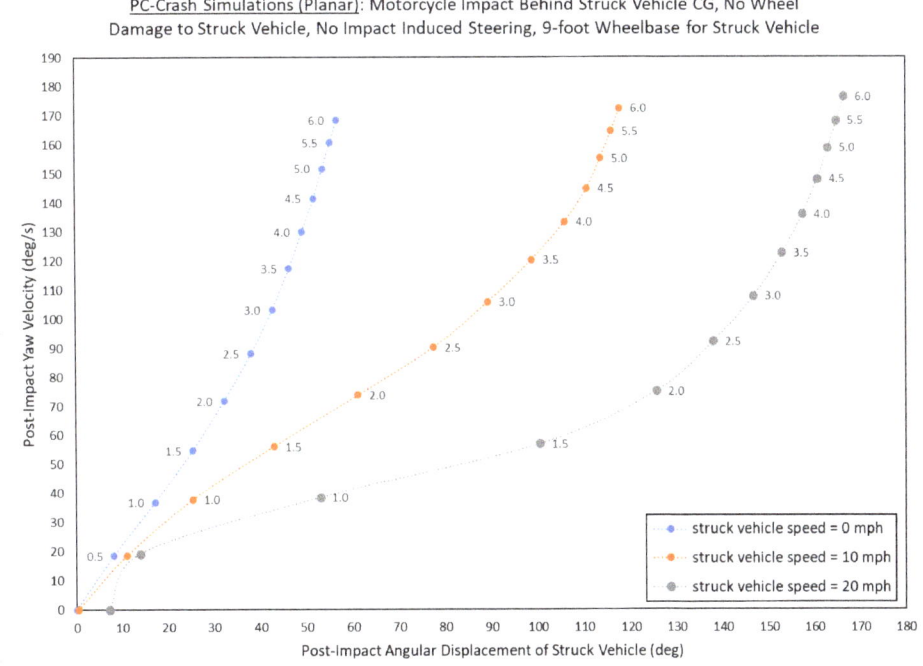

©2022, SAE International

FIGURE 8.12 Collision force moment arm versus postimpact yaw velocity.

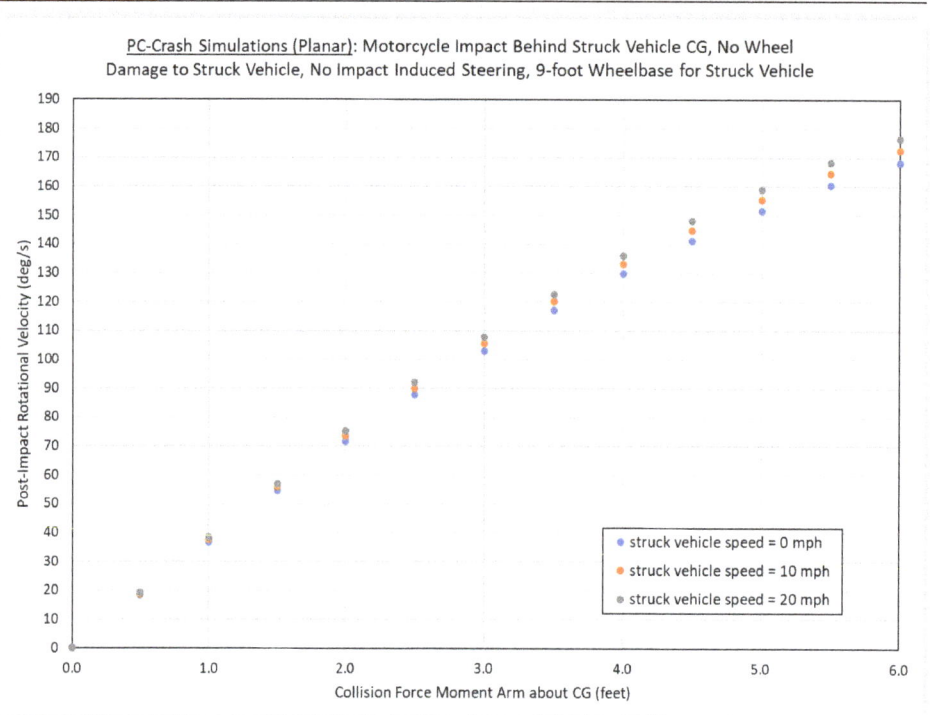

Three additional sets of simulations were run. For the first, the wheelbase of the struck vehicle was increased from 108 in. (9 ft) to 120 in. (10 ft). The struck vehicle was again assumed to have a weight of 4385 lb, a yaw moment of inertia of 2408 lb-ft-s², a length of 185.7 in., a width of 75.8 in., and the distance from the CG to the front axle of the struck vehicle was maintained at 46.9 in. With these assumptions, the static load on the front axle was approximately 2671 lb and the static load on the rear axle was approximately 1714 lb. Next, a set of simulations was run with these same parameters, with the exception that the yaw moment of inertia of the struck vehicle was increased by 25% to 3010 lb-ft-s². And finally, a set of simulations was run with the same parameters, with the exception that the yaw moment of inertia was reduced by 25% to 1806 lb-ft-s².

These sets of additional simulations were run with the initial speed of the struck vehicle at zero and the impact speed of the motorcycle at 44.7 mph. The moment arm of the collision force about the CG of the struck vehicle was varied in ½-ft increments between 0 and 6 ft behind the CG. These simulations again assumed no wheel damage or impact induced steering occurred to the struck vehicle. The brake factors on all four wheels of the struck vehicle were set at 5%. The coefficient of restitution for the collisions was held constant at 0.1 and the impulse ratio was held constant at 0.6.

The results from these simulations are plotted in Figure 8.13. As this graph shows, the relationship between the postimpact angular displacement of the struck vehicle and the postimpact yaw velocity was not significantly altered by changing the wheelbase. This is contrary to what Equation (8.40) predicts. Increasing or decreasing the yaw moment of inertia did, however, alter the relationship between the angular displacement and the

FIGURE 8.13 Angular displacement versus postimpact yaw velocity.

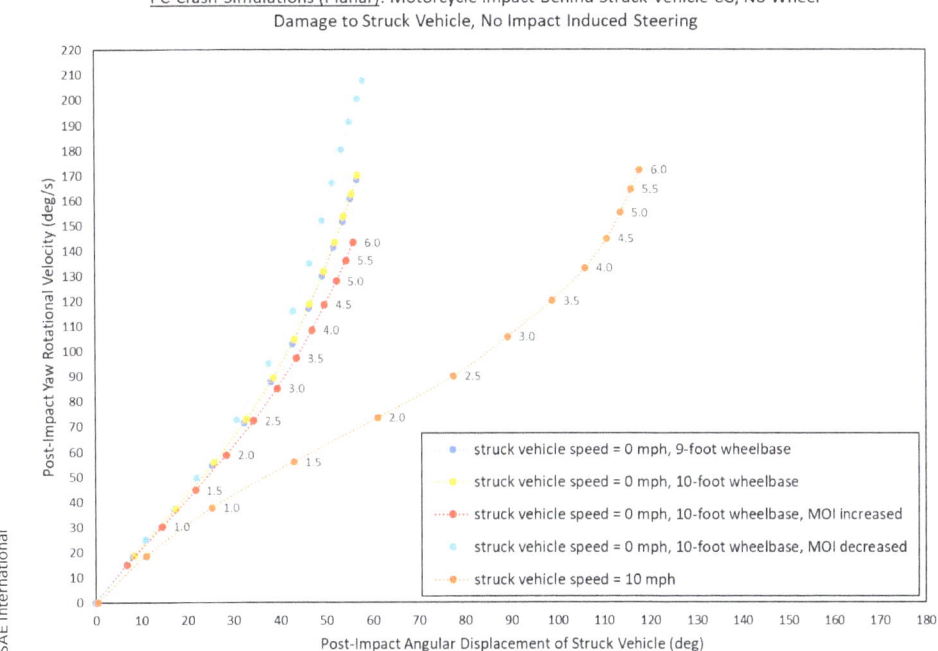

angular velocity. The effect was less substantial than the influence of the struck vehicle speed, but still potentially significant within speed analysis.

Ogden and Kloberdanz Equation

Ogden and Kloberdanz [39] presented the following equation for relating the postimpact angular displacement of the struck vehicle to the impact speed of the motorcycle:

$$V_{mc} = \left(\frac{g}{W_{mc}d_i} + \frac{g}{W_{car}d_i} + \frac{d_i}{I_{y,car}} \right) \sqrt{I_{y,car} \theta \cdot \eta \mu W_{car} l_{w,car}} \qquad (8.45)$$

In this equation, V_{mc} is the impact speed of the motorcycle in feet per second, g is the gravitational constant, W_{mc} is the weight of the motorcycle in lbs (including whatever portion of the rider's weight is appropriate for the circumstances), W_{car} is the weight of the struck car in pounds, $I_{y,car}$ is the yaw moment of inertia of the car in lb-ft-s², $l_{w,car}$ is the wheelbase of the struck car in feet, d_i is the moment arm of the collision force about the CG of the struck car in feet, θ is the postimpact angular displacement of the struck car in radians, η is an efficiency factor, and μ is the roadway coefficient of friction. The term $\eta\mu$ can be referred to as the drag factor. Reasonable values for most of these variables would be straightforward to establish. The efficiency factor η is the exception, and so, application of Equation (8.45) would depend on having a source of data for this parameter.

Keifer et al. [35, 36] referred to the term $\eta\mu$ as the rotational coefficient of friction and the variable η as the rotational friction factor. They observed that "the rotational coefficient of friction is dependent on the vehicle size parameters and is likely to change between various classes of vehicles." They also noted that the rotational coefficient of friction will always be less than the roadway coefficient of friction, so η will take on values less than 1. Finally, these authors noted that the rotational coefficient of friction will depend on the preimpact velocity of the struck vehicle and on the total magnitude of rotation experienced by the vehicle following the collision.

Keifer et al. used EDSMAC to perform a study of rotational friction factors. They ran simulations in which a large bullet vehicle (1998 Ford Expedition) struck the side of a target vehicle. Target vehicles of three sizes were utilized: large (1998 Ford Expedition), medium (1998 Honda Accord), and small (1998 Geo Metro). The collision in each simulation was set up to occur at the passenger side rear wheel of the target vehicle and the speed of the bullet vehicle was varied to obtain a variety of angular displacements for the target vehicle. In the first set of simulations, the target vehicles were initially stationary. The roadway coefficients of friction were varied between four different levels (0.25, 0.50, 0.75, and 1.00). All four wheels were assumed unbraked following the collision (either from driver input or damage). For each simulation the total angular displacement and rotational friction factor were tabulated and plotted. For the discussion here, the tabular data from Keifer et al. were plotted and it is included in Figure 8.14.

For the medium vehicle configuration, Keifer et al. ran additional simulations with the target vehicle stopped, varying additional conditions from the first set of simulations. In some of these, all four wheels were locked. In others, a single wheel was locked. And, in others, no wheels were locked, but the front of the target vehicle was struck rather than the

FIGURE 8.14 Rotational friction factors for initially stopped struck vehicle.

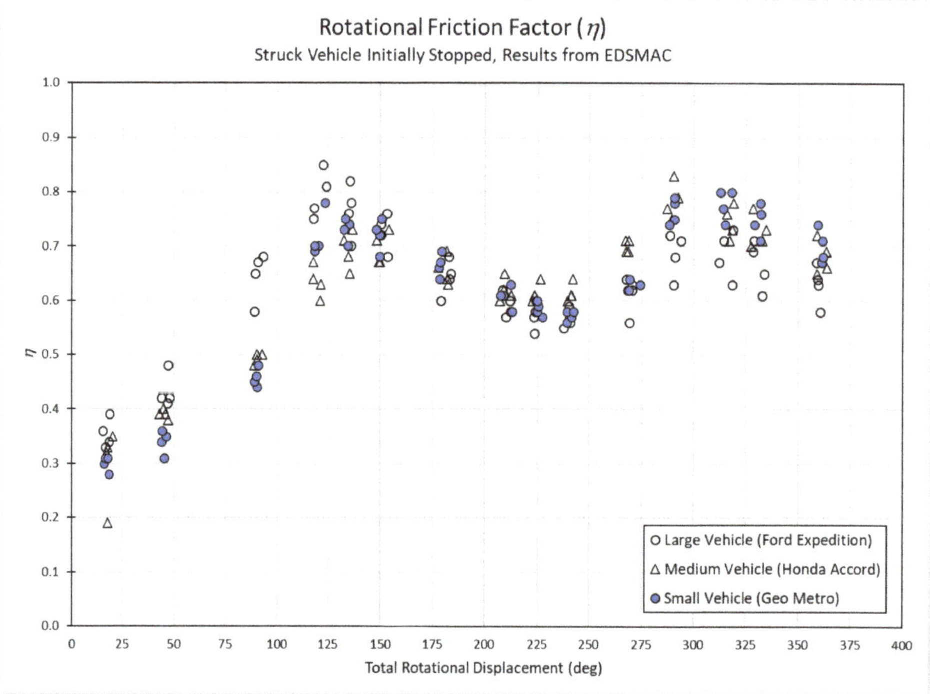

rear. Each of these variations produced different trends from the baseline simulations. The results from these simulations are plotted in Figure 8.15. For both the simulations with four wheels locked and those with one wheel locked, the rotational friction factors started out at approximately 0.4 and rose to between 0.7 and 0.8 by the time the rotational displacement of the vehicle reached 150°. The rotational friction factor remained generally in that range for the rotational displacements up to 360°.

When the vehicle was struck at the front, the trend of the rotational friction factor was significantly different from that for when the vehicle was struck at the rear. For rotational displacements between 0° and 50°, the rotational friction factors were slightly higher for the vehicles struck at the front. For rotational displacements between 75° and 125°, the rotational friction factors were significantly higher for the vehicle struck at the front than for those struck at the rear. For rotational displacements between 125° and 175°, the rotational friction factors were comparable for the vehicle struck at the front or the rear. Beyond 175°, the rotational friction factors were significantly lower for the vehicles struck in the front than for those struck in the rear. The data in these graphs could potentially be applied in instances when the struck vehicle was initially stopped. Many staged collisions between motorcycles and passenger vehicles will meet this criterion. Some real-world collisions will also involve a stationary struck vehicle, but most will not. Before extending this approach to instance where the struck vehicle was moving, it should be mentioned that Ogden and Kloberdanz [39] applied Equation (8.45) in conjunction with the data in Figure 8.14 to analyze eight of the WREX 2000 collisions and four collisions reported by Craig [40]. The struck vehicles in these tests were initially stopped. They reported errors between −2.5 and +2.6 mph (−9.9% and +6.31%).

FIGURE 8.15 Rotational friction factors for initially stopped struck medium vehicle under varying conditions.

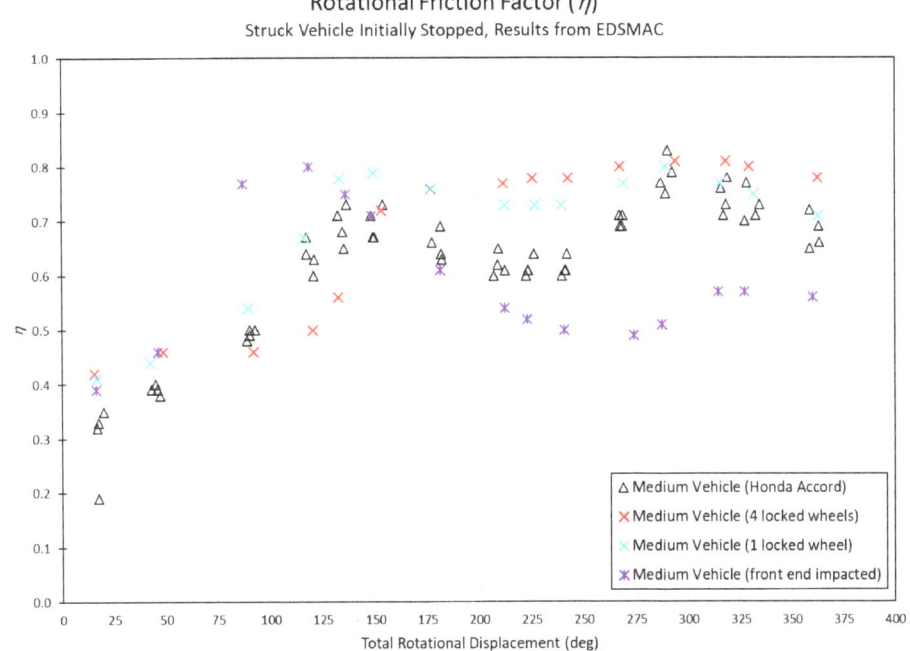

Keifer et al. reported additional simulations for instances where the target vehicle was initially moving between 5 and 30 mph. These are plotted in Figure 8.16. These data show that as the initial speed of the target vehicle increases, the η decreases for any given magnitude of angular displacement. This means that the faster the struck vehicle is initially traveling, the lower the motorcycle speed necessary to cause any particular level of angular displacement. The data in this graph also reveal limitations to the use of Equation (8.45). For example, for instances where the struck vehicle is initially traveling 30 mph, η is equal to zero for angular displacements between 0° and 135°. Equation (8.45) would yield zero for the motorcycle collision speed in these instances, which is not physically realistic.

Additional limitations should be mentioned. First, all the simulation data reported by Keifer et al. assumed a 90° impact configuration. This again is a condition likely to be met in many staged collisions, but not in many real-world collisions. How is the rotational friction and the speed calculation influenced when the collision occurs with something other than a 90° angle between the vehicles—an intersection collision, for instance, where a component of the speed of the struck vehicle would be likely to lie along the same line, though in the opposite direction, of the speed of the motorcycle? Second, when a vehicle rotates significantly following a collision, particularly if it is struck at the rear, steering angles can develop at the front tires. These steering angles develop, in part, because the frictional forces between the tires and ground cause the yaw rotation of the tires to lag the yaw rotation of the vehicle. The development of these steering angles also depends, in part, on moments generated because of the caster angle, which would cause some steering to occur even for a crash test where a significant yaw velocity did not develop. Rose [28]

FIGURE 8.16 Rotational friction factors for initially moving struck vehicle.

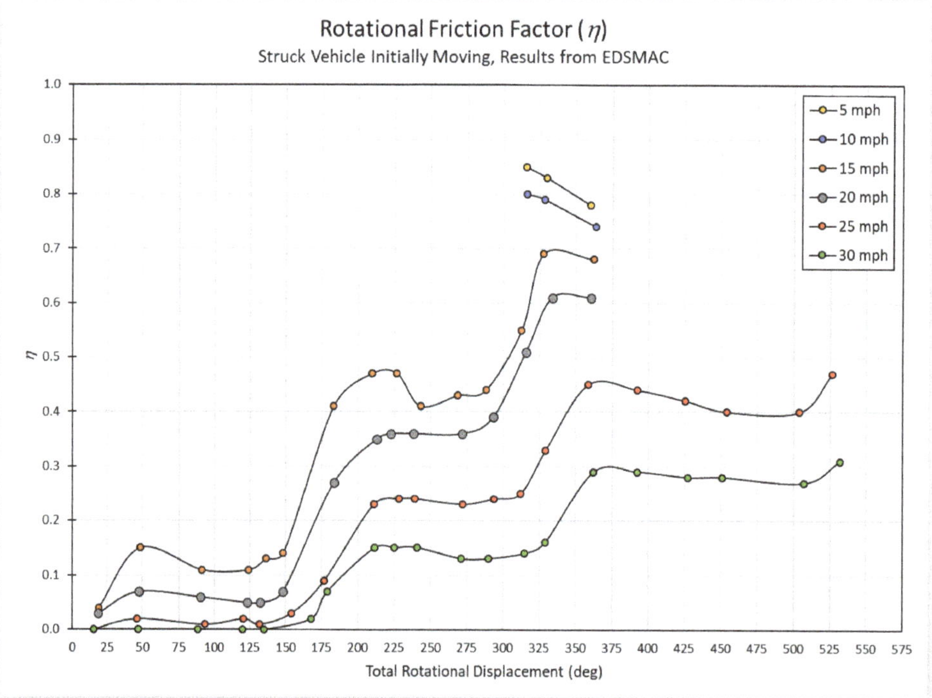

demonstrated that simulations can be improved by including this steering. Such steering did occur in some of the WREX collisions discussed below and can also occur in real-world crashes. Including this steering in the simulation can improve the match with the known evidence. It is not apparent from Keifer et al.'s descriptions of their results whether or not such steering inputs were considered, but it is likely they were not.

Keifer et al. observed that their data "does not examine all possibilities of initial conditions and therefore must be used with care and good judgement. It is necessary to evaluate the effect of the particulars of the case under investigation ..." Unfortunately, one would need to use simulation to "evaluate the effect of the particulars of the case under investigation," and this whole approach seems intended to avoid the need for simulation. There is not a simple, analytical way to incorporate many potential factors. One is consigned either to endless tabulation of scenarios for graphs and tables or to the use of simulation. Simulation provides the best opportunity to incorporate all the unique features of the postimpact movement of the struck vehicle for any particular case.

Deyerl and Cheng [25] and Rose and Carter [26] illustrated this type of analysis using EDSMAC4 simulation. In their study, Deyerl and Cheng simulated the WREX 2000 motorcycle-car collisions and assessed the degree to which the simulations accurately predicted the vehicle rest positions. They concluded that "the use of EDSMAC4 in the simulation of motorcycle-into-vehicle collisions provides a valid means for analyzing collisions with configurations similar to these crash tests ... The simulation effort described resulted in good to excellent correlation of post-impact translations and heading angle changes of the vehicles struck by the motorcycles." Other software packages commonly used by accident reconstructionists can be used for similar analysis (PC-Crash or V-Crash, for example). Several examples of analysis with PC-Crash of WREX 2016 collisions follow.

WREX 2016 Crash Tests

This section uses seven staged motorcycle-to-car collisions from the World Reconstruction Exposition in 2016 (WREX 2016) [41] to illustrate and test the following methods for modeling these collisions:

Method 1—Equation (8.39), implemented using Equation (8.40) to estimate the postimpact yaw velocity of the struck vehicle.

Method 2—Equation (8.39), implemented using the data in Figure 8.11 and Figure 8.13 to estimate the postimpact yaw velocity of the struck vehicle.

Method 3—Equations (8.40) through (8.44).

Method 4—Equation (8.45), using the data in Figure 8.14 and Figure 8.15 to estimate the value of η.

Method 5—PC-Crash Simulation.

Method 6—Wheelbase Reduction and Crush.

Each of these methods is implemented first with the yaw moment of inertia of the struck vehicle obtained from the internal calculations of PC-Crash, and second, with the yaw moment of inertia determined from the equations in reference [42]. Other methods of estimating the struck vehicle yaw moment of inertia are available in the literature, but these combinations are sufficient to illustrate the expected accuracy and the limitations of these equations.

The seven WREX 2016 tests considered here involved Harley-Davidson (H-D) motorcycles and various passenger vehicles. The passenger vehicles were stationary prior to the collisions and the motorcycle impact speeds varied between 30.3 and 46.3 mph. One of the primary limitations in these tests is that the struck vehicles were stationary prior to the collision. This will not necessarily be true of motorcycle collisions in the real world. These tests were conducted on an asphalt surface near the Orlando International Airport with a reported coefficient of friction of 0.7.

Case Study: WREX 2016, Test #3

Test #3 involved a 2013 H-D FXSB 103B Softail Breakout motorcycle (667 lb) impacting the center of the passenger's side front door of a stationary 2006 Nissan Maxima (3449 lb), at an approximate 90° angle, and at a speed of 43.0 mph. Figure 8.17 shows the damage to these vehicles, their rest positions, tire marks, and chalk marks identifying the passenger's side wheel positions both pre- and postcollisions. The Nissan moved 2.75 ft laterally as a result of the collision. The front wheels of the Nissan were steered to the left at rest. Review of the video for this test confirmed that these wheels were oriented straight ahead prior to the collision and that they steered to the left during the postimpact motion of the Nissan. The motorcycle experienced a wheelbase reduction of 11.3 in. and the car a maximum crush of 18.7 in.

FIGURE 8.17 Impact damage and rest positions for WREX 2016 test #3.

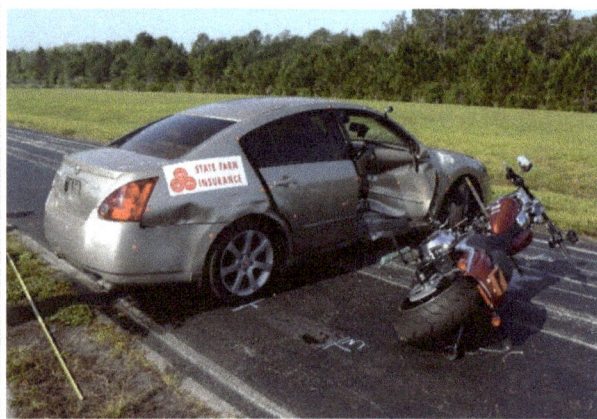

Methods 1 to 4

The resultant collision force in this instance was directed through the CG of the Nissan, and the Nissan experienced negligible rotation because of the collision. Given this, equations relating struck vehicle rotation to the motorcycle impact speed will not be useful. In fact, Equations (8.40) through (8.44) and Equation (8.45) would calculate that the motorcycle impact speed was zero for this case, which is not correct. This demonstrates that these sets of equations do not represent general cases, but special cases that neglect the translational motion of the struck vehicle.

This is also a limitation of Equation (8.39), when it is applied in isolation from Equation (8.38). However, Equations (8.38) and (8.39) could be applied in combination. To illustrate this, first consider the following equation, which is derived with the principle of work and energy. This equation yields the postimpact speed of the Nissan. The variable d_{post} is the postimpact travel distance of the Nissan. The variable η_{trans} specifies the percentage of the roadway friction that is recruited in decelerating the vehicle. In this instance, the vehicle is simply pushed sideways, and it can be assumed that all the available friction will be recruited. Thus, η_{trans} can be set equal to 1. The roadway friction coefficient has been set to 0.7.

$$v_{post,\,nissan} = \sqrt{2\eta_{trans}\mu g d_{post}} \tag{8.46}$$

$$v_{post,nissan} = \sqrt{2 \times 0.7 \times 32.2 \times 2.75} = 11.1\ fps = 7.59\ mph$$

Thus, the calculated postimpact speed of the Nissan is 7.59 mph. Because the Nissan was stopped prior to the collision, this is also the calculated ΔV for the Nissan. Equation (8.38) can now be applied with $h = 0$. With the moment arm set to zero, Equation (8.38) reduces to the following:

$$v_{2n,i} = v_{1n,f}\frac{1}{1+e}\frac{W_1+W_2}{W_2}$$

$$v_{2n,i} = 7.59\frac{1}{1+0.09}\frac{3,449+667}{667} = 43.0\ mph$$

Thus, when combined with reasonable estimates of the postimpact speed and coefficient of restitution (0.09, in this instance), Equation (23) yields the actual impact speed of the motorcycle.

Note that the postimpact distance used above was measured from the pretest position of the Nissan to the posttest position of the Nissan. Thus, this calculation inherently assumes no movement of the Nissan during the collision itself, which is consistent with the assumptions typically employed with conservation of momentum impact models. In this approach, any movement of the struck vehicle that occurs during the collision is treated as if it occurred following the collision. This approach works well in the present context, but issues with it will arise in the next chapter when we take up the topic of applying crash data from event data recorders (EDRs).

Method 5

This test was next simulated with PC-Crash. The roadway coefficient of friction was set at 0.7 and the TM-Easy tire model was used with default values for both vehicles. To have a direct comparison with the prior calculations, the simulation was initially run in 2-D. A leftward steering input of 15°, occurring over 250 ms, was used in modeling the postimpact motion of the Nissan. Brake factors of 100% were used for the front drive wheels of the Nissan and 1% for the rear wheels. Brake factors of 100% were applied to the front wheel of the motorcycle, though the simulation was not sensitive to this input. An integration time step of 1 ms was used.

FIGURE 8.18 Optimized two-dimensional PC-Crash simulation for WREX 2016 test #3.

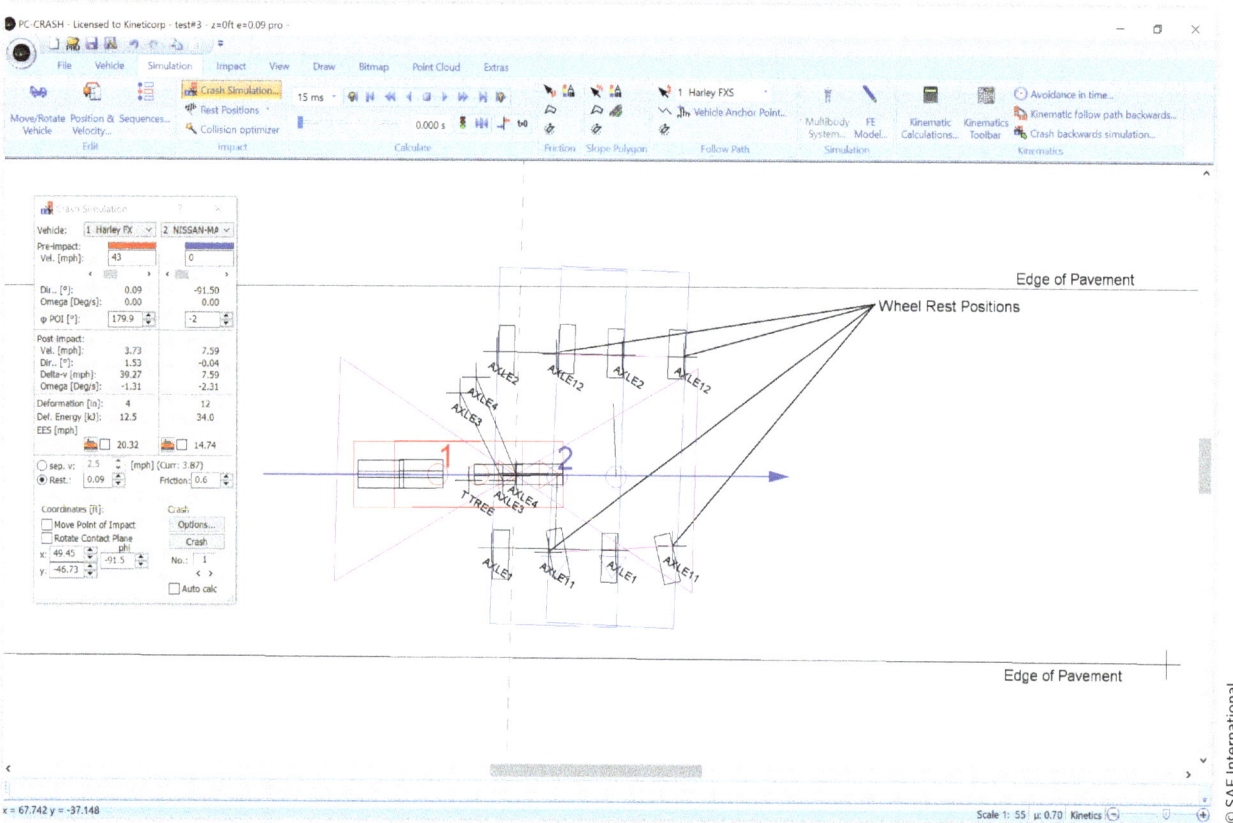

The actual impact speed was entered and the simulation was optimized using the coefficient of restitution. A high-quality match with the rest position of the Nissan was obtained with the actual motorcycle impact speed and a coefficient of restitution of 0.09, which is in agreement with the 2-D conservation of momentum solution above. Figure 8.18 is a screen capture from this PC-Crash simulation, which shows the impact and rest positions of the vehicles and the crash simulation dialog box showing the impact parameter inputs. This dialog box also shows that PC-Crash calculated a postimpact speed for the Nissan of 7.59 mph, in agreement with the 2-D conservation of energy and momentum solution.

Simulations were also run in 3-D. The CG height of the Nissan was set at 22.7 in., which is approximately 39% of the total vehicle height [22, 23]. The motorcycle CG height was estimated at 18.7 in., which was just fewer than 28% of the motorcycle wheelbase. The coefficient of restitution needed to match the actual motorcycle impact speed and struck vehicle rest position was found to be dependent on the impact center height. With an impact center height of 1.0 ft, a high-quality match with the actual rest position of the Nissan was obtained with the actual motorcycle impact speed and a coefficient of restitution of 0.25 (Figure 8.19). With an impact center height of 1.25 ft, a coefficient of restitution of 0.17 was required. With an impact center height of 1.5 ft, a high-quality match was obtained with a

CHAPTER 8 Motorcycle Collisions with Vehicles and Roadside Barriers 211

FIGURE 8.19 Optimized PC-Crash simulation for WREX 2016 test #3 (impact center height = 1.0 ft).

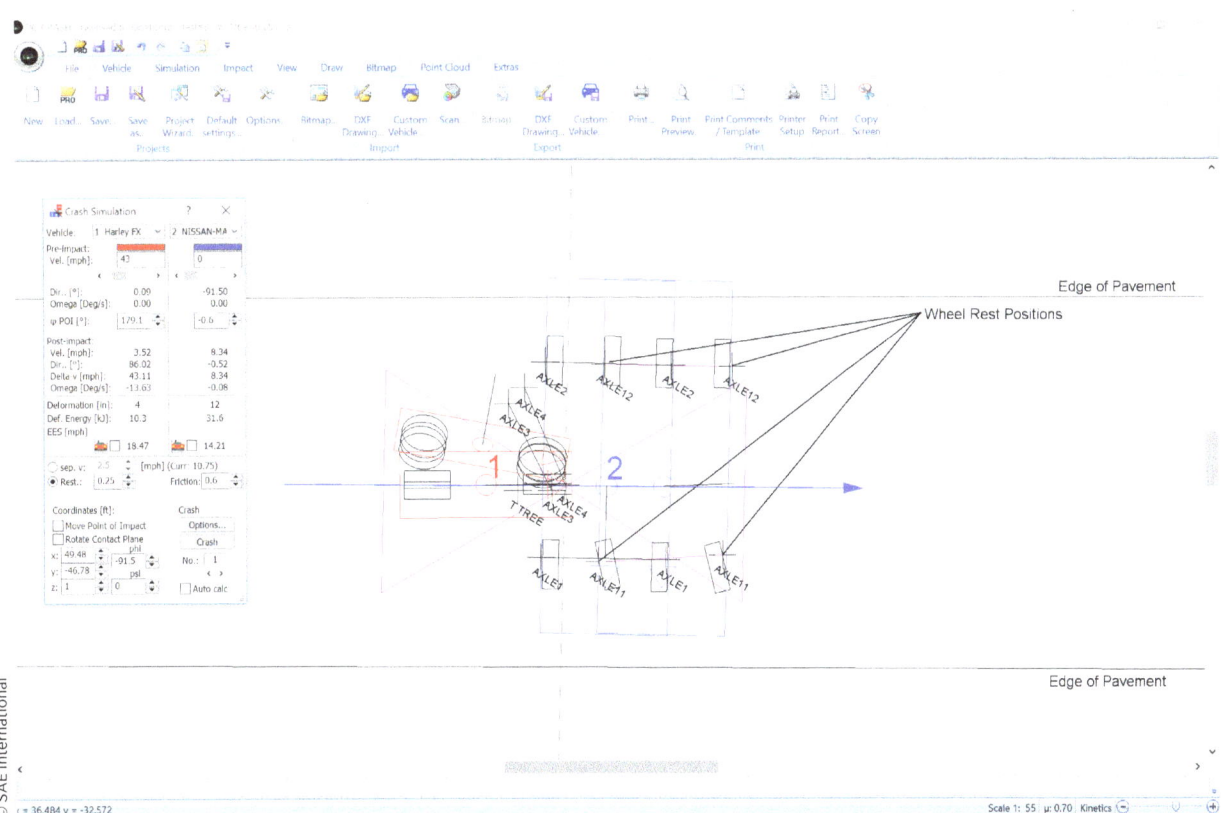

coefficient of restitution of 0.12. With an impact center height of 1.75 ft, a high-quality match was obtained with a coefficient of restitution of 0.09 (Figure 8.20), back to being consistent with the 2-D analysis. Review of the test video and photographs of the deformation to the Nissan revealed that the impact center height of 1.75 ft made physical sense, though there is no way to precisely make this determination.

Method 6

This collision is also conducive to analysis using the Nissan crush and motorcycle wheelbase shortening. This collision can be classified as a door impact, and thus, Equation (8.7) can be applied, as follows:

$$S = 1.50(L + C) + 9.27$$

$$S = 1.50(11.3 + 18.7) + 9.27 = 54.3 \; mph$$

In this instance, the crush plus wheelbase reduction method yields a speed that is 11.3 mph higher than the actual motorcycle impact speed. This falls within the 95% confidence interval for this equation, but not the 68% confidence interval.

FIGURE 8.20 Optimized PC-Crash simulation for WREX 2016 test #3 (impact center height = 1.75 ft).

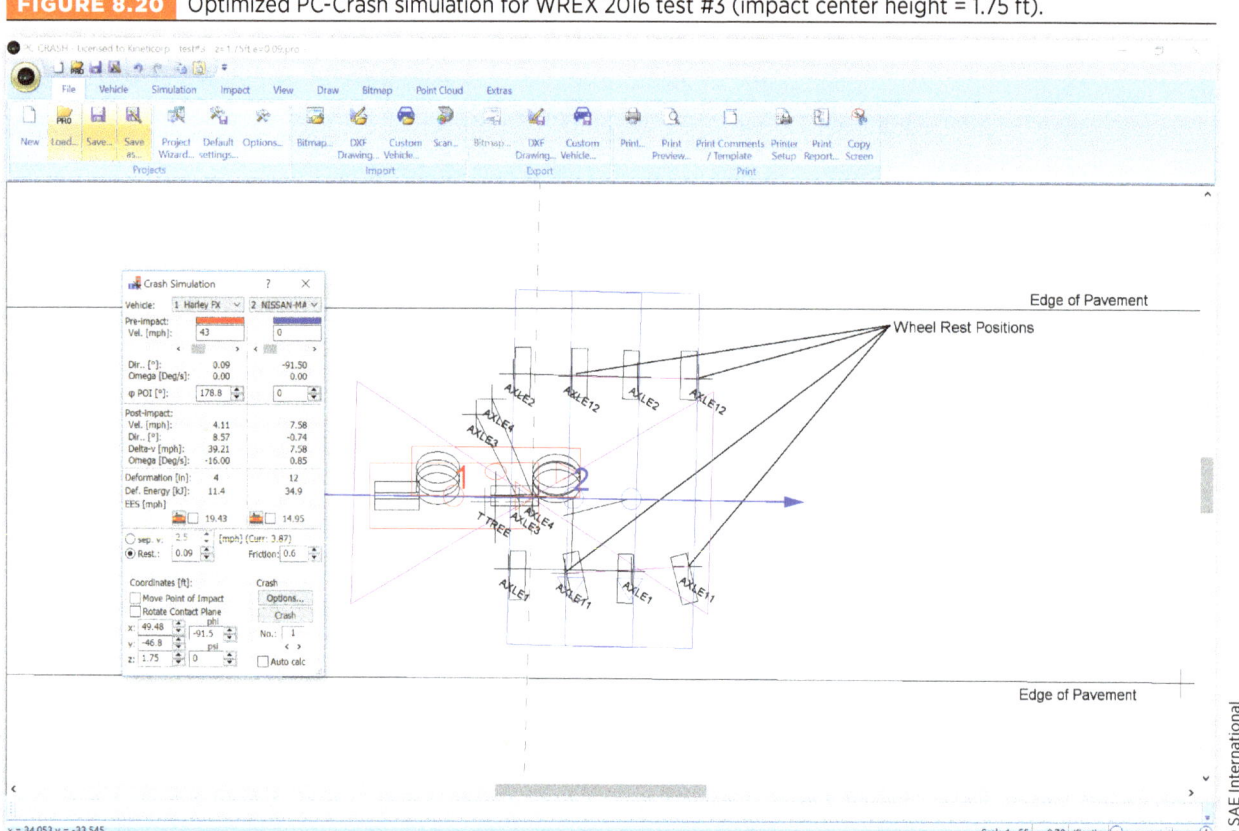

Case Study: WREX 2016, Test #5

Test #5 involved a 2013 H-D Dyna Street Bob FXDBA motorcycle impacting the driver's side rear door of stationary 2006 Nissan Maxima at a 90° angle and at a speed of 36.9 mph. Figure 8.21 shows the damage to these vehicles, their rest positions, tire marks, and chalk marks identifying the driver's side wheel positions both pre- and postcollisions. The front wheels of the Nissan were steered to the right when the vehicle came to rest following the test. Review of the video for this test confirmed that these wheels were oriented straight ahead prior to the collision and that they steered to the right during the postimpact motion of the Nissan. Review of the test video also revealed that it would be reasonable to assume the center of rotation for the struck vehicle was the center of the front axle.

The motorcycle weighed 633 lb and the Nissan weighed 3449 lb. The Nissan had a wheelbase of 111.2 in. (9.27 ft) and approximately 61% of its weight was on the front axle. Thus, the CG was 43.4 in. (3.62 ft) behind the front axle and 67.8 in. (5.65 ft) in front of the rear axle. The static load on the front axle was approximately 2104 lb, and the static load on the rear axle was approximately 1345 lb. The motorcycle impacted the Nissan 3.25 ft behind its CG and 6.86 ft behind its front axle. The collision resulted in 26.7° (0.466 rad) of counterclockwise rotation of the Nissan. The motorcycle in this test experienced a wheelbase reduction of 12.8 in. and the car a maximum crush of 8.6 in.

FIGURE 8.21 Impact damage and rest positions for WREX 2016 test #5.

Method 1

Using the internal calculations of PC-Crash, the yaw moment of inertia of the Nissan was calculated at 2032 lb-ft-s². Adjusting this to be about the center of the front axle with Equation (8.42) resulted in a value of approximately 3436 lb-ft-s². Using Equation (8.41) to calculate the torque resulted in a value of 8728 lb-ft. Equation (8.40) can be applied to calculate the postimpact yaw velocity, as follows:

$$\omega = \sqrt{\frac{2 \cdot 8,728 \cdot 0.466}{3,436}} = 1.54 \frac{rad}{s} = 88.2 \frac{deg}{s}$$

This estimate of the postimpact yaw velocity from Equation (8.40) can be used in conjunction with Equation (8.39), as follows:

$$\overline{m} = \frac{1}{32.2} \cdot \frac{3,449 \cdot 633}{3,449 + 633} = 16.6$$

$$v_{2n,i} = 1.54 \frac{1}{1+0.12} \frac{2032 + 16.6 \cdot 3.25^2}{16.6 \cdot 3.25} = 56.26 \frac{ft}{s} = 38.4 \ mph$$

Thus, with the yaw moment of inertia of the struck vehicle calculated from PC-Crash and the angular velocity of the struck vehicle estimated with Equation (8.40), Equation (8.39) yields a speed that is 1.5 mph higher than actual. The coefficient of restitution in this calculation was set at 0.12 based on analysis of this test with PC-Crash described below.

Calculations of the impact speed of a motorcycle based on the postimpact rotation of the struck vehicle will be sensitive to the calculated yaw moment of inertia of the struck vehicle. There are several methods in the literature for making this calculation. As an example, Neptune [42] proposed the following equation for estimating the yaw moment of inertia of a passenger vehicle:

$$I_{yaw} = \frac{m}{K}(L^2 + w^2) \tag{8.47}$$

In Equation (8.47), L is the length of the vehicle and w is the width of the vehicle. K is an empirically determined coefficient, defined for each vehicle type in Table 8.7.

TABLE 8.7 Empirical coefficients for Equation (8.47)

Vehicle type	K
All	13.1
Car	13.8
Pickup	13.4
Utility	12.2
Van	12.3

© SAE International

For the Nissan Maxima in this test, Equation (8.47) yields a yaw moment of inertia of 2295 lb-ft-s², a value that is approximately 13% higher than that calculated by PC-Crash. When this higher moment of inertia was incorporated into PC-Crash, the coefficient of restitution increased to 0.15. Further, when this yaw moment of inertia was adjusted to be about the center of the front axle, it became 3699 lb-ft-s². With the moment of inertia calculated with Equation (8.47) and a coefficient of restitution of 0.15, Equation (8.40) yields an angular velocity of 85 deg/s (1.48 rad/s). Equation (8.39) yields a speed of 40.2 mph (3.3 mph higher than actual).

Method 2
The data in Figure 8.13 yield an estimate of the postimpact yaw velocity of approximately 60 deg/s (1.05 rad/s). This estimate can then be used in Equation (8.39), as follows:

$$v_{2n,i} = 1.05 \frac{1}{1+0.12} \frac{2032 + 16.6 \cdot 3.25^2}{16.6 \cdot 3.25} = 38.25 \frac{ft}{s} = 26.1 \; mph$$

Thus, with the yaw moment of inertia of the struck vehicle calculated from PC-Crash and the angular velocity of the struck vehicle estimated with Figure 8.13, Equation (8.39) yields a speed that is 10.8 mph lower than actual. With the moment of inertia calculated with Equation (8.47) and the angular velocity estimated with Figure 8.13, Equation (8.39) yields a speed of 28.5 mph (8.4 mph lower than actual), as follows:

$$v_{2n,i} = 1.05 \frac{1}{1+0.15} \frac{2{,}295 + 16.6 \cdot 3.25^2}{16.6 \cdot 3.25} = 41.8 \frac{ft}{s} = 28.5 \; mph$$

Method 3
The motorcycle velocity change can be calculated with Equation (8.43), as follows:

$$\Delta v_{mc} = \frac{3436 \cdot 32.2 \cdot 1.54}{633 \cdot 6.86} = 39.2 \frac{ft}{s} = 26.8 \; mph$$

Finally, Equation (8.44) will yield the impact speed of the motorcycle, as follows:

$$v_{mc} = \frac{6.86 \cdot 1.54 + 39.2}{1+0.12} = 44.4 \frac{ft}{s} = 30.3 \; mph$$

Thus, with the yaw moment of inertia of the struck vehicle calculated from PC-Crash, Equations (8.40) through (8.44) result in a speed that is 6.6 mph lower than actual. With the moment of inertia calculated with Equation (8.47) and a coefficient of restitution of 0.15, Equations (8.40) through (8.44) yield a speed of 30.1 mph (6.8 mph lower than actual).

Method 4

Using the medium vehicle points in the graph of Figure 8.14, the value for η would be approximately 0.36. With this value, Equation (8.45) yields an impact speed of 37.9 mph, as follows. This speed is 1.0 mph higher than actual (2.7%).

$$V_{mc} = \left(\frac{32.2}{633 \times 3.25} + \frac{32.2}{3449 \times 3.25} + \frac{3.25}{2032}\right) \times \sqrt{2032 \cdot 0.466 \cdot 0.36 \cdot 0.7 \cdot 3449 \cdot \left(\frac{111.2}{12}\right)} = 55.6 \; fps$$
$$= 37.9 \; mph$$

With the moment of inertia calculated with Equation (8.47), Equation (8.45) yields a speed of 39.9 mph (3.0 mph higher than actual). The coefficient of restitution is not an input into this equation; thus, no adjustment is needed for the increased coefficient of restitution. This is a limitation of this equation.

Method 5

In simulating this test with PC-Crash, the roadway coefficient of friction was set at 0.7 and the TM-Easy tire model was used with default values for both vehicles. The CG height of the Nissan was set at 22.7 in. and the motorcycle CG height was set at 17.8 inches. For the first simulation run, the yaw moment of inertia of the Nissan was set at 2032 lb-ft-s² which was the value obtained from the internal calculations of PC-Crash. The impact height was set at 1.75 ft. A rightward steering input of 10° occurring over 250 ms was used for the postimpact motion of the Nissan. Brake factors of 100% were used for the front drive wheels of the Nissan and 1% for the rear wheels. Brake factors of 100% were applied to the front wheel of the motorcycle, though the simulation was not sensitive to this input. An integration time step of 1 ms was used.

The actual impact speed was entered for the motorcycle and the simulation was optimized using the coefficient of restitution. Figure 8.22 shows the optimized simulation results for this test. The postimpact rotational motion of the Nissan was well matched with a coefficient of restitution of 0.12. There was a small translational discrepancy between the simulated and actual rest positions of the vehicle. An additional simulation was run for Test #5 that utilized the struck vehicle moment of inertia calculated with Equation (8.47). This increase in the moment of inertia resulted in an increase in the coefficient of restitution needed to optimize the simulation, from 0.12 to 0.15.

Method 6

This collision is also conducive to analysis using the Nissan crush and motorcycle wheelbase shortening with Equation (8.7), as follows:

$$S = 1.50(L + C) + 9.27$$
$$S = 1.50(12.8 + 8.6) + 9.27 = 41.4 \; mph$$

FIGURE 8.22 Optimized PC-Crash simulation for WREX 2016 test #5.

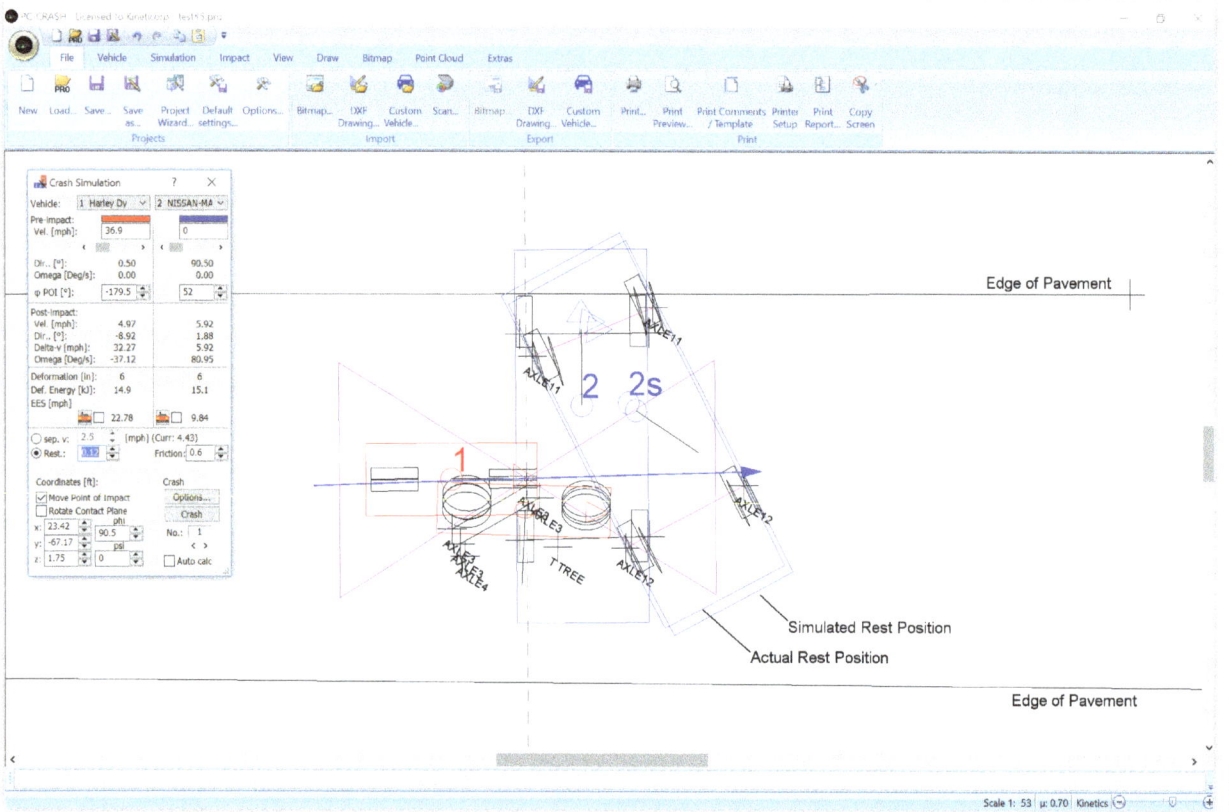

In this instance, the crush plus wheelbase reduction method yields a speed that is 4.5 mph higher than the actual motorcycle impact speed. This falls within the 68% confidence interval for this equation.

Case Study: WREX 2016, Test #8

Test #8 involved a 2012 H-D FXDF Dyna Fat Boy motorcycle impacting the driver's side rear of a stationary 2005 Dodge Durango at a 90° angle and at a speed of 46.3 mph. Figure 8.23 contains photographs showing the damage to these vehicles, their rest positions, tire marks, and chalk marks identifying the driver's side wheel positions both pre- and postcollisions. The test video revealed that it would be reasonable to assume the center of rotation for the struck vehicle during the postimpact rotation was the center of the front axle.

The motorcycle in this test weighed 678 lb and the Dodge weighed 4740 lb. The Dodge had a wheelbase of 119.2 in. (9.93 ft) and approximately 53.3% of its weight was on the front axle. Thus, the CG was approximately 55.7 in. (4.64 ft) behind the front axle. The static load on the front axle was approximately 2514 lb, and the static load on the rear axle was approximately 2226 lb. The motorcycle impacted the Dodge approximately 7 ft behind its CG and the collision resulted in approximately 44.5° (0.78 rad) of counterclockwise rotation of the Dodge. The motorcycle in this test experienced a wheelbase reduction of 3.3 in. and the Dodge a maximum crush of 9.1 in.

FIGURE 8.23 Impact damage and rest positions for WREX 2016 test #8.

Photo courtesy of Louis Peck

Method 1

Using the internal calculations of PC-Crash, the yaw moment of inertia of the Dodge was estimated at 3106 lb-ft-s². Adjusting this to be about the center of the front axle with Equation (8.42) resulted in a value of approximately 6275 lb-ft-s². Using Equation (8.41) to calculate the torque results in 15,473 lb-ft. Equation (8.40) can now be applied to calculate the postimpact yaw velocity, as follows:

$$\omega = \sqrt{\frac{2 \cdot 15{,}473 \cdot 0.78}{6{,}275}} = 1.96 \frac{rad}{s} = 112.4 \frac{deg}{s}$$

This estimate of the postimpact yaw velocity can be used in conjunction with Equation (8.39) as follows:

$$\bar{m} = \frac{1}{32.2} \cdot \frac{4{,}740 \cdot 678}{4{,}740 + 678} = 18.4$$

$$v_{2n,i} = 1.96 \frac{1}{1+0.08} \frac{3106 + 18.4 \cdot 7^2}{18.4 \cdot 7} = 56.46 \frac{ft}{s} = 38.5 \; mph$$

Thus, with the yaw moment of inertia of the struck vehicle calculated with PC-Crash and the yaw velocity calculated with Equation (8.40), Equation (8.39) yields a speed that is 7.8 mph lower than actual. The coefficient of restitution in this calculation was set at 0.08 based on analysis of this test with PC-Crash described below.

For the Dodge Durango in this test, Equation (8.47) yields a yaw moment of inertia of 3863 lb-ft-s², a value that is approximately 24% higher than that calculated by PC-Crash. When this value was incorporated into PC-Crash the coefficient of restitution required to optimize the simulation increased from 0.08 to 0.10. When the increased yaw moment of inertia was adjusted to be about the center of the front axle it became 7032 lb-ft-s². With the moment of inertia calculated with Equation (8.47), Equation (8.40) yields the following angular velocity:

$$\omega = \sqrt{\frac{2 \cdot 15{,}473 \cdot 0.78}{7{,}032}} = 1.85 \frac{rad}{s} = 106.2 \frac{deg}{s}$$

With this angular velocity and a coefficient of restitution of 0.10, Equation (8.39) yields a speed of 42.4 mph (3.9 mph lower than actual), as follows:

$$v_{2n,i} = 1.85 \frac{1}{1+0.10} \frac{3{,}863 + 18.4 \cdot 7^2}{18.4 \cdot 7} = 62.2 \frac{ft}{s} = 42.4 \; mph$$

Method 2

The data in Figure 8.13 would yield an estimate of the postimpact yaw velocity of approximately 100 deg/s (1.75 rad/s). This estimate can then be used in Equation (8.39), as follows:

$$v_{2n,i} = 1.75 \frac{1}{1+0.08} \frac{3{,}106 + 18.4 \cdot 7^2}{18.4 \cdot 7} = 50.4 \frac{ft}{s} = 34.4 \; mph$$

Thus, with the yaw moment of inertia of the struck vehicle calculated from PC-Crash and the angular velocity of the struck vehicle estimated with Figure 8.13, Equation (8.39) yields a speed that is 11.9 mph lower than actual. With the moment of inertia calculated with Equation (8.47), the angular velocity estimated with Figure 8.13, and a coefficient of restitution of 0.10, Equation (8.39) yields a speed of 40.1 mph (6.2 mph lower than actual), as follows:

$$v_{2n,i} = 1.75 \frac{1}{1+0.10} \frac{3{,}863 + 18.4 \cdot 7^2}{18.4 \cdot 7} = 58.8 \frac{ft}{s} = 40.1 \; mph$$

Method 3

The motorcycle velocity change can be calculated with Equation (8.43), with the rotation rate of the struck vehicle calculated with Equation (8.40), as follows:

$$\Delta v_{mc} = \frac{6{,}275 \cdot 32.2 \cdot 1.96}{678 \cdot 11.64} = 50.2 \frac{ft}{s} = 34.2 \; mph$$

Equation (8.44) will yield the impact speed of the motorcycle, as follows.

$$v_{mc} = \frac{11.64 \cdot 1.96 + 50.2}{1+0.08} = 67.6 \frac{ft}{s} = 46.1 \; mph$$

Thus, with the yaw moment of inertia of the struck vehicle calculated from PC-Crash, Equations (8.40) through (8.44) result in a speed that is 0.2 mph lower than actual. With the moment of inertia calculated with Equation (8.47) and a coefficient of restitution of 0.10, Equations (8.40) through (8.44) yield a motorcycle impact speed of 46.3 mph, which is the actual speed.

Method 4

Using the large vehicle points in the graph of Figure 8.14, the value for η would be approximately 0.45. With this value and the PC-Crash moment of inertia, Equation (8.45) yields an impact speed of 40.8 mph (5.5 mph lower than actual), as follows:

$$V_{mc} = \left(\frac{32.2}{678 \times 7} + \frac{32.2}{4740 \times 7} + \frac{7}{3,106}\right) \times \sqrt{3,106 \cdot 0.776 \cdot 0.45 \cdot 0.7 \cdot 4740 \cdot 9.93} = 59.8 \ fps$$
$$= 40.8 \ mph$$

With the moment of inertia calculated with Equation (8.47), Equation (8.45) yields an impact speed of 43.1 mph (3.2 mph lower than actual) as follows:

$$V_{mc} = \left(\frac{32.2}{678 \times 7} + \frac{32.2}{4740 \times 7} + \frac{7}{3,863}\right) \times \sqrt{3,863 \cdot 0.776 \cdot 0.45 \cdot 0.7 \cdot 4740 \cdot 9.93}$$
$$V_{mc} = 63.2 \ fps = 43.1 \ mph$$

Method 5

In analyzing this test with PC-Crash, the roadway coefficient of friction was set at 0.7 and the TM-Easy tire model was used with default values for both vehicles. The CG height of the Dodge was set at 29 in., which was 39% of the overall height of the vehicle. The motorcycle CG height was set at 17.7 in. The impact height was set at 2.0 ft. Brake factors of 1% were used for the front wheels of the Dodge and 100% for the rear drive wheels. An integration time step of 1 ms was used. The simulation was optimized using the coefficient of restitution. Figure 8.24 shows the optimized simulation for this test. The postimpact rotational motion of the Dodge was well matched with a coefficient of restitution of 0.08. There was a small translational discrepancy between the simulated and actual rest positions of the vehicle. An additional simulation was run for Test #8 that utilized the struck vehicle moment of inertia calculated with Equation (8.47). This increase in the moment of inertia resulted in an increase in the coefficient of restitution needed to optimize the simulation, from 0.08 to 0.10.

Method 6

This collision is conducive to analysis using the Dodge crush and motorcycle wheelbase shortening. The collision could be classified as a fender impact and analyzed with Equation (8.8), as follows:

$$S = 1.26(L + C) + 22.95$$
$$S = 1.26(3.3 + 9.1) + 22.95 = 38.6 \ mph$$

This yields a speed that is 7.7 mph lower than the actual motorcycle impact speed. This falls within the 95% confidence interval for this equation, but not the 68% confidence

FIGURE 8.24 Optimized PC-Crash simulation for WREX2016 test #8.

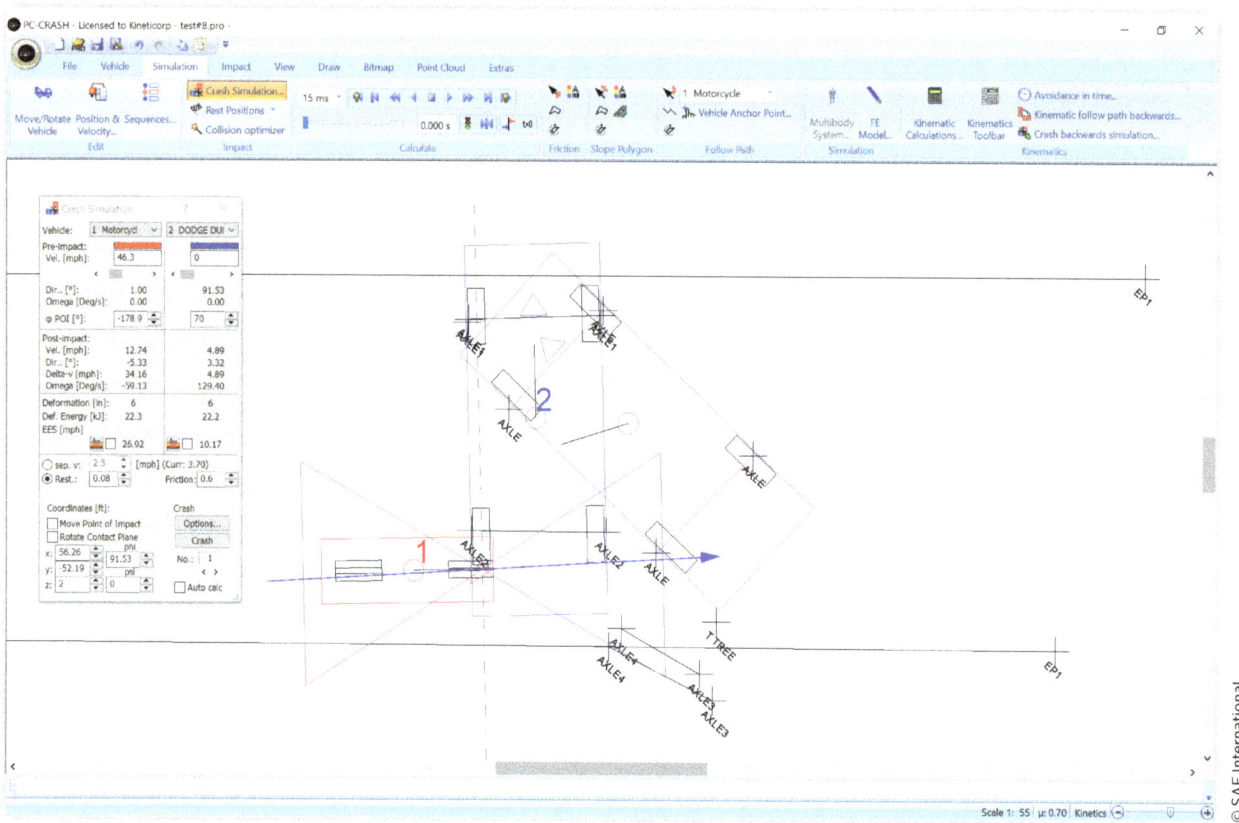

interval. On the other hand, this collision could be classified as an axle impact because the rear wheel was contacted, in which case Equation (8.5) could be used, as follows:

$$S = 2.16(L + C) + 17.33$$

$$S = 2.16(3.3 + 9.1) + 17.33 = 44.1 \; mph$$

This is 2.1 mph lower than the actual impact speed, well within the 68% confidence interval of Equation (8.5).

Case Study: WREX 2016, Test #11

Test #11 involved a 2012 H-D Dyna Low Rider FXDL motorcycle impacting the passenger's side front door of a stationary 2005 Dodge Durango at a 90° angle and at a speed of 42.9 mph. The rear tires of the Dodge Durango were initially positioned on moist dirt and grass and the front tires were positioned on asphalt. Figure 8.25 shows the damage to these vehicles, their rest positions, tire marks, and chalk marks identifying the passenger's side wheel positions both pre- and postcollisions. The motorcycle in this test weighed 649 lb and the Dodge weighed 4740 lb. The Dodge had a wheelbase of 119.2 in. (9.93 ft) and approximately

FIGURE 8.25 Impact damage and rest positions for WREX 2016 test #11.

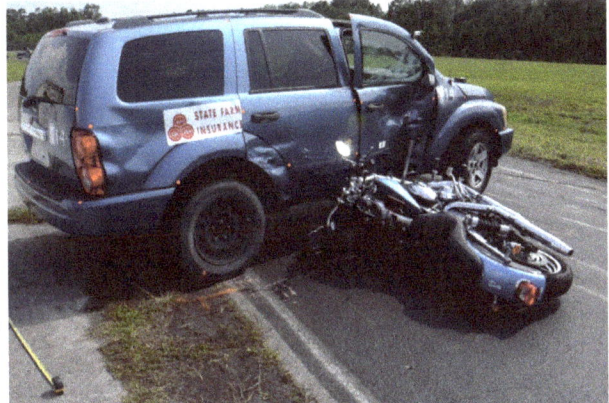

Photo courtesy of Louis Peck

53% of its weight was on the front axle. The CG was approximately 55.7 in. (4.64 ft) behind the front axle and 63.5 in. (5.29 ft) in front of the rear axle. The static load on the front axle was approximately 2514 lb, and the static load on the rear axle was approximately 2226 lb. The motorcycle impacted the Dodge approximately 1.8 ft in front of its CG and approximately 7 ft in front of the rear axle. The collision resulted in approximately 15° (0.26 rad) of counterclockwise rotation of the Dodge. The front tires of the Dodge were steered slightly to the left at rest. The motorcycle in this test experienced a wheelbase reduction of 10.7 in. and the Dodge a maximum crush of 10.3 in.

Method 1

Review of the test video revealed that it would be reasonable to assume the center of rotation for the struck vehicle was the center of the rear axle. Using the internal calculations of PC-Crash, the yaw moment of inertia of the Dodge was estimated at 3106 lb-ft-s². Adjusting this to be about the center of the rear axle with Equation (8.42) resulted in a value of approximately 7225 lb-ft-s². Using Equation (8.41) to calculate the torque resulted in a value of 17,475 lb-ft. Equation (8.40) can now be applied to calculate the postimpact yaw velocity, as follows:

$$\omega = \sqrt{\frac{2 \cdot 17{,}475 \cdot 0.26}{7{,}225}} = 1.12 \frac{rad}{s} = 64.3 \frac{deg}{s}$$

This estimate of the postimpact yaw velocity can be used in conjunction with Equation (8.39), as follows:

$$\bar{m} = \frac{1}{32.2} \cdot \frac{4{,}740 \cdot 649}{4{,}740 + 649} = 17.7$$

$$v_{2n,i} = 1.12 \frac{1}{1+0.18} \frac{3106 + 17.7 \cdot 1.8^2}{17.7 \cdot 1.8} = 94.2 \frac{ft}{s} = 64.3 \ mph$$

©2022, SAE International

Thus, with the yaw moment of inertia of the struck vehicle calculated with PC-Crash and the yaw velocity calculated with Equation (8.40), Equation (8.39) yields a speed that is 21.4 mph higher than actual. The coefficient of restitution in this calculation was set at 0.18 based on analysis of this test with PC-Crash described below.

For the Dodge Durango in this test, Equation (8.47) yields a yaw moment of inertia of 3863 lb-ft-s^2, a value that is approximately 24% higher than that calculated by PC-Crash. When this value was incorporated into PC-Crash, the coefficient of restitution needed to optimize the simulation increased to 0.26. When the increased moment of inertia was adjusted to be about the center of the rear axle a value of 7982 lb-ft-s^2 resulted. With this moment of inertia, Equation (8.40) yields an angular velocity of 61.1 deg/s (1.07 rad/s), as follows:

$$\omega = \sqrt{\frac{2 \cdot 17{,}475 \cdot 0.26}{7{,}982}} = 1.07 \frac{rad}{s} = 61.1 \frac{deg}{s}$$

With this angular velocity and the coefficient of restitution of 0.26, Equation (8.39) yields a speed of 80.6 mph (37.7 mph higher than actual), as follows:

$$v_{2n,i} = 1.07 \frac{1}{1+0.18} \frac{3863 + 17.7 \cdot 1.8^2}{17.7 \cdot 1.8} = 118.3 \frac{ft}{s} = 80.6 \; mph$$

Method 2

The data in Figure 8.13 was for collisions in which the motorcycle impacted the struck vehicle behind the CG, and so, this data should be applied with caution to this collision where motorcycle struck the vehicle in front of the CG. That said, the data in this figure would yield an estimate of the postimpact yaw velocity of approximately 32 deg/s (0.56 rad/s). This estimate can then be used in Equation (8.39) as follows:

$$v_{2n,i} = 0.56 \frac{1}{1+0.18} \frac{3106 + 17.7 \cdot 1.8^2}{17.7 \cdot 1.8} = 47.1 \frac{ft}{s} = 32.1 \; mph$$

Thus, with the yaw moment of inertia of the struck vehicle calculated from PC-Crash and the angular velocity of the struck vehicle estimated with Figure 8.13, Equation (8.39) yields a speed that is 10.8 mph lower than actual. With the moment of inertia calculated with Equation (8.47) and the angular velocity estimated with Figure 8.13, Equation (8.39) yields a speed of 37.3 mph (5.6 mph lower than actual), as follows:

$$v_{2n,i} = 0.56 \frac{1}{1+0.26} \frac{3{,}863 + 17.7 \cdot 1.8^2}{17.7 \cdot 1.8} = 54.7 \frac{ft}{s} = 37.3 \; mph$$

Method 3

The motorcycle velocity change can be calculated with Equation (8.43), with the rotation rate of the struck vehicle calculated with Equation (8.40), as follows:

$$\Delta v_{mc} = \frac{7{,}228 \cdot 32.2 \cdot 1.05}{649 \cdot 7} = 53.8 \frac{ft}{s} = 36.7 \; mph$$

Equation (8.44) will yield the impact speed of the motorcycle, as follows.

$$v_{mc} = \frac{7 \cdot 1.05 + 53.8}{1 + 0.18} = 51.8 \frac{ft}{s} = 35.3 \; mph$$

Thus, with the yaw moment of inertia of the struck vehicle calculated from PC-Crash, Equations (8.40) through (8.44) result in a speed that is 7.6 mph lower than actual. With the moment of inertia calculated with Equation (8.47), Equations (8.40) through (8.44) yield a motorcycle impact speed of 36.8 mph (6.1 mph lower than actual), as follows:

$$\Delta v_{mc} = \frac{7,982 \cdot 32.2 \cdot 1.07}{649 \cdot 7} = 60.5 \frac{ft}{s} = 41.3 \; mph$$

$$v_{mc} = \frac{7 \cdot 1.07 + 60.5}{1 + 0.26} = 54.0 \frac{ft}{s} = 36.8 \; mph$$

Method 4

Using the large vehicle points in the graph of Figure 8.14, the value for η would be approximately 0.36. With these values, Equation (8.45) yields an impact speed of 62.5 mph (19.6 mph higher than actual), as follows:

$$V_{mc} = \left(\frac{32.2}{649 \times 1.8} + \frac{32.2}{4740 \times 1.8} + \frac{1.8}{3106} \right) \times \sqrt{3106 \times 0.26 \times 0.36 \times 0.6 \times 4740 \times 9.93} = 91.7 \frac{ft}{s}$$
$$= 62.5 \; mph$$

This overestimation is significant and outside of the range of errors reported by Ogden for application of this equation to the WREX 2000 collisions. If Equation (8.47) is used for calculating the moment of inertia of the Dodge, Equation (8.45) yields a speed of 69.3 mph (26.4 mph higher than actual), as follows:

$$V_{mc} = \left(\frac{32.2}{649 \times 1.8} + \frac{32.2}{4740 \times 1.8} + \frac{1.8}{3863} \right) \times \sqrt{3863 \times 0.26 \times 0.36 \times 0.6 \times 4740 \times 9.93} = 101.6 \frac{ft}{s}$$
$$= 69.3 \; mph$$

Method 5

In the simulation for this test, a coefficient of friction of 0.5 was used for the dirt and grass surface. The CG height of the Dodge was set at 29 in., which was 39% of the overall height of the vehicle. The motorcycle CG height was set at 17.8 in. The impact height was set at 2.55 ft. A leftward steering input of 5T, occurring over 250 ms, was used for the postimpact motion of the Dodge. Brake factors of 1% were used for the front wheels of the Dodge and 100% for the rear drive wheels. Brake factors of 100% were applied to the front wheel of the motorcycle, though the simulation was not sensitive to this input. An integration time step of 1 ms was used. The simulation was optimized using the coefficient of restitution (Figure 8.26). The postimpact motion of the Dodge was well matched with a coefficient of restitution of 0.18.

224 Motorcycle Accident Reconstruction

FIGURE 8.26 Optimized PC-Crash simulation for WREX 2016 test #11.

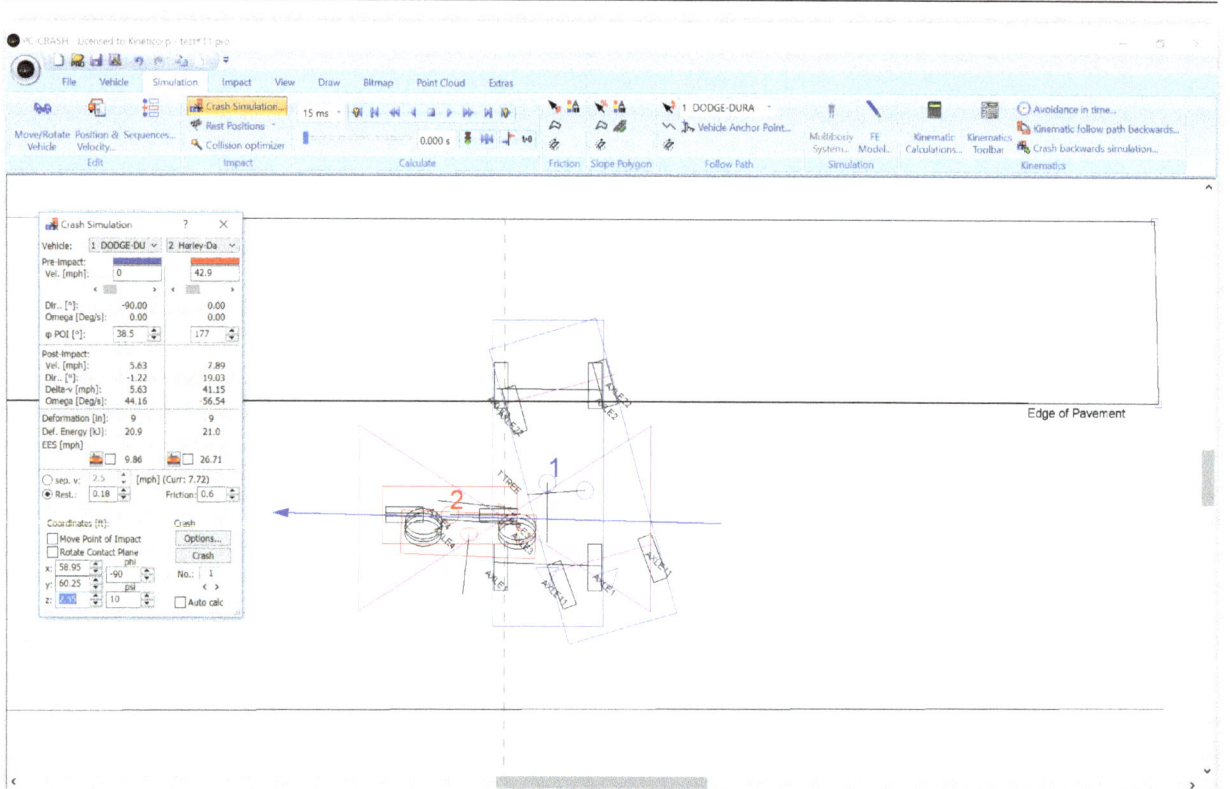

An additional simulation was run for Test #11 that utilized the struck vehicle moment of inertia calculated with Equation (8.47). This increase in the moment of inertia resulted in an increase in the coefficient of restitution needed to optimize the simulation, from 0.18 to 0.26.

Method 6

This collision is conducive to analysis using the Dodge crush and motorcycle wheelbase shortening. In this instance, the collision would be classified as a door impact and analyzed with Equation (8.7) as follows:

$$S = 1.50(L + C) + 9.27$$

$$S = 1.50(10.7 + 10.3) + 9.27 = 40.8 \; mph$$

This yields a speed that is 2.1 mph lower than the actual motorcycle impact speed.

Case Study: WREX 2016, Test #22

Test #22 involved a 1997 H-D Sportster 883 motorcycle impacting the passenger's side rear wheel of a stationary 2006 Hyundai Sonata at a 90° angle and at a speed of 30.3 mph. Figure 8.27 shows the damage to these vehicles and their rest positions. The motorcycle in

FIGURE 8.27 Impact damage and rest positions for WREX 2016 test #22.

this test weighed 498 lb and the Hyundai weighed 3547 lb. The Hyundai had a wheelbase of 107.4 in. (8.95 ft) and approximately 63% of its weight was on the front axle. The static load on the front axle was approximately 2220 lb, and the static load on the rear axle was approximately 1329 lb. The CG was approximately 40.2 in. (3.35 ft) behind the front axle, and the motorcycle impacted the Hyundai approximately 6.5 ft behind its CG and 9.85 ft behind the front axle. The collision resulted in approximately 20° of clockwise rotation of the Hyundai. The front tires of the Dodge were steered slightly to the left at rest. The motorcycle in this test experienced a wheelbase reduction of 4.3 in. and the car a maximum crush of 2.1 in.

Method 1

Review of the test video revealed that it would be reasonable to assume the center of rotation for the struck vehicle was the center of the front axle. Using the internal calculations of PC-Crash, the yaw moment of inertia of the Hyundai was estimated at 1982 lb-ft-s². Adjusting this to be about the center of the rear axle with Equation (8.42) resulted in a value of approximately 3218 lb-ft-s². Using Equation (8.41) to calculate the torque resulted in a value of 8326 lb-ft. Equation (8.40) can now be applied to calculate the postimpact yaw velocity as follows:

$$\omega = \sqrt{\frac{2 \cdot 8{,}326 \cdot 0.349}{3{,}218}} = 1.34 \frac{rad}{s} = 77 \frac{deg}{s}$$

This estimate of the postimpact yaw velocity can be used in conjunction with Equation (8.39) as follows:

$$\bar{m} = \frac{1}{32.2} \cdot \frac{3549 \cdot 498}{3549 + 498} = 13.56$$

$$v_{2n,i} = 1.34 \frac{1}{1+0.20} \frac{1982 + 13.56 \cdot 6.5^2}{13.56 \cdot 6.5} = 32.4 \frac{ft}{s} = 22.1 \; mph$$

Thus, with the yaw moment of inertia of the struck vehicle calculated with PC-Crash and the yaw velocity calculated with Equation (8.40), Equation (8.39) yields a speed that is

8.2 mph lower than actual. The coefficient of restitution in this calculation was set at 0.20 based on analysis of this test with PC-Crash described below.

For the Hyundai Sonata in this test, Equation (8.47) yields a yaw moment of inertia of 2288 lb-ft-s², a value that is approximately 15% higher than that calculated by PC-Crash. When this moment of inertia was incorporated into PC-Crash, the coefficient of restitution increased to 0.25. Adjusting the increased moment of inertia to be about the center of the front axle with Equation (8.42) resulted in a value of approximately 3524 lb-ft-s². With this moment of inertia, Equation (8.40) yields an angular velocity of 73.6 deg/s (1.28 rad/s). With this angular velocity and a coefficient of restitution of 0.25, Equation (8.39) yields a speed of 22.7 mph (7.6 mph lower than actual) as follows:

$$v_{2n,i} = 1.28 \frac{1}{1+0.25} \frac{2288 + 13.56 \cdot 6.5^2}{13.56 \cdot 6.5} = 33.2 \frac{ft}{s} = 22.7\ mph$$

Method 2

The data in Figure 8.13 would yield an estimate of the postimpact yaw velocity of approximately 48 deg/s (0.84 rad/s). This estimate can then be used in Equation (8.39) as follows:

$$v_{2n,i} = 0.84 \frac{1}{1+0.20} \frac{1982 + 13.56 \cdot 6.5^2}{13.56 \cdot 6.5} = 20.3 \frac{ft}{s} = 13.8\ mph$$

Thus, with the yaw moment of inertia of the struck vehicle calculated from PC-Crash and the angular velocity of the struck vehicle estimated with Figure 8.13, Equation (8.39) yields a speed that is 16.5 mph lower than actual. With the moment of inertia calculated with Equation (8.47) and the angular velocity estimated with Figure 8.13, Equation (8.39) yields a speed of 14.9 mph (15.4 mph lower than actual) as follows:

$$v_{2n,i} = 0.84 \frac{1}{1+0.25} \frac{2288 + 13.56 \cdot 6.5^2}{13.56 \cdot 6.5} = 21.9 \frac{ft}{s} = 14.9\ mph$$

Method 3

The motorcycle velocity change can be calculated with Equation (8.43), with the rotation rate of the struck vehicle calculated with Equation (8.40) as follows:

$$\Delta v_{mc} = \frac{3,218 \cdot 32.2 \cdot 1.34}{498 \cdot 9.85} = 28.3 \frac{ft}{s} = 19.3\ mph$$

Equation (8.44) will yield the impact speed of the motorcycle as follows:

$$v_{mc} = \frac{9.85 \cdot 1.34 + 28.3}{1 + 0.20} = 34.6 \frac{ft}{s} = 23.6\ mph$$

Thus, with the yaw moment of inertia of the struck vehicle calculated from PC-Crash, Equations (8.40) through (8.44) result in a speed that is 6.7 mph lower than actual. With the moment of inertia calculated with Equation (8.47), Equations (8.40) through (8.44) yield a motorcycle impact speed of 23.0 mph (7.3 mph lower than actual) as follows:

$$\Delta v_{mc} = \frac{3{,}524 \cdot 32.2 \cdot 1.28}{498 \cdot 9.85} = 29.6 \frac{ft}{s} = 20.2 \ mph$$

$$v_{mc} = \frac{9.85 \cdot 1.28 + 29.6}{1 + 0.25} = 33.8 \frac{ft}{s} = 23.0 \ mph$$

Method 4

Using the medium vehicle points in the graph of Figure 8.14, the value for η would be approximately 0.325. With these values, Equation (8.45) yields an impact speed of 22.3 mph (8.0 mph lower than actual) as follows:

$$V_{mc} = \left(\frac{32.2}{498 \times 6.5} + \frac{32.2}{3549 \times 6.5} + \frac{6.5}{1982} \right) \times \sqrt{1982 \times 0.349 \times 0.325 \times 0.7 \times 3547 \times 8.95} = 32.7 \frac{ft}{s}$$
$$= 22.3 \ mph$$

With the moment of inertia from Equation (8.47), Equation (8.45) yields an impact speed of 23.2 mph (7.1 mph lower than actual).

Method 5

In the PC-Crash simulation for this test, the impact center height was set at 1.5 ft. Brake factors of 100% were used for the front drive wheels of the Hyundai and 1% for the rear wheels. An integration time step of 1 ms was used. Brake factors of 100% were applied to the front wheel of the motorcycle. The simulation was optimized using the coefficient of restitution (Figure 8.28). The postimpact motion of the Dodge was well matched with a coefficient of restitution of 0.2. An additional simulation was run for Test #22 that utilized the struck vehicle moment of inertia calculated with Equation (8.47). This increase in the moment of inertia resulted in an increase in the coefficient of restitution needed to optimize the simulation, from 0.20 to 0.25.

Method 6

This collision is conducive to analysis using the car crush and motorcycle wheelbase shortening. In this instance, the collision would be classified as a fender impact and analyzed with Equation (8.8) as follows:

$$S = 1.26(L + C) + 22.95$$
$$S = 1.26(4.3 + 2.1) + 22.95 = 31.0 \ mph$$

This yields a speed that is 0.7 mph higher than the actual motorcycle impact speed.

Case Study: WREX 2016, Test #23

Test #23 involved a 2002 H-D Sportster 883 motorcycle impacting the passenger's side rear door of a stationary 2006 Hyundai Sonata, just in front of the rear wheel, at a 90° angle and at a speed of 42.7 mph. Figure 8.29 shows the vehicle rest positions and postcollision damage.

228 Motorcycle Accident Reconstruction

FIGURE 8.28 PC-Crash simulation for WREX 2016 test #22.

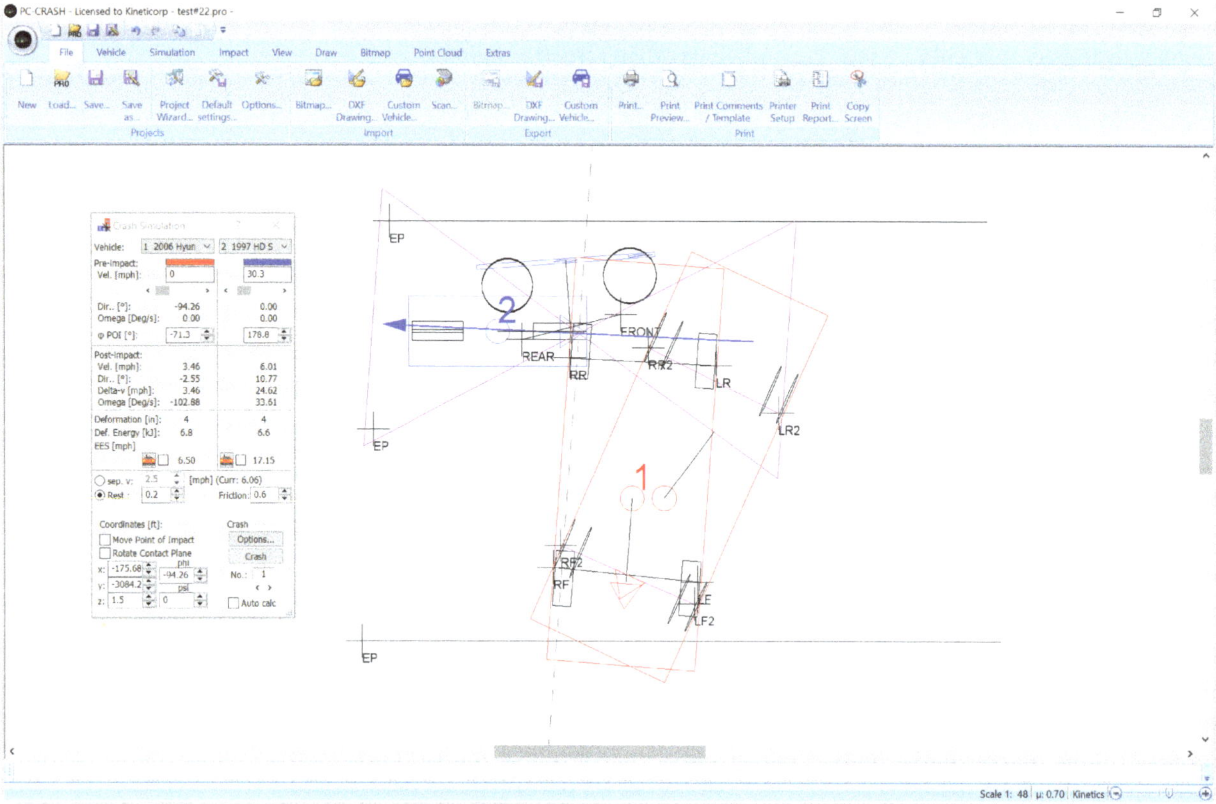

FIGURE 8.29 Impact damage and rest positions for WREX 2016 test #23.

The motorcycle in this test weighed 500 lb and the Hyundai weighed 3547 lb. The Hyundai had a wheelbase of 107.4 in. (8.95 ft) and approximately 63% of its weight was on the front axle. The static load on the front axle was approximately 2220 lb, and the static load on the rear axle was approximately 1329 lb. The CG was approximately 40.2 in. (3.35 ft) behind the front axle. The motorcycle impacted the Hyundai approximately 3.8 ft behind its CG and approximately 7.15 ft behind the front axle. The collision resulted in approximately 26° (0.454 rad) of clockwise rotation of the Hyundai. The front tires of the Dodge were steered slightly to the left at rest. The motorcycle in this test experienced a wheelbase reduction of 8.8 in. and the car a maximum crush of 7.1 in.

Method 1

Review of the test video revealed that it would be reasonable to assume the center of rotation for the struck vehicle was the center of the front axle. Using the internal calculations of PC-Crash, the yaw moment of inertia of the Hyundai was estimated at 1982 lb-ft-s². Adjusting this to be about the center of the front axle with Equation (8.42) resulted in a value of approximately 3218 lb-ft-s². Using Equation (8.41) to calculate the torque resulted in a value of 8326 lb-ft. Equation (8.40) can now be applied to calculate the postimpact yaw velocity as follows:

$$\omega = \sqrt{\frac{2 \cdot 8{,}326 \cdot 0.454}{3{,}218}} = 1.53 \frac{rad}{s} = 87.8 \frac{deg}{s}$$

This estimate of the postimpact yaw velocity can be used in conjunction with Equation (8.39) as follows:

$$\bar{m} = \frac{1}{32.2} \cdot \frac{3549 \cdot 500}{3549 + 500} = 13.56$$

$$v_{2n,i} = 1.53 \frac{1}{1+0.14} \frac{1982 + 13.56 \cdot 3.8^2}{13.56 \cdot 3.8} = 56.7 \frac{ft}{s} = 38.7 \; mph$$

Thus, with the yaw moment of inertia of the struck vehicle calculated with PC-Crash and the yaw velocity calculated with Equation (8.40), Equation (8.39) yields a speed that is 4.0 mph lower than actual. The coefficient of restitution in this calculation was set at 0.14 based on analysis of this test with PC-Crash described below.

For the Hyundai Sonata in this test, Equation (8.47) yields a yaw moment of inertia of 2288 lb-ft-s², a value that is approximately 15% higher than that calculated by PC-Crash. When this value was incorporated into PC-Crash, the coefficient of restitution increased to 0.18. Adjusting the increased moment of inertia to be about the center of the front axle with Equation (8.42) resulted in a value of 3524 lb-ft-s². With these moments of inertia, Equation (8.40) yields an angular velocity of 83.9 deg/s (1.46 rad/s). With this angular velocity, Equation (8.39) yields a speed of 40.6 mph (2.1 mph lower than actual) as follows:

$$v_{2n,i} = 1.46 \frac{1}{1+0.18} \frac{2288 + 13.56 \cdot 3.8^2}{13.56 \cdot 3.8} = 59.6 \frac{ft}{s} = 40.6 \; mph$$

Method 2

The data in Figure 8.13 would yield an estimate of the postimpact yaw velocity of approximately 60 deg/s (1.05 rad/s). This estimate can then be used in Equation (8.39) as follows:

©2022, SAE International

$$v_{2n,i} = 1.05 \frac{1}{1+0.14} \frac{1982 + 13.56 \cdot 3.8^2}{13.56 \cdot 3.8} = 38.9 \frac{ft}{s} = 26.5 \; mph$$

Thus, with the yaw moment of inertia of the struck vehicle calculated from PC-Crash and the angular velocity of the struck vehicle estimated with Figure 8.13, Equation (8.39) yields a speed that is 16.2 mph lower than actual. With the moment of inertia calculated with Equation (8.47) and the angular velocity estimated with Figure 8.13, Equation (8.39) yields a speed of 29.2 mph (13.5 mph lower than actual) as follows:

$$v_{2n,i} = 1.05 \frac{1}{1+0.18} \frac{2288 + 13.56 \cdot 3.8^2}{13.56 \cdot 3.8} = 42.9 \frac{ft}{s} = 29.2 \; mph$$

Method 3

The motorcycle velocity change can be calculated with Equation (8.43), with the rotation rate of the struck vehicle calculated with Equation (8.40) as follows:

$$\Delta v_{mc} = \frac{3{,}218 \cdot 32.2 \cdot 1.53}{500 \cdot 7.15} = 44.3 \frac{ft}{s} = 30.2 \; mph$$

Equation (8.44) will yield the impact speed of the motorcycle as follows:

$$v_{mc} = \frac{7.15 \cdot 1.53 + 44.3}{1 + 0.14} = 48.5 \frac{ft}{s} = 33.1 \; mph$$

Thus, with the yaw moment of inertia of the struck vehicle calculated from PC-Crash, Equations (8.40) through (8.44) result in a speed that is 9.6 mph lower than actual. With the moment of inertia calculated with Equation (8.47), Equations (8.40) through (8.44) yield a motorcycle impact speed of 32.8 mph (9.9 mph lower than actual) as follows:

$$\Delta v_{mc} = \frac{3{,}524 \cdot 32.2 \cdot 1.46}{500 \cdot 7.15} = 46.3 \frac{ft}{s} = 31.6 \; mph$$

$$v_{mc} = \frac{7.15 \cdot 1.46 + 46.3}{1 + 0.18} = 48.1 \frac{ft}{s} = 32.8 \; mph$$

Method 4

Using the medium vehicle points in the graph of Figure 8.14, the value for η would be approximately 0.35. With these values, Equation (8.45) yields an impact speed of 38.5 mph (4.2 mph lower than actual).

$$V_{mc} = \left(\frac{32.2}{500 \times 3.8} + \frac{32.2}{3547 \times 3.8} + \frac{3.8}{1982} \right) \times \sqrt{1982 \times 0.454 \times 0.35 \times 0.7 \times 3547 \times 8.95} = 56.4 \frac{ft}{s}$$
$$= 38.5 \; mph$$

With the moment of inertia calculated with Equation (8.47), Equation (8.45) yields an impact speed of 40.7 mph (2.0 mph lower than actual).

$$V_{mc} = \left(\frac{32.2}{500 \times 3.8} + \frac{32.2}{3549 \times 3.8} + \frac{3.8}{2288}\right) \times \sqrt{2288 \times 0.454 \times 0.35 \times 0.7 \times 3547 \times 8.95} = 59.7 \frac{ft}{s}$$

$$= 40.7 \; mph$$

Method 5

In the simulation for this test, the impact center height was set at 1.5 ft. Brake factors of 100% were used for the front wheels of the Hyundai and 1% for the rear wheels. An integration time step of 1 ms was used. The simulation was optimized using the coefficient of restitution. Figure 8.30 shows the optimized simulation for this test. The postimpact translation and rotation of the Dodge was well matched with a coefficient of restitution of 0.14. An additional simulation was run for Test #23 that utilized the struck vehicle moment of inertia calculated with Equation (8.47). This increase in the moment of inertia resulted in an increase in the coefficient of restitution needed to optimize the simulation, from 0.14 to 0.18.

Method 6

This collision is conducive to analysis using the car crush and motorcycle wheelbase shortening. The collision could be classified as a door impact and analyzed with Equation (8.7) as follows:

$$S = 1.50(L + C) + 9.27$$

$$S = 1.50(8.8 + 7.1) + 9.27 = 33.1 \; mph$$

FIGURE 8.30 Optimized PC-Crash simulation for WREX 2016 test #23.

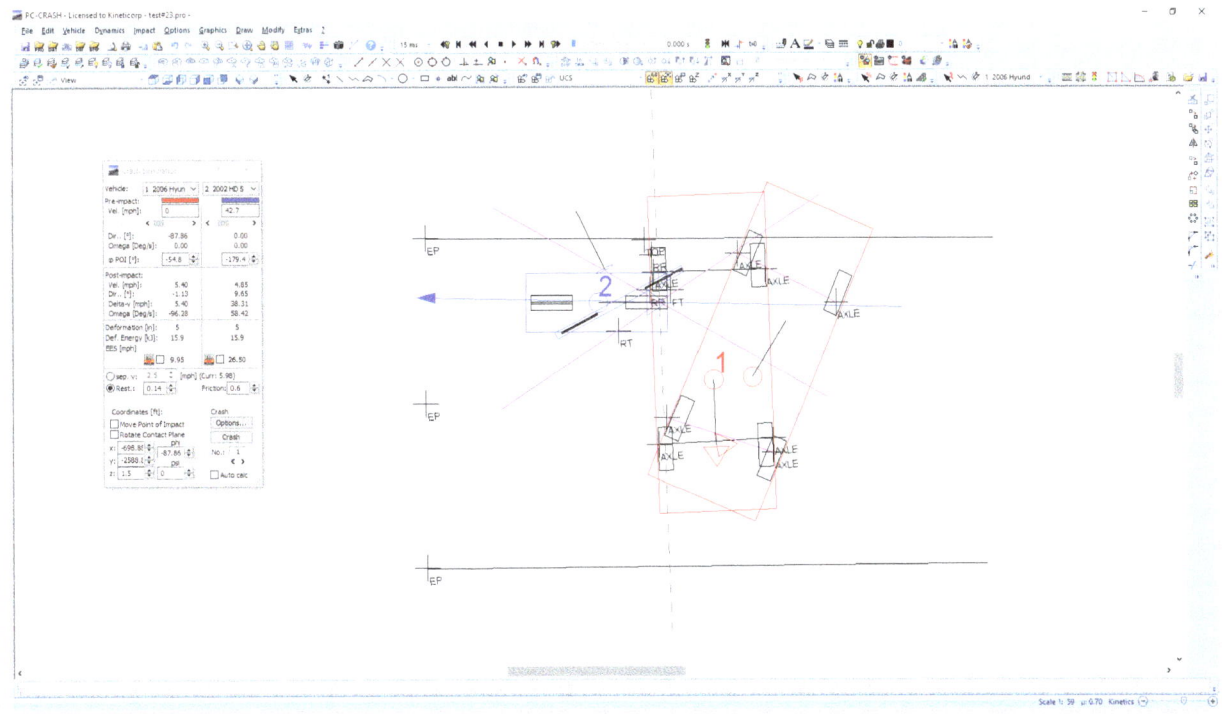

This yields a speed that is 9.6 mph lower than the actual motorcycle impact speed. On the other hand, this collision could be classified as a pillar impact, in which case Equation (8.6) could be used as follows:

$$S = 1.36(L + C) + 19.50$$

$$S = 1.36(8.8 + 7.1) + 19.50 = 41.1 \; mph$$

This result is 1.6 mph lower than the actual impact speed.

Case Study: WREX 2016, Test #24

Test #24 involved a 2003 H-D Sportster 883 motorcycle impacting the driver's side rear door of a stationary 2006 Hyundai Sonata at a 90° angle and at a speed of 35.5 mph. Figure 8.31 shows the damage to the vehicles along with their rest positions. The motorcycle in this test weighed 483 lb and the Hyundai weighed 3547 lb. The Hyundai had a wheelbase of 107.4 in. (8.95 ft) and approximately 63% of its weight was on the front axle. The static load on the front axle was approximately 2220 lb, and the static load on the rear axle was approximately 1329 lb. The CG was approximately 40.2 in. (3.35 ft) behind the front axle. The motorcycle impacted the Hyundai approximately 3.5 ft behind its CG and 6.85 ft behind the front axle. The collision resulted in approximately 15° (0.262 rad) of counterclockwise rotation of the Hyundai. The motorcycle in this test experienced a wheelbase reduction of 8.9 in. and the car a maximum crush of 4.2 in.

Method 1

Review of the test video revealed that it would be reasonable to assume the center of rotation for the struck vehicle was the center of the front axle. Using the internal calculations of PC-Crash, the yaw moment of inertia of the Hyundai was estimated at 1982 lb-ft-s^2. Adjusting this to be about the center of the rear axle with Equation (8.42) resulted in a value of approximately 3218 lb-ft-s^2. Using Equation (8.41) to calculate the torque resulted in a value of 8326 lb-ft. Equation (8.40) can now be applied to calculate the postimpact yaw velocity as follows:

FIGURE 8.31 Impact damage and rest positions for WREX 2016 test #24.

$$\omega = \sqrt{\frac{2 \cdot 8,326 \cdot 0.262}{3,218}} = 1.16 \frac{rad}{s} = 66.7 \frac{deg}{s}$$

This estimate of the postimpact yaw velocity can be used in conjunction with Equation (8.39) as follows:

$$\bar{m} = \frac{1}{32.2} \cdot \frac{3549 \cdot 483}{3549 + 483} = 13.2$$

$$v_{2n,i} = 1.16 \frac{1}{1+0.07} \frac{1982 + 13.2 \cdot 3.5^2}{13.2 \cdot 3.5} = 50.3 \frac{ft}{s} = 34.3 \; mph$$

Thus, with the yaw moment of inertia of the struck vehicle calculated with PC-Crash and the yaw velocity calculated with Equation (8.40), Equation (8.39) yields a speed that is 1.2 mph lower than actual. The coefficient of restitution in this calculation was set at 0.07 based on analysis of this test with PC-Crash described below.

For the Hyundai Sonata, Equation (8.47) yields a yaw moment of inertia of 2288 lb-ft-s², a value that is approximately 15% higher than that calculated by PC-Crash. When this moment of inertia was incorporated into PC-Crash, the coefficient of restitution increased to 0.12. Adjusting the increased moment of inertia to be about the center of the rear axle with Equation (8.42) resulted in a value of approximately 3524 lb-ft-s². With this moment of inertia, Equation (8.40) yields an angular velocity of 63.8 deg/s (1.11 rad/s). With this angular velocity, Equation (8.39) yields a speed of 35.8 mph (0.3 mph higher than actual) as follows:

$$v_{2n,i} = 1.11 \frac{1}{1+0.12} \frac{2288 + 13.2 \cdot 3.5^2}{13.2 \cdot 3.5} = 52.5 \frac{ft}{s} = 35.8 \; mph$$

Method 2

The data in Figure 8.13 would yield an estimate of the postimpact yaw velocity of approximately 35 deg/s (0.61 rad/s). This estimate can then be used in Equation (8.39) as follows:

$$v_{2n,i} = 0.61 \frac{1}{1+0.07} \frac{1982 + 13.2 \cdot 3.5^2}{13.2 \cdot 3.5} = 26.5 \frac{ft}{s} = 18.0 \; mph$$

Thus, with the yaw moment of inertia of the struck vehicle calculated from PC-Crash and the angular velocity of the struck vehicle estimated with Figure 8.13, Equation (8.39) yields a speed that is 17.5 mph lower than actual. With the moment of inertia calculated with Equation (8.47) and the angular velocity estimated with Figure 8.13, Equation (8.39) yields a speed of 19.7 mph (15.8 mph lower than actual) as follows:

$$v_{2n,i} = 0.61 \frac{1}{1+0.12} \frac{2288 + 13.2 \cdot 3.5^2}{13.2 \cdot 3.5} = 28.9 \frac{ft}{s} = 19.7 \; mph$$

Method 3

The motorcycle velocity change can be calculated with Equation (8.43), with the rotation rate of the struck vehicle calculated with Equation (8.40) as follows:

$$\Delta v_{mc} = \frac{3{,}128 \cdot 32.2 \cdot 1.16}{483 \cdot 6.85} = 35.3\frac{ft}{s} = 24.1 \; mph$$

Equation (8.44) will yield the impact speed of the motorcycle as follows.

$$v_{mc} = \frac{6.85 \cdot 1.16 + 35.3}{1 + 0.07} = 40.4\frac{ft}{s} = 27.6 \; mph$$

Thus, with the yaw moment of inertia of the struck vehicle calculated from PC-Crash, Equations (8.40) through (8.44) result in a speed that is 7.9 mph lower than actual. With the moment of inertia calculated with Equation (8.47), Equations (8.40) through (8.44) yield a motorcycle impact speed of 27.8 mph (7.7 mph lower than actual) as follows:

$$\Delta v_{mc} = \frac{3524 \cdot 32.2 \cdot 1.11}{483 \cdot 6.85} = 38.1\frac{ft}{s} = 26.0 \; mph$$

$$v_{mc} = \frac{6.85 \cdot 1.11 + 38.1}{1 + 0.12} = 40.8\frac{ft}{s} = 27.8 \; mph$$

Method 4

Using the medium vehicle points in the graph of Figure 8.14, the value for η would be approximately 0.35. With these values, Equation (8.45) yields an impact speed of 32.1 mph (3.4 mph lower than actual) as follows:

$$V_{mc} = \left(\frac{32.2}{483 \times 3.5} + \frac{32.2}{3549 \times 3.5} + \frac{3.5}{1982}\right) \times \sqrt{1982 \times 0.262 \times 0.35 \times 0.7 \times 3547 \times 8.95} = 47.0\frac{ft}{s}$$
$$= 32.1 \; mph$$

Using the moment of inertia obtained with Equation (8.47), Equation (8.45) yields an impact speed of 34.1 mph (1.4 mph lower than actual) as follows:

$$V_{mc} = \left(\frac{32.2}{483 \times 3.5} + \frac{32.2}{3549 \times 3.5} + \frac{3.5}{2288}\right) \times \sqrt{2288 \times 0.262 \times 0.35 \times 0.7 \times 3547 \times 8.95} = 50.0\frac{ft}{s}$$
$$= 34.1 \; mph$$

Method 5

In the simulation for this test, the impact center height was set at 1.5 ft. Brake factors of 100% were used for the front wheels of the Hyundai and 1% for the rear wheels. An integration time step of 1 ms was used. The simulation was optimized using the coefficient of restitution. Figure 8.32 shows the optimized simulation for this test. The postimpact translation and rotation of the Dodge was well matched with a coefficient of restitution of 0.07. An additional simulation was run for Test #24 that utilized the struck vehicle moment of inertia calculated with Equation (31). This increase in the moment of inertia resulted in an increase in the coefficient of restitution needed to optimize the simulation, from 0.07 to 0.12.

CHAPTER 8 Motorcycle Collisions with Vehicles and Roadside Barriers **235**

FIGURE 8.32 Optimized PC-Crash simulation for WREX 2016 test #24.

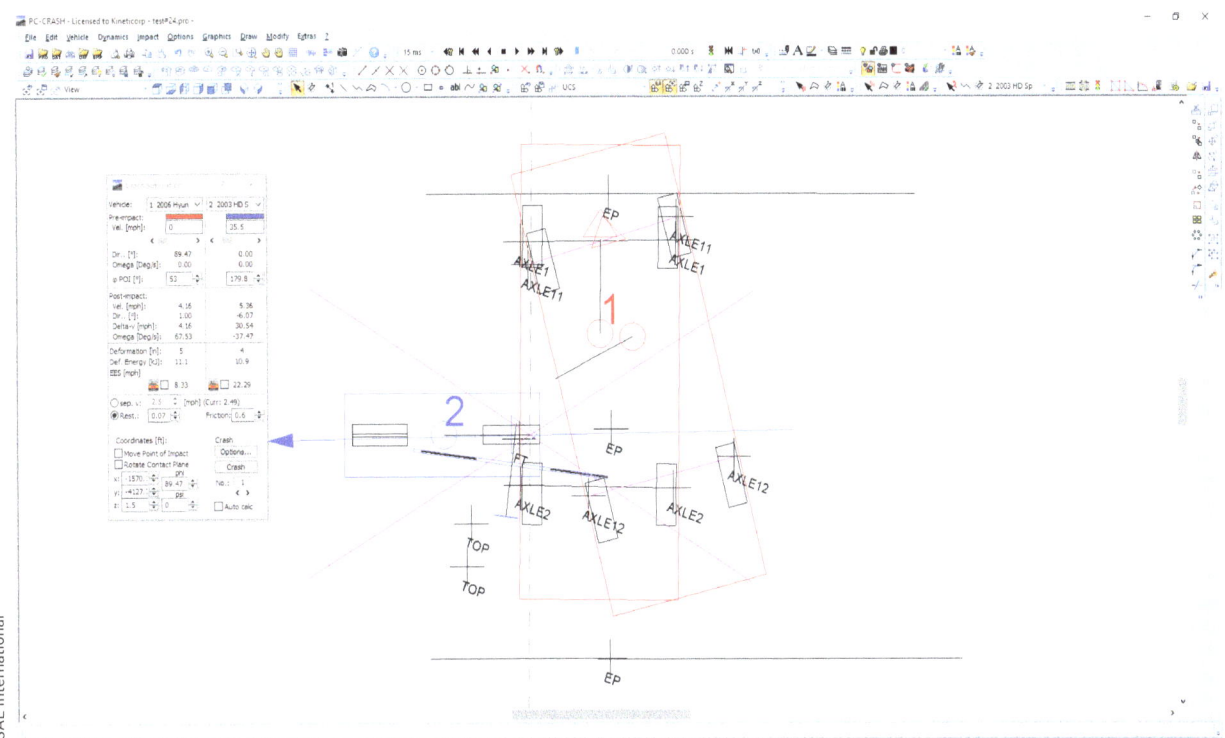

© SAE International

Method 6

This collision is conducive to analysis using the car crush and motorcycle wheelbase shortening. The collision would be classified as a door impact and analyzed with Equation (8.7) as follows:

$$S = 1.50(L + C) + 9.27$$

$$S = 1.50(8.9 + 4.2) + 9.27 = 28.9 \; mph$$

This yields a speed that is 6.6 mph lower than the actual motorcycle impact speed.

Discussion of WREX 2016 Tests

Table 8.8 and Table 8.9 summarize the results of applying Methods 1 through 4 to the WREX 2016 collisions with two methods of obtaining the struck vehicle yaw moment of inertia. Table 8.8 reports the speeds calculated with each method and the absolute difference from the actual motorcycle impact speed. Table 8.9 reports the calculated speeds and the percent difference in those speeds from the actual impact speeds. The four methods in these tables were as follows:

Method 1—This method was Equation (8.39), implemented using Equation (8.40) to estimate the postimpact yaw velocity of the struck vehicle. Equation (8.39) was developed

TABLE 8.8 Results of applying the various methods to the WREX2016 collisions (absolute difference in calculated speed and actual)

Test	Actual motorcycle (MC) impact speed (mph)	Impact location relative to center of gravity (CG)	Method 1 with PC-Crash moments of inertia (MOI) (mph)	Method 1 with Neptune moments of inertia (MOI) (mph)	Method 2 with PC-Crash moments of inertia (MOI) (mph)	Method 2 with Neptune moments of inertia (MOI) (mph)	Method 3 with PC-Crash moments of inertia (MOI) (mph)	Method 3 with Neptune moments of inertia (MOI) (mph)	Method 4 with PC-Crash moments of inertia (MOI) (mph)	Method 4 with Neptune moments of inertia (MOI) (mph)
3	43.0	At center of gravity (CG)	-	-	-	-	-	-	-	-
5	36.9	Behind center of gravity (CG)	38.4 (+1.5)	40.2 (+3.3)	26.1 (−10.8)	28.5 (−8.4)	30.3 (−6.6)	30.1 (−6.8)	37.9 (+1.0)	39.9 (+3.0)
8	46.3	Behind center of gravity (CG)	38.5 (−7.8)	42.4 (−3.9)	34.4 (−11.9)	40.1 (−6.2)	46.1 (−0.2)	46.3 (0.0)	40.8 (−5.5)	43.1 (−3.2)
11	42.9	In front of center of gravity (CG)	64.3 (+21.4)	80.6 (+37.7)	32.1 (−10.8)	37.3 (−5.6)	35.3 (−7.6)	36.8 (−6.1)	62.5 (+19.6)	69.3 (+26.4)
22	30.3	Behind center of gravity (CG)	22.1 (−8.2)	22.7 (−7.6)	13.8 (−16.5)	14.9 (−15.4)	23.6 (−6.7)	23.0 (−7.3)	22.3 (−8.0)	23.2 (−7.1)
23	42.7	Behind center of gravity (CG)	38.7 (−4.0)	42.0 (−0.7)	26.5 (−16.2)	29.2 (−13.5)	33.1 (−9.6)	32.8 (−9.9)	38.5 (−4.2)	40.7 (−2.0)
24	35.5	Behind center of gravity (CG)	34.3 (−1.2)	35.8 (+0.3)	18.0 (−17.5)	19.7 (−15.8)	27.6 (−7.9)	27.8 (−7.7)	32.1 (−3.4)	34.1 (−1.4)
Avg		All	+0.3 ± 11.0	+4.8 ± 16.5	−14.0 ± 3.1	−10.8 ± 4.6	−6.4 ± 3.2	−6.3 ± 3.3	−0.1 ± 10.1	2.6 ± 12.1
Avg		Behind center of gravity (CG)	−4.0 ± 4.2	−1.7 ± 4.2	−14.6 ± 3.0	−11.9 ± 4.3	−6.2 ± 3.6	−6.3 ± 3.7	−4.0 ± 3.5	−2.1 ± 3.6

© SAE International

with PIM, but its development invoked several simplifying assumptions to arrive at a single equation relating the postimpact rotations speed of the struck vehicle to the impact speed of the motorcycle. One of the simplifying assumptions inherent in this equation is that the struck vehicle is initially stopped. This assumption is accurate for the WREX collisions, but this equation would not be applicable to many real-world collisions.

In this first method, Equation (8.40) was implemented to estimate the postimpact yaw velocity of the struck vehicle. This equation was developed assuming that the translational and angular kinetic energies could be partitioned from each other and equated separately to translational and angular work terms. This is not an accurate assumption, and this equation generally does not yield accurate estimates of the angular velocity.

The most significant error produced by this method occurred for the test where the collision force was applied in front of the CG (Test #11). If Test #11 was included, then Method 1 utilizing the PC-Crash moment of inertia yielded an average difference in speed from actual of +0.3 mph with a standard deviation of 11.0 mph. If this test was excluded, then Method 1 utilizing the PC-Crash moment of inertia yielded an average difference in speed from actual of −4.0 mph with a standard deviation of 4.2 mph. The maximum error (Test #11) was an overestimate of 21.4 mph (49.9%). If Test #11 was included, then Method 1 utilizing the moment of inertia from Equation (8.47) yielded an average difference in speed from actual of +4.8 mph with a standard deviation of 16.5 mph. If this test was excluded, then Method 1 utilizing the

TABLE 8.9 Results of applying the various methods to the WREX2016 collisions (percent difference in calculated speed and actual)

Test	Actual motorcycle (MC) impact speed (mph)	Impact location relative to center of gravity (CG)	Method 1 with PC-Crash moments of inertia (MOI) (mph)	Method 1 with Neptune moments of inertia (MOI) (mph)	Method 2 with PC-Crash moments of inertia (MOI) (mph)	Method 2 with Neptune moments of inertia (MOI) (mph)	Method 3 with PC-Crash moments of inertia (MOI) (mph)	Method 3 with Neptune moments of inertia (MOI) (mph)	Method 4 with PC-Crash moments of inertia (MOI) (mph)	Method 4 with Neptune (MOI) (mph)
3	43.0	At center of gravity (CG)	-	-	-	-	-	-	-	-
5	36.9	Behind center of gravity (CG)	38.4 (−4.1%)	40.2 (+8.9%)	26.1 (−29.3%)	28.5 (−22.8%)	30.3 (−17.9%)	30.1 (−18.4%)	37.9 (+2.7%)	39.9 (+8.1%)
8	46.3	Behind center of gravity (CG)	38.5 (−16.8%)	42.4 (−8.4%)	34.4 (−25.7%)	40.1 (−13.4%)	46.1 (−0.4%)	46.3 (0.0%)	40.4 (−11.9%)	43.1 (−6.9%)
11	42.9	In front of center of gravity (CG)	64.3 (+49.9%)	80.6 (+87.9%)	32.1 (−25.2%)	37.3 (−13.1%)	35.3 (−17.7%)	36.8 (−14.2%)	62.5 (+45.7%)	69.3 (+61.5%)
22	30.3	Behind center of gravity (CG)	22.1 (−27.1%)	22.7 (−25.1%)	13.8 (−54.1%)	14.9 (−50.8%)	23.6 (−22.1%)	23.0 (−24.1%)	22.3 (−26.4%)	23.2 (−23.4%)
23	42.7	Behind center of gravity (CG)	38.7 (−9.4%)	42.0 (−1.6%)	26.5 (−37.9%)	29.2 (−31.6%)	33.1 (−22.5%)	32.8 (−23.2%)	38.5 (−9.8%)	40.7 (−4.7%)
24	35.5	Behind center of gravity (CG)	34.3 (−3.4%)	35.8 (+0.8%)	18.0 (−49.3%)	19.7 (−44.5%)	27.6 (22.3%)	27.8 (−21.7%)	32.1 (−9.6%)	34.1 (−3.9%)

moment of inertia from Equation (8.47) yielded an average difference in speed from actual of −1.7 mph with a standard deviation of 4.2 mph. The maximum error was 37.7 mph (87.9%).

Method 2—This method was Equation (8.39), implemented using the PC-Crash data in Figure 8.13 to estimate the postimpact yaw velocity of the struck vehicle. In applying this method, Test #11 was not such an outlier as it was for Method 1. Using the PC-Crash data to estimate the postimpact rotational velocity of the struck vehicle reduced the standard deviation of the errors significantly. However, despite less spread in the results, this method consistently underestimated the actual motorcycle impact speed by significant amounts. Method 2 utilizing the PC-Crash moment of inertia yielded an average difference in speed from actual of −14.0 mph with a standard deviation of 3.1 mph. Method 2 utilizing the moment of inertia from Equation (8.47) yielded an average difference in speed from actual of −10.8 mph with a standard deviation of 4.6 mph. Keep in mind that the WREX collisions represent the only conditions under which Methods 1 or 2 could be applied because the struck vehicles in these tests were stationary at impact and oriented at approximately 90° to the motorcycle. Initial velocity of the struck vehicle, as would typically be present in real-world crashes, would further degrade the accuracy of both methods.

Method 3—This method was Equations (8.40) through (8.44) applied as a unit. This method utilizing the PC-Crash moment of inertia yielded an average difference in speed from actual of −6.4 mph with a standard deviation of 3.2 mph. Utilizing the moment of inertia from Equation (8.47), this method yielded an average difference in speed from actual of −6.3 mph with a standard deviation of 3.3 mph. Thus, this method consistently underestimated the motorcycle impact speed. It is important to recognize, though, that this method does not incorporate the fact that, as the initial speed of the struck vehicle increases, less motorcycle speed is needed to cause any given level of struck vehicle rotation. Thus, under real-world conditions, in which the struck vehicle is often moving, the trend in errors within this method would reverse itself and the method would begin to overestimate the motorcycle impact speed.

Method 4—This method was Equation (8.45), using the data in Figure 8.14 to estimate the value of η. If Test #11 was included, then Method 4 utilizing the PC-Crash moment of inertia yielded an average difference in speed from actual of −0.1 mph with a standard deviation of 10.1 mph. If this test was excluded, then Method 1 utilizing the PC-Crash moment of inertia yielded an average difference in speed from actual of −4.0 mph with a standard deviation of 3.5 mph. If Test #11 was included, then Method 1 utilizing the moment of inertia from Equation (8.47) yielded an average difference in speed from actual of +2.6 mph with a standard deviation of 12.1 mph. If this test was excluded, then Method 1 utilizing the moment of inertia from Equation (8.47) yielded an average difference in speed from actual of −2.1 mph with a standard deviation of 3.6 mph. Thus, if Test #11 is excluded, then this method was the best overall performer of the four simple analytical methods. It is important to keep in mind though, that this method does not explicitly account for the effects of increasing struck vehicle speed and does not include the coefficient of restitution as an input. The accuracy of this method may deteriorate significantly as the initial speed of the struck vehicle increases.

In reflecting on these results, it is important to state that all four of the analytical methods considered here used the rotational displacement or rotational velocity of the struck vehicle *alone* to calculate the motorcycle impact speed. These methods do not consider the translational displacement or velocity of the vehicle, nor do they consider the sequencing and coordination of the translation and rotation experienced by the struck vehicle after the collision (something that can be considered with simulation software). Given that Method 1 only accounts for the postimpact rotational velocity, and not the translational, it is not representative of the results that could be achieved with PIM generally. If the analyst uses a reliable method for calculating the postimpact translational *and* rotational velocities and uses them within a fuller version of the PIM equations, a higher level of accuracy would be achievable. On the other hand, simulation will provide the greatest potential for an accurate solution because simulation software enables a reconstructionist to consider wheel-by-wheel steering or braking because of damage and the possibility of struck vehicle moving on multiple surfaces of varying friction. Wheel-by-wheel steering and braking can have a significant influence on the postimpact rotation experienced by a vehicle. A simulation can also be run in 3-D, and weight transfer or wheel lift can be modeled. These software packages also typically provide a visualization of the postimpact vehicle motion that can be used to assess its consistency with the physical evidence.

How Much of the Rider's Weight Should be Included?

The WREX 2016 collisions did not utilize anthropomorphic test devices (ATDs) to represent riders. Therefore, the discussion so far leaves an important issue unaddressed. When analyzing real-world collisions, the rider will need to be considered. As Niederer [43] has observed: "A major difficulty in the reconstruction of motorcycle-vehicle collisions derives from the fact that the motorcycle-rider and the motorcycle itself execute in general different trajectories after a collision with another vehicle." The previous section noted that, when using damage based methods to analyze the energy dissipation during the collision, the weight of the rider should generally not be included in the motorcycle weight. This is because the rider will typically continue moving forward as the motorcycle is decelerated

CHAPTER 8 Motorcycle Collisions with Vehicles and Roadside Barriers 237

out from under them. Any collision between the rider and the passenger vehicle will typically occur later, after the collision between the motorcycle and the car is complete. When using simulation to model the collision, this situation is different. Even though there is a time delay between the motorcycle and rider collisions with the vehicle, the rider collision to the vehicle could influence the amount of translation and rotation the passenger vehicle experiences following the collision. In general, if both the motorcycle and rider's full weight participate in the collision, the motorcycle and the rider should come to rest near each other. If none of the rider's weight participates in the collision, then the rider should be thrown a distance commensurate with the precollision speed of the motorcycle.

In PC-Crash and HVE, a multibody motorcycle-rider model could be used to automatically assess the percentage of the rider's weight that participates in the collision. Often, though, the simplest way to analyze the collision will be to collapse the collisions of the motorcycle and the rider into one event, with the option to include or not include all or part of the weight of the rider. This is an issue that will have to be dealt with on a case-by-case basis, and the reconstructionist will need to determine the degree to which the rider's body engages with the vehicle. In some instances, there will be significant engagement, and, in others, there will be little engagement. In the former case, the weight of the rider would be included, and, in the later, it would not. There might also be some warrant for including part, but not all, of the rider's weight. Describing the damage to the motorcycle fuel tank damage observed in a crash test, Smith et al. [44] observed: "As the rider moved forward on the motorcycle and engaged the fuel tank, the loading to the fuel tank created by the rider's legs and pelvis was symmetric and produced a balanced damage pattern. This damage is consistent with fuel tank damage observed by the authors in similar 'real world' accidents." Such loading of the motorcycle by the rider would indicate that at least part of the rider's weight is participating with the motorcycle in the collision with the car. Ogden and Kloberdanz [39] observed, "if there is evidence of handle bar or tank deformation due to motorcycle rider contact with the motorcycle, but no contact between the rider and the vehicle, then adding 1/3 to 2/3 of the rider's weight to the weight of the motorcycle at impact is appropriate, otherwise, the weight of the rider is omitted if the rider is ejected without contacting the vehicle or the lack of the rider engaging with the components of the motorcycle during ejection."

Frank et al. [45] applied such an approach in the EDSMAC4 and SIMON simulations they reported, noting that "the initial weight used was the as-tested weight of the [motorcycle], ignoring the weight of the [anthropomorphic test device]. However, after performing some initial simulations it was determined that the ATD was a factor in the impact. Review of the test video showed interaction between the ATD and the [car] during the impact; however, it appeared that the interaction involved only a portion of the ATD's mass. Currently, EDSMAC4 and SIMON are not capable of simulating the free body movement of the ATD traveling forward off of a motorcycle and impacting the side of a car. So, after some experimentation, it was determined that adding half of the ATD's weight to the motorcycle improved the results of the simulations significantly."

Simulation software does exist that can simulate the motion of the motorcycle and rider independently, along with their interaction with the vehicle. For example, Nieboer et al. [46] presented simulation of motorcycle collisions using MADYMO. Nieboer noted that "computer simulation of motorcycle rider behavior during a collision event is far more difficult than the simulation of passenger car occupants. This is due to the complex way in which motorcycles and their riders behave after impact and the numerous contact interactions required in a mathematical model associated with this ... The motorcycle rider interacts with the motorcycle, the motorcycle interacts with the passenger car (or another collision partner), but the rider interacts with the car directly as well. Considering these complex interactions, a step-by-step approach to specify more complete mechanical properties for each model component is often needed to improve the accuracy of the simulation." This type of simulation is often used in safety system research and development. However, because of the detailed component-level knowledge necessary for the application of this type of modeling, it is not customarily used in a collision reconstruction context.

Impacts into Moving Vehicles

In the crash tests discussed so far, the motorcycle struck a stationary barrier or vehicle. In real-world crashes, a struck vehicle could be moving, and this will influence the collision forces and the resulting motion of the vehicles. As Smith et al. [44] has observed, "For two-moving-vehicle accidents, the rotation of the motorcycle front wheel and the resulting asymmetric compression and deformation of the front forks affects the post-impact motion of the motorcycle rider and the deformation of both the motorcycle and the other vehicle." Smith reported four crash tests where both the motorcycle and the car were moving. These tests are summarized in Table 8.10. The motorcycles in the first two tests were ridden by Hybrid III, 50th-percentile male adult ATDs with a full-face helmet. The motorcycles in the third and fourth tests were ridden by Hybrid II, 5th-percentile female adult ATDs.

TABLE 8.10 Summary of the crash tests reported by Smith et al. [44]

Test	Bullet vehicle	Bullet vehicle speed (mph)	Target vehicle	Target vehicle speed (mph)	Test description
1	1981 Kawasaki KX750E	30.0	1986 Ford Mustang	14.0	The Kawasaki (449 lb + 172-lb ATD) impacted the driver's side of the Mustang (2771 lb), near the rear of the driver's door; 90° impact angle
2	1981 Kawasaki KZ750E	29.9	1986 Ford Mustang	30.1	The Kawasaki (456 lb + 172-lb ATD) impacted the passenger's side of the Mustang (2771 lb), near the rear of the passenger's door; 90° impact angle
3	2007 Kawasaki Ninja ZX-10R	63.2	1986 Jaguar XJ6	41.9	The Kawasaki (407 lb + 103-lb ATD) impacted the passenger's side of the Mustang (4360 lb), near the rear of the passenger's door; 60° impact angle
4	2006 Kawasaki Ninja ZX-10R	68.1	1986 Jaguar XJ6	46.6	The Kawasaki (409 lb + 103-lb ATD) impacted the passenger's side of the Mustang (4362 lb), near the rear of the passenger's door; 60° impact angle

© SAE International

For these tests, Smith observed that "the post-impact motion of the motorcycle and the rider … was significantly influenced by the velocity of the car … the area of direct contact is typically wider on the side of the car than the comparable direct contact area for a stationary car crash test. The direct contact is not symmetric around the first point of contact and there is typically no direct contact damage from the motorcycle on the car forward of the initial point of contact … After contact, the front tire and wheel assembly of the motorcycle turns to align with the travel direction of the moving car … As a result of this rotation, the wheel and tire do not become trapped between the forward-most components of the motorcycle body and the car. Therefore, symmetric damage to the wheel is typically not evident … As a result of the impact, the motorcycle develops rotational motion primarily in yaw … Fork tube damage was also related to the speed of the moving car …" These observations imply that the extent of the damage on the struck car will be one indicator of the direction and magnitude of that car's motion at the time of the collision. They also imply that, as the speed of the struck vehicle increases, wheelbase reduction methods for determining the motorcycle impact speed become less and less applicable. Smith observes that for the lowest car speed (14.0 mph), the fork tubes "bent in a manner similar to an impact into a stationary car." This was not the case for the higher car impact speeds.

Frank et al. [45] presented analysis of Smith's tests using EDSMAC4 and SIMON. They concluded that "the results demonstrate that it is indeed possible to simulate a motorcycle in these packages and that both packages can simulate two-moving motorcycle-to-car crashes reasonably well." There are several details of these simulations that will be useful to a reconstructionist carrying out simulation analysis of a motorcycle-to-vehicle collision. First, "in both EDSMAC4 and SIMON a two-wheeled motorcycle must be modeled as a narrow-track four-wheeled vehicle. In all of the simulations discussed in this article the rear track width was set to 2 inches and the front track width was set to 4 inches. This configuration kept the contact patches of the front and rear tires sufficiently close to the simulate a single tire, while also providing the capacity to input front wheel steering without generating a software error code." Second, Frank shortened the distance from the motorcycle CG to the front of the motorcycle by 6 in. in each simulation. He stated that "a thorough review of the crash test footage and data traces revealed that the motorcycles did not experience any substantial deceleration until they had traveled some distance beyond initial contact. On closer examination it was determined that immediately after the initial contact steering inputs produced by the relative motion across the engagement interface, coupled with bending of the front forks, allowed the motorcycle to move approximately six inches forward before sustaining significant deceleration." Third, Frank noted that "it was expected that the post-impact motion of the motorcycle would be difficult to match. This expectation was borne out by the results of the simulations. In reality the simulation packages cannot be held accountable for this because they do not claim the ability to model a motorcycle."

Frank also observed that "initially there was some difficulty achieving the [car's] measured rest position in the simulation. A review of the test photos and video showed that the front tires of the [car] experienced caster steer post-impact. This was due to the fact that the steering gear was free to respond to tire forces once the vehicle was released from tow. To mimic the steering response observed in this test a fixed post-impact clockwise steering wheel input of 270° was added. With this simulated caster steer applied it became possible to produce a rest position for the [car] that closely approximated the measured rest position."

Motorcycle Collisions with Roadside Barriers

In 2007, Gabler [47] examined US accident statistics related to fatal motorcycle collisions with guardrails. He observed that "in 2005 for the first time, motorcycle riders suffered more fatalities (224) than the passengers of cars (171) … involved in a guardrail collision. In terms of fatalities per registered vehicle, motorcycle riders are dramatically overrepresented in number of fatalities resulting from guardrail impacts … Motorcycle-guardrail crash fatalities are a growing problem. From 2000–2005, the number of car occupants who were fatally injured in guardrail collisions declined by 31% from 251 to 171 deaths. In contrast, the number of motorcyclists fatally-injured in guardrail crashes increased by 73% from 129 to 224 fatalities during the same period. Over two-thirds of motorcycle riders who were fatally injured in a guardrail crash were wearing a helmet."

Also in 2007, Ibitoye et al. [48] reported MADYMO simulations of upright motorcyclists colliding with guardrails next to lanes designated exclusively for motorcycles. They also reported a crash test that they used to validate the MADYMO finite element modeling of the guardrail system and the multibody model of the motorcycle and rider. This study was focused on improving the safety of motorcyclists in Malaysia where motorcycles constitute 49% of registered vehicles and "about 68% of all road accident injuries involved motorcyclists with overall relative risks of about 20 times higher than that of passenger cars." The simulations utilized a 110-kg motorcycle impacting the guardrail at speeds of 32, 48, and 60 km/h at angles of 15°, 30°, and 45°. Ibitoye et al. reported that "the kinematics of [the] rider for all impact conditions are similar during the initial stage with the dummy having leg contact with the guardrail surface and projecting with head forward. But, the dynamics of [the] rider towards the landing depend on the impact speeds and angles." Further, the guardrail "causes the rider to slide and tumble along the top of [the] guardrail before landing on the ground with [their] head."

In 2010, Bambach et al. [49] reported a study of 78 crashes in which motorcyclists were fatally injured in Australia and New Zealand following a collision with a roadside barrier. The study period was 2001 to 2006. "Of particular note were the findings that 97% of the motorcyclists were wearing a helmet prior to the crash, 86% of the crashes were single vehicle run-off crashes, 80% occurred on a corner, 92% of motorcyclists were male with a mean age of 34.2 years, 72% were less than 40 years and 81% of motorcyclists died at the crash scene. Motorcyclist behavior such as speeding and alcohol/drug use were identified as common causal factors … In Australia and New Zealand the main barrier types installed are steel W beam barriers … followed by concrete and wire rope (steel cable) barriers. Amongst motorcyclists fatally injured in barrier crashes, 77% involved W beams, 10% involved concrete barriers, 8% involved wire rope barriers and 5% involved other barriers … In the 78 cases the majority of motorcycles were sports motorcycles (n = 51), followed by touring motorcycles (n = 17) and off-road motorcycles (n = 3), with insufficient information to classify the motorcycle in seven cases."

Bambach et al. found that in 37 cases the motorcyclist was upright at impact with the barrier and in 34 cases the motorcycle and the rider were sliding on the ground into the barrier. They noted that in the upright crashes, "the motorcycle is typically redirected along the barrier. Due to the impact trajectory angle of the motorcycle relative to the barrier, momentum causes the upper body of the motorcyclist to want to continue over the barrier. In nine cases the motorcyclist was ejected over the barrier upon impact. In 20 cases this momentum and the redirection of the motorcycle along the barrier resulted in the motorcyclists scraping/tumbling/skidding along the top of the barrier. After scraping along the top of the barrier for some distance the motorcyclist was then ejected from the barrier, and in 15 of the 20 cases this occurred as a result of the motorcyclist impacting a barrier post."

Rizzi et al. [50] examined the influence of the barrier type and the orientation of the motorcycle (upright or capsized) at the time of the collision on the injury outcome for the riders. They utilized police-reported crashes in Sweden from between 2003 and 2010. They also conducted in-depth interviews with 55 Swedish motorcyclists who had been involved in collisions with roadside barriers. Rizzi et al. found no statistically significant difference in the injury outcomes among wire rope barriers, Kohlswa-beam (similar to W-beam) barriers, and W-beam barriers. They noted that "the small number of in-depth case findings, however, showed that injury severity was lower in crashes in which the motorcyclists were in an upright position during the collision."

Maza et al. [51] noted that "a particularly severe type of motorcycle collision is the crash into road barriers due to losing control of the motorcycle." Noting the past research efforts that have examined this issue, these researchers state that "some of these efforts resulted in the adoption of the European Technical Specification CEN/TS 1317-8 in which a modified Hybrid III test dummy with a helmet is launched sliding on the ground in a head-on impact at 30° and 60 km/h against different locations of [a] Motorcycle Protective Systems (MPS). [An] MPS consist[s] of a lower continuous rail that is installed in existing W-beam barriers to prevent the sliding motorcyclist from passing under the upper W-beam and impact into other rigid roadside obstacles. However, there is evidence that a significant share of motorcyclist impacts against roadside barriers happen in upright position with the rider still attached to the motorcycle." Maza et al. also observed that "the International Federation of Motorcycles (FIM) proposed an internal regulation of roadside protection used in the racing tracks during competition. This regulation requires barriers to provide a certain level of deceleration of a body-block in vertical position, impacting the barrier at different speeds in free flight." These researchers sought to determine the most common orientations of riders impacting roadside barriers on normal roads and on racing tracks. Ultimately, they sought to assess the realism of the test standards.

In conducting their research, Maza et al. examined 110 motorcycle collisions with roadside barriers that were captured on video and posted to YouTube. These researchers concluded that "For race track crashes, the pilot spine—ground angle distribution found in the video analysis shows mostly low angles, which is not well represented in the barrier test procedure applied, consisting of a body-block vertically impacting the barrier at different speeds in free flight. For normal road crashes, the motorcyclist spine-ground angle distribution found is consistent with the barrier test procedure defined by CEN/TS 1317-8, consisting of a full dummy sliding and impacting the barrier in supine position."

Motorcycle-to-Roadside Barrier Crash Tests

Berg et al. [52, 53] reported crash tests of motorcycles (Kawasaki ER 5 Twisters) and rider dummies colliding with three different barrier systems—a conventional steel barrier, a concrete barrier, and a modified steel guardrail design. Tests were conducted with the motorcycle and rider initially upright and with the motorcycle and rider capsized prior to colliding with the barrier. The nominal collision speed for each test was 60 km/h (37.3 mph). For the upright tests, the impact angle was approximately 12°, and for the capsized tests, the impact angle was approximately 25°. The motorcycles in these tests had a mass of approximately 180 kg (397 lb) and the rider dummies had a mass of approximately 92 kg (198 lb).

Steel Guardrail: Upright

In the upright test into the steel guardrail, the motorcycle exited the sled at 60 km/h (37.3 mph) and collided with the steel barrier at a speed of approximately 58 km/h (36.0 mph). From first contact to rest, the motorcycle traveled 28 m (91.9 ft). The dummy traveled 21 m (68.8 ft). Thus, from first contact to rest, the motorcycle decelerated at an average rate of 0.47 g and the rider decelerated at an average rate of 0.63 g. Figure 8.33 is a series of frames from the video of this test showing the motion of the motorcycle and rider from the time of contact with barrier until the motorcycle is nearly capsized. As these images show, the motorcycle collided with the barrier with its right side leading. The motorcycle quickly began rebounding, leaning, and steering to the left. The upper body of the rider continued to the right and folded over the barrier. Eventually, the front tire of the motorcycle steered back to the right as the motorcycle continued to be in contact with the lower extremities of the rider. The motorcycle began leaning right and eventually capsized onto its right side.

Figure 8.34 is another series of frames from the test video showing the motion of the motorcycle and the rider following capsize of the motorcycle. As these images show, the

FIGURE 8.33 Upright, steel barrier, frames from the test video showing the motion of the motorcycle and rider from the time of contact with barrier until the motorcycle is nearly capsized.

FIGURE 8.34 Upright, steel barrier, frames from the test video showing the motion of the motorcycle and rider after capsize.

motorcycle capsized onto its right side and remained on its right side as it slid and spun to rest. The upper body of the rider dummy continued interacting with and sliding along the barrier, while the dummy's lower body began sliding on the pavement.

Concrete Barrier: Upright

In the upright test into the concrete barrier, the motorcycle again exited the sled at 60 km/h (37.3 mph) in an upright position. After the motorcycle collided with the barrier, the dummy separated from the motorcycle and traveled over the barrier, coming to rest on the opposite side of the barrier. The motorcycle traveled 38 m (124.7 ft) from first contact with the barrier to rest. The dummy traveled 26 m (85.3 ft) from first contact to rest. Thus, from first contact to rest, the motorcycle decelerated at an average rate of 0.37 g and the rider decelerated at an average rate of 0.54 g. Figure 8.35 is a series of frames from the test video showing the motion of the motorcycle and rider as it contacts the barrier. The motorcycle begins sliding along the barrier and the rider is thrown over the barrier.

FIGURE 8.35 Upright, concrete barrier, frames from the test video showing the motion of the motorcycle and rider.

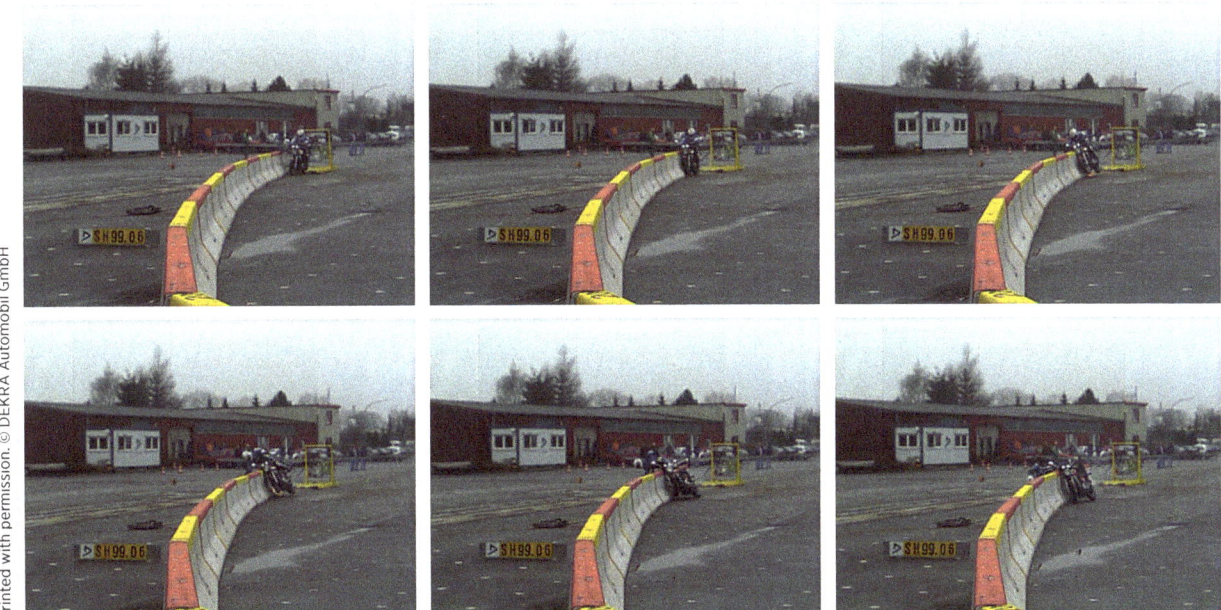

Reprinted with permission. © DEKRA Automobil GmbH

Modified Steel Guardrail: Upright

Berg et al. designed a modification to the steel guardrail system in which "an additional underrun protection board was mounted near to the ground to prevent both the direct impact onto a post and movement of the motorcyclist underneath the barrier protection system." In relationship to the upright motorcycle collision into this modified barrier, again at a speed of 60 km/h (37.3 mph), Berg et al. observed that "after first contact into the barrier the motorcycle was redirected away from the barrier. The dummy separated from the motorcycle and fell onto the protection system. After sliding for a short distance on the guard rail the dummy fell to the ground on the opposite side. Because of the closed

FIGURE 8.36 Upright, modified steel guardrail, frames from the test video showing the motion of the motorcycle and rider.

shape of the box-type profile, snagging did not occur and injury risk from impact was low as observed from the analysis of the film." Figure 8.36 is a series of frames from the video of this test. The motorcycle traveled 23 m (75.5 ft) from first contact to rest and the dummy traveled 22 m (72.2 ft). Thus, from first contact to rest, the motorcycle decelerated at an average rate of 0.61 g and the rider decelerated at an average rate of 0.64 g.

Steel Guardrail: Capsized

In this test, the motorcycle exited the sled at 60 km/h (37.3 mph). As it slid on its side, the motorcycle decelerated and, at the time it collided with the barrier, it was traveling a speed of 47 km/h (29.2 mph). Because it was on its side, the motorcycle bypassed the metal rail and directly impacted one of the posts of the barrier system. This impact brought the motorcycle to a stop quickly and the motorcycle came to rest underneath the guardrail. The dummy slid toward the barrier behind the motorcycle. When the motorcycle struck the barrier, the dummy separated from the motorcycle and struck a post of the barrier system. The motorcycle traveled 2 m during its interaction with the barrier and the dummy traveled 5 m. Figure 8.37 is a series of frames from the video of this test.

Concrete Barrier: Capsized

In this test, the motorcycle exited the sled at 59 km/h (36.7 mph). At the time it collided with the barrier, it was traveling 46 km/h (28.6 mph). The collision with the barrier system redirected the motorcycle and the rider along the barrier, and they both slid to rest traveling parallel with the barrier. Figure 8.38 is a series of frames from the video of this test. In relationship to this test, the authors observed that the "deceleration of the motorcycle and

CHAPTER 8 Motorcycle Collisions with Vehicles and Roadside Barriers 247

FIGURE 8.37 Capsized, steel barrier, frames from the test video showing the motion of the motorcycle and rider.

Reprinted with permission. © DEKRA Automobil GmbH

FIGURE 8.38 Capsized, concrete barrier, frames from the test video showing the motion of the motorcycle and rider.

Reprinted with permission. © DEKRA Automobil GmbH

©2022, SAE International

dummy were not as rapid as during the impact where the motorcycle slid into the guard rail made from steel. Nevertheless, the measured dummy decelerations for the primary impact were high, indicating a risk of severe and life-threatening injuries."

Modified Steel Guardrail: Capsized

In this test, the motorcycle exited the sled at a speed of 60 km/h and impacted the barrier at a speed of 54 km/h. The authors observed that "due to the impact the underrun protection board broke and the motorcycle struck a Sigma post. The dummy separated from the motorcycle immediately after the initial primary impact and then the helmeted head struck the underrun protection board." Following first contact with the barrier, the motorcycle traveled 1 m and the dummy traveled 7 m.

References

[1] Severy, D., Brink, H., and Blaisdell, D., "Motorcycle Collision Experiments," SAE Technical Paper 700897, 1970, doi:10.4271/700897.

[2] Adamson, K., Alexander, P., Robinson, E., Johnson, G. et al., "Seventeen Motorcycle Crash Tests into Vehicles and a Barrier," SAE Technical Paper 2002-01-0551, 2002, doi:10.4271/2002-01-0551.

[3] Baxter, A.T., *Motorcycle Crash Investigation*, Institute of Police Technology and Management, Jacksonville, FL, 2017, ISBN 978-1-934807-18-7.

[4] Bartlett, W., "Motorcycle Crush Analysis," *Accident Reconstruction Journal*, 19(2): 25–28, 2009, ISSN 1057-8153.

[5] Eubanks, J., *Motorcycle Speed-from-Damage Estimates Update*, Society of Accident Reconstruction, 22, 1991.

[6] Bartlett, W., Focha, B., and Kauderer, C., "25 Moving Motorcycle into Stationary Car Tests: CA²RS 2009 Data," *Accident Reconstruction Journal*, 23(4): 23–27, 64, 2013.

[7] Peck, L., Bartlett, W., Manning, J., Dickerson, C. et al., "Eleven Instrumented Motorcycle Crash Tests and Development of Updated Motorcycle Impact-Speed Equations," SAE Technical Paper 2018-01-0517, 2018, doi:10.4271/2018-01-0517.

[8] Campbell, K., "Energy Basis for Collision Severity," SAE Technical Paper 740565, 1974, doi:10.4271/740565.

[9] McHenry, R.R., "A Comparison of Results Obtained with Different Analytical Techniques for Reconstruction of Highway Accidents," SAE Technical Paper 750893, 1975, doi:10.4271/750893.

[10] McHenry, R.R., "Extensions and Refinements of the CRASH Computer Program Part II," DOT HS-801 838, February 1976.

[11] McHenry, R.R. and Jones, I.S., "Extensions and Refinements of the CRASH Computer Program Part III, Evaluation of the Accuracy Reconstruction Techniques for Highway Accidents," DOT HS-801 839, February 1976.

[12] McHenry, R.R., "Computer Aids for Accident Investigation," SAE Technical Paper 760776, 1976, doi:10.4271/760776.

[13] McHenry, R.R. and McHenry, B.G., "A Revised Damage Analysis Procedure for the CRASH Computer Program," SAE Technical Paper Number 861894, 1986, doi:10.4271/861894.

[14] Rose, N. and Ziernicki, R., "An Examination of the CRASH3 Effective Mass Concept," SAE Technical Paper 2004-01-1181, 2004, doi:10.4271/2004-01-1181.

[15] Rose, N. and Ziernicki, R., "Crush and Conservation of Energy Analysis: Toward a Consistent Methodology," SAE Technical Paper 2005-01-1200, 2005, doi:10.4271/2005-01-1200.

[16] Rose, N., "Restitution Modeling for Crush Analysis: Theory and Validation," SAE Technical Paper 2006-01-0908, 2006, doi:10.4271/2006-01-0908.

[17] Rose, N. and Carter, N., "Further Assessment of the Uncertainty of CRASH3 ΔV and Energy Loss Calculations," SAE Technical Paper 2014-01-0477, 2014, doi:10.4271/2014-01-0477.

[18] Searle, J., "The Reconstruction of Speed in Motorcycle Collisions from the Extent of Damage," *IMPACT: Journal of the ITAI*, Kent, UK, 2010.

[19] Wood, D.P., Glynn, C., and Walsh, D., "Estimation of the Collision Speed in a Collision of a Motorcycle or Scooter with a Car from Individual Vehicle Deformation," *Proc IMechE Part D: J Automobile Engineering*, 228(3): 295–309, 2014a, doi:10.1177/0954407012471272.

[20] Wood, D.P., Glynn, C., O'Dea, C., and Walsh, D., "Physical and Empirical Models for Motorcycle Speed Estimation from Crush," *International Journal of Crashworthiness*, 19(5): 540–554, 2014b, doi:10.1080/13588265.2014.918300.

[21] Wood DP, Glynn C, Walsh D. Motorcycle-to-car and scooter-to-car collisions: speed estimation from permanent deformation. *Proceedings of the Institution of Mechanical Engineers, Part D: Journal of Automobile Engineering*, 223(6): 737–756, 2009. doi:10.1243/09544070JAUTO1069.

[22] MacInnis, D., Cliff, W., and Ising, K., "A Comparison of Moment of Inertia Estimation Techniques for Vehicle Dynamics Simulation," SAE Technical Paper 970951, 1997, doi:10.4271/970951.

[23] Allen, R., Klyde, D., Rosenthal, T., and Smith, D., "Estimation of Passenger Vehicle Inertial Properties and Their Effect on Stability and Handling," SAE Technical Paper 2003-01-0966, 2003, doi:10.4271/2003-01-0966.

[24] Deyerl, E. and Cheng, L., "Computer Simulation of Staged Motorcycle-Vehicle Collisions Using EDSMAC4," white paper presented at the *2008 HVE Forum, HVE-WP2008-3*, https://edccorp.com/index.php/support-learning/library/white-papers, accessed on November 6, 2021.

[25] Deyerl, E. and Cheng, L., "Computer Simulation of Staged Motorcycle-Vehicle Collisions Using EDSMAC4," *Accident Reconstruction Journal*, 17(4): 33–45, 2007.

[26] Rose, N. and Carter, N., "An Analytical Review and Extension of Two Decades of Research Related to PC-Crash Simulation Software," SAE Technical Paper 2018-01-0523, 2018, doi:10.4271/2018-01-0523.

[27] Day, T. and Hargens, R., "An Overview of the Way EDSMAC Computes Delta-V," SAE Technical Paper 880069, 1988, doi:10.4271/880069.

[28] https://www.vcrashusa.com/vc-validation-vc, accessed on November 6, 2021.

[29] Brach, R.M., Manuel, E.J., Bailey, R., Rogers, J. et al., "Sensitivity Analysis of Various Vehicle Dynamic Simulation Software Packages Using Design of Experiments (DOE)," SAE Technical Paper 2020-01-0639, 2020, doi:10.4271/2020-01-0639.

[30] Brach, R.M., *Mechanical Impact Dynamics: Rigid Body Collisions*, Revised edition, Brach Engineering, South Bend, IN, 2007, ISBN 978-1-4028-9462-6.

[31] Brach, R.M. and Brach, R.M., "A Review of Impact Models for Vehicle Collision," SAE Technical Paper Number 870048, 1987a, doi:10.4271/870048.

[32] Brach, R.M. and Brach, R.M., "Energy Loss in Vehicle Collision," SAE Technical Paper Number 871993, 1987b, doi:10.4271/871993.

[33] Brach, R.M. and Brach, R.M., *Vehicle Accident Analysis and Reconstruction Methods*, 2nd ed., SAE International, Warrendale, PA, 2005, ISBN 978-0-7680-3437-0.

[34] Marine, M.C., "On the Concept of Inter-Vehicle Friction and Its Application in Automobile Accident Reconstruction," SAE Technical Paper Number 2007-01-0744, 2007, doi:10.4271/2007-01-0744.

[35] Keifer, O., Reckamp, B., Heilmann, T., and Layson, P., "A Parametric Study of Frictional Resistance to Vehicular Rotation Resulting from a Motor Vehicle Impact," SAE Technical Paper 2005-01-1203, 2005, doi:10.4271/2005-01-1203.

[36] Keifer, O., Conte, R. and Reckamp, B., *Linear and Rotational Motion Analysis in Traffic Crash Reconstruction*, Institute of Police Technology and Management, Jacksonville, FL, 2007.

[37] McNally, B.F. and Bartlett, W., "Motorcycle Speed Estimates Using Conservation of Linear and Rotational Momentum," *20th Annual Special Problems in Traffic Crash Reconstruction at the Institute of Police Technology and Management*, University of North Florida, Jacksonville, FL, April 15–19, 2002.

[38] DiTallo, M., "2019 IATAI Conference, Motorcycle Case Study," *Presentation, Motorcycle Crash Investigation Seminar*, Illinois Association of Technical Accident Investigators (IATAI), Springfield, IL, October 8, 2019.

[39] Ogden, J.S. and Kloberdanz, K.M., "Forensic Engineering Analysis of Motorcycle Impacts Using Rotational Mechanics and Fork/Vehicle Deformation," NAFE 561F/430C, National Academy of Forensic Engineers, June 2012.

[40] Craig, V., "Motorcycle High-Speed Crash Tests," *Accident Reconstruction Journal*, 20(1): 25–32, 2010.

[41] Peck, L., Bartlett, W., Manning, J., Dickerson, C., et al., "Eleven Instrumented Motorcycle Crash Tests and Development of Updated Motorcycle Impact-Speed Equations," SAE Technical Paper 2018-01-0517, 2018, doi:10.4271/2018-01-0517.

[42] Neptune, J., "Overview of an HVE Vehicle Database," SAE Technical Paper 960896, 1996, doi:10.4271/960896.

[43] Niederer, P., "Some Aspects of Motorcycle-Vehicle Collision Reconstruction," SAE Technical Paper 900750, 1990, doi:10.4271/900750.

[44] Smith, J., Frank, T., Bosch, K., Fowler, G. et al., "Full-Scale Moving Motorcycle into Moving Car Crash Testing for Use in Safety Design and Accident Reconstruction," SAE Technical Paper 2012-01-0103, 2012, doi:10.4271/2012-01-0103.

[45] Frank, T., Smith, J., Fowler, G., Carter, J. et al., "Simulating Moving Motorcycle to Moving Car Crashes," SAE Technical Paper 2012-01-0621, 2012, doi:10.4271/2012-01-0621.

[46] Nieboer, J., Wismans, J., Versmissen, A., van Slagmaat, M. et al., "Motorcycle Crash Test Modelling," SAE Technical Paper 933133, 1993, doi:10.4271/933133.

[47] Gabler, H.C., *The Emerging Risk of Fatal Motorcycle Crashes with Guardrails*, January 2007, https://www.sbes.vt.edu/gabler/publications/TRB-07-3456-Motorcycles-Final.pdf, accessed on November 6, 2021.

[48] Ibitoye, A.B., Radin, R.S., and Hamouda, A.M.S., "Roadside Barrier and Passive Safety of Motorcyclists Along Exclusive Motorcycle Lanes," *Journal of Engineering Science and Technology*, 2(1): 1–20, 2007.

[49] Bambach, M.R., Grzebieta, R.H., and McIntosh, A.S., "Crash Characteristics of Motorcyclists Impacting Road Side Barriers," *2010 Australasian Road Safety Research, Policing and Education Conference*, August 31–September 3, 2010, Canberra, Australian Capital Territory.

[50] Rizzi, M., Strandroth, J., Sternlund, S., Tingvall, C. et al., "Motorcycle Crashes into Road Barriers: The Role of Stability and Different Types of Barriers for Injury Outcome," *IRCOBI Conference 2012*, Dublin, Ireland, IRC-12-41.

[51] Maza, M., Larriba, J., Juste-Lorente, O., and Lopez-Valdes, F.J., "Motorcyclists Crashes into Race Tracks and Normal Road Barriers: Kinematic Analysis and Correlation with Test Procedures," *IRCOBI Conference 2016*, Seoul, South Korea, IRC-16-103.

[52] Berg, F.A., Rücker, P., Gärtner, M., König, J. et al., "Motorcycle Impacts into Roadside Barriers – Real World Accident Studies, Crash Tests and Simulations Carried Out in Germany and Australia," *International Technical Conference on the Enhanced Safety of Vehicles*, Washington, DC, June 2005, Paper Number 05-0095.

[53] Berg, F.A., Rücker, P., and König, J., "Motorcycle Crash Tests – An Overview," *International Journal of Crashworthiness*, 10(4): 327–339, 2005, doi:10.1533/ijcr.2005.0349.

9

Event Data Recorders and Speedometers in Motorcycle Accidents

Event Data from the Struck or Striking Vehicle

A common motorcycle crash scenario occurs when a passenger vehicle equipped with an event data recorder (EDR) turns left across the path of a motorcycle and a collision occurs. The EDR data on the passenger vehicle will often be accessible with either the Bosch Crash Data Retrieval (CDR) system or the Global Information Technology (GIT) system. Precrash EDR data can be useful for establishing the specific characteristics of the left turn. This data may include speed, throttle percentage, brake status (ON or OFF), and steering angles. In addition, an EDR-reported change in velocity (ΔV) from the passenger vehicle can potentially be used to infer the ΔV and impact speed of the motorcycle.

Precrash Data

Reference [1] reported that the EDR-reported precrash speeds are "typically measured by sensors monitoring the output of the transmission or an average of the speed of the drive wheels. These sensors can accurately report wheel speed but … may not represent the true over-the-ground speed of the vehicle … factors [causing this discrepancy] may include longitudinal wheel slip due to acceleration or braking, wheel slip due to rotation of the vehicle about the vertical axis, significant changes in the tire's rolling radius as compared to the vehicle's original equipment, and changes to final drive ratio compared to the vehicle's original equipment." This reference reported that under heavy braking, the reported precrash speed will typically be lower than the actual speed but within 4 mph of the actual speed. For steady driving, the EDR-reported precrash speed will typically be lower than the actual speed but within 1 or 2 mph of the actual speed.

One challenge in using EDR-reported precrash speed data to reconstruct a collision is to determine precisely how the data syncs to the crash. Suppose, for example, that an EDR reports 5 sec of precrash data at 0.5-sec intervals. How should one interpret the last reported speed (time 0.0 for some systems, −0.5 sec for others)? This depends on the specific system, but for some systems this reported speed at time zero will simply be the last speed reading collected by the airbag control module (ACM) prior to the collision, which could be anytime between 0 and 0.499 sec prior to impact. As Ruth and Wright [2] note, "all of this [pre-crash data] … is collected at regular intervals and saved (buffered) to an intermediate memory location. Once the buffer is full—usually with 5 seconds of data—the most recent sample displaces and replaces the oldest data sample. This process, one of writing the new and discarding the old, continues over and over, with the result that the buffer always contains the most recent 5 seconds of information. We can characterize the metronome-like sampling of the pre-crash data as 'blind' because the data is unsynchronized to the crash; it is unsynchronized because the EDR has no way of knowing ahead of time when the crash will occur … Essentially, the unsynchronized, independent collection of pre-crash data means that the last reported sample has been recorded sometime prior to the crash. It follows, then, that the last recorded speed is not necessarily the actual [speed at impact]. It also follows that there will usually be some time delay between the last speed sampling and the time of the crash." In some instances, the best way to handle this uncertainty will be to consider both ends of the range, considering the degree to which the vehicle is accelerating or decelerating over this ½-sec interval. In other instances, the reconstruction will reveal when the final reading was likely captured. For example, the reconstructionist may be able to determine what impact speed is most compatible with the EDR-reported ΔV. Some ACMs do sample the precrash data elements at the time of the triggering event, and for these ACMs, this consideration would not be necessary.

Another issue with precrash data is that the multiple data elements reported may be asynchronous. This asynchronicity arises from the fact that not every sensor can be using the CAN bus to transmit their message at the same time, so the CAN messages get staggered. The sensors take their measurements and send them, but a controller decides what messages get to use the CAN based on the priority assigned to that sensor and a message priority from within the sensor. As a result, the level of asynchronicity is related to the volume of CAN traffic at any given time and the priority of those competing messages. Finally, sensors that measure and report the precrash data elements can have full-scale values that are too limited to capture the full value of a data element. For example, some systems have a limit on the highest speed they can report and others a limit on the magnitude of the steering input they will report. These limitations are often listed in the data limitations section of the CDR report.

Crash Data (ΔV)

There are also issues that can arise when using the passenger vehicle EDR-reported ΔV to infer the ΔV and impact speed of the motorcycle. One of the primary issues relates to the large weight ratio that often exists between the motorcycle and the passenger vehicle. Conservation of momentum dictates that, during a collision between two vehicles, the ratio of the mass of Vehicle #1 (m_1) to the mass of Vehicle #2 (m_2) is equal to the ratio of the change in velocity experienced by Vehicle #2 (ΔV_2) to the change in velocity experienced by Vehicle #1 (ΔV_1), as follows:

$$\frac{m_1}{m_2} = \frac{\Delta V_2}{\Delta V_1} \tag{9.1}$$

As an example, one of the case studies discussed later in this chapter had a ratio between the weight of the passenger vehicle and the weight of the motorcycle (including the rider) of approximately 7.15. The EDR-reported resultant ΔV of the passenger vehicle was 12.1 mph. Applying Equation (9.1) yields a ΔV for the motorcycle of 85.8 mph, as follows:

$$\Delta V_{motorcycle} = \frac{W_{struck\ vehicle}}{W_{motorcycle}} \Delta V_{struck\ vehicle} = 7.15 \times 12.1\ mph = 86.5\ mph$$

There is some level of error that could be present in the EDR-reported ΔV. Given the weight ratio, every 1 mph of potential error in the ΔV would produce 7.15 mph of uncertainty in the calculated ΔV for the motorcycle. Thus, ±1 mph of uncertainty would produce a range in the calculated ΔV for the motorcycle of 79.4 to 93.7 mph.

Potential Error Sources

The following systematic error sources can affect the accuracy of the EDR-reported speed change (ΔV) [3]. This list can be treated as a checklist of error sources that can be ruled out or accounted for through analysis.

- *Error Source #1*: During an impact, there can be a ΔV that occurs prior to the accelerations exceeding the event-triggering threshold (algorithm enable or AE). This is an error source that contributes to under-reporting of the ΔV. Prior studies have indicated that the triggering threshold for most passenger vehicle EDRs is in 1 to 2 g range.

- *Error Source #2*: Some EDRs reportedly have a built-in positive offset in the measurement of longitudinal accelerations that can contribute to under-reporting of the ΔV for frontal impacts and over-reporting for rear impacts. Ruth [4] notes that this offset can vary from vehicle to vehicle and EDR generation to generation. A positive longitudinal acceleration bias is documented for Toyotas in references [5–7].

- *Error Source #3*: The recording window for the ΔV may be too short to capture the full ΔV, an error source that contributes to under-reporting of the ΔV. This error source can be ruled out if the EDR-reported ΔV reaches a maximum and begins to decrease prior to the end of the recording window. A similar, but rarer, error source is that the recording window for the ΔV may be too long and a ΔV experienced by the vehicle because of postimpact tire and dragging forces could be recorded. This error source could contribute to over-reporting of the ΔV. This error source can be ruled out if the EDR-reported ΔV reaches a maximum prior to the end of the recording window and then decreases for the remainder of the recording window. This error source would be recognizable if the EDR-reported ΔV reached a local maximum, then began to decrease, but eventually began to increase again prior to the end of the recording window.[1]

---

[1] This error source will become less of an issue with time now that the Code of Federal Regulations (49 CFR 563) requires passenger vehicle EDRs to report cumulative ΔV over a duration of 250 ms and to monitor and report the maximum ΔV over a period of 300 ms [8]. Part 563 also contains criteria under which the recording can be terminated short of these times, but these criteria are designed to ensure full capture of the collision. This requirement has been in place since September 1, 2012.

- *Error Source #4*: The peak accelerations during a collision can exceed the capabilities of the accelerometer in the ACM where the EDR resides, in which case the system will not capture the peak accelerations. This is referred to as clipping. This error source contributes to under-reporting of the ΔV. This error source will be recognizable through a flatline portion of either the EDR-reported ΔV or acceleration curve. Because of the significant weight discrepancy that is typically present in motorcycle-to-passenger vehicle collisions, error source #4 will not be common in motorcycle crashes.

- *Error Source #5*: The ACM may reside some distance from the center of gravity (CG) of the vehicle. This is relevant because accident reconstruction calculations often calculate and utilize the CG ΔV. In instances where the collision induces significant rotation, the ΔV of the passenger vehicle—which is measured at the ACM location—may need to be adjusted to accurately reflect the ΔV at the CG of the passenger vehicle. Bundorf [9], Marine and Werner [10, 11], Rose [12], and Haight and Haight [13] describe methods for making this adjustment.

- *Error Source #6*: Damage to or displacement of the ACM can affect the accuracy of the reported ΔV. A vehicle involved in a collision can be split into two regions—a deforming or crushing region and a nondeforming region [14]. When a collision is severe enough that the ACM ends up being within the crushing region of the vehicle, the accelerations and ΔV reported by this module will not be representative of the accelerations and ΔV experienced by the nondeforming region [15]. This error source can contribute to significant under-reporting or over-reporting of the ΔV. However, it will not be common for motorcycle collisions with passenger vehicles, and this error source can be ruled out through physical examination of the vehicle and ACM.

- *Error Source #7*: The EDR could lose power before the collision is complete and some of the data may not get recorded. When this occurs, the system would under-report the ΔV. Typically, the EDR report will indicate whether recording of a reported event was complete. If the report states that the recording was complete, then this error source can be ruled out.

- *Error Source #8*: The accelerations measured by the ACM, and the ΔV that is calculated from those accelerations, include the accelerations not only because of the collision forces, but also because of external forces present during the collision. Tire forces are a common external force that would be present. Many of the physics models that reconstructionists utilize assume that these tire forces are not present. When using a physics model that neglects these tire forces during the collision, the ΔV generated by the tire forces may need to be added or subtracted to the EDR-reported ΔV to get to the ΔV that would have been produced by the collision force alone. This is to accommodate the physics model. Alternatively, a physics model that incorporates the tire forces could be utilized.

Potential Magnitude of the Error—Longitudinal ΔV

Figure 9.1 plots the error in the EDR-reported ΔV for crash tests reported in the literature [16–30]. The error was quantified in the cited studies by comparing the EDR-reported ΔV to the ΔV calculated from onboard, laboratory grade accelerometers with which the vehicles were instrumented. The data in this figure are for full-overlap, front or rear collisions where the vehicles experienced insignificant yaw rotation following the collision. The ΔV calculated from the laboratory accelerometers is plotted on the horizontal axis, and the error in the EDR-reported ΔV is plotted on the vertical axis. A negative ΔV on the horizontal axis

FIGURE 9.1 Error in the event data recorder (EDR)-reported ΔV for full-overlap, frontal impacts.

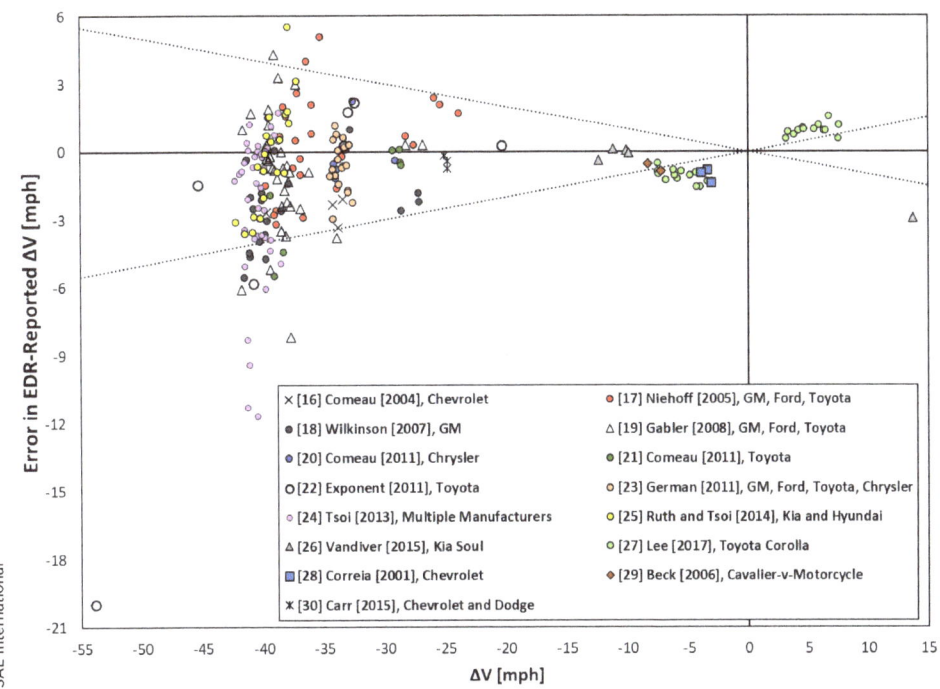

is a frontal impact and a positive ΔV is a rear impact. The errors reported on the vertical axis are calculated from the ΔV magnitudes, such that a negative error is always an under-reporting of the *magnitude* of the ΔV and a positive error is always an over-reporting of the *magnitude* of the ΔV.

The dashed black lines represent a window of ±10% error, the required accuracy per 49 CFR part 563.8 and the rule of thumb reported in a study by Chidester et al. [31] examining early General Motors Sensing and Diagnostic Modules. Figure 9.1 shows generally increasing absolute error in the EDR-reported ΔVs with increasing ΔV, consistent with the underlying implication of the 10% error rule. Much of the data do lie with the 10% error band, but there are also points that fall outside of this window. For the frontal collisions with ΔVs around 40 mph there are points with error higher than 10%. The higher errors associated with these higher severity tests are related to the proximity of the ACM to the crushing region of the vehicle. The more severe the collision and the closer the ACM is to the collision, the greater the likelihood that the ACM or surrounding structure will be damaged or deformed. If such deformation occurs, increased error in the EDR-reported ΔV will be the result. The highest errors occurred around a ΔV of 40 mph and above, a severity level unlikely to be experienced by the passenger vehicle during a collision with a motorcycle. If damage or deformation to the ACM or surrounding structure can be ruled out, then these higher error rates would not be applicable.

The error rates related to lower severity collisions are more likely to apply to motorcycle collisions. For frontal collisions with ΔVs lower than 10 mph, almost all the points lie outside of the 10% error window. For these lower speed collisions, it is more useful to think in terms of absolute error, rather than percentage error. For example, Correia et al. [28] reported 12 low-severity, front-to-rear vehicle-to-vehicle staged collisions using two different General Motors bullet vehicles—a 2000 Chevrolet Malibu and a 1997 Chevrolet Cavalier. These vehicles passenger one of two target vehicles—a 1984 VW Rabbit or a 1994 Honda Accord. The sensing and diagnostic module (SDM)[2] on the Cavalier would report the ΔV, but the SDM on the Malibu was not equipped to do so. Of the seven tests with the Cavalier, four did not cause an event to be recorded on the SDM. In three tests where a nondeployment event was recorded, the EDR-reported ΔVs were between 1.3 and 2.2 km/h (0.8 to 1.4 mph) low. The authors attributed the under-reporting to a recording window that was too short to capture the full collisions.

Lawrence et al. [32] examined the accuracy of EDR-reported ΔVs from the SDMs on MY 1996-1999 General Motors vehicles for low-severity collisions. Two SDM-equipped vehicles were subjected to 260 staged frontal collisions with speed changes below 11 km/h (6.8 mph). Of the 260 collisions, 105 of them were vehicle-to-barrier tests and 155 of them were vehicle-to-vehicles tests, with the front bumper of the SDM-equipped vehicle striking the rear bumper of a 1984 Volvo GL. Airbags did not deploy in any of the tests. The authors reported that "in all of the vehicle tests, the speed change reported by the SDM underestimated the actual speed change of the vehicle … The difference between the SDM-reported speed change and the actual speed change was as high as 4 km/h [2.5 mph] at low speed changes and decreased to a maximum of 2.6 km/h [1.6 mph] at a speed change of 10 km/h [6.2 mph]." They attributed this difference to the portion of the ΔV that occurred prior to the threshold acceleration necessary to wake up the module. They determined that this threshold acceleration was between 1.2 and 1.4 g for the modules in their study. The authors also found that the error in the SDM-reported ΔV was sensitive to pulse shape and duration. For the same ΔV, shorter collisions pulses generated more accurate SDM-reported ΔVs, because more of the pulse was above the threshold acceleration. The authors observed that "long duration and low peak acceleration pulses have larger areas excluded from the integration and will result in larger errors in the SDM-reported speed change than short duration or high peak acceleration pulses of the same general shape." These authors did not report tabular data for their tests, and so these tests are not included in Figure 9.1. However, the results of these tests are consistent with other low-severity tests that are included in the figure.

Wilkinson et al. [33] examined the accuracy of the EDR-reported ΔVs from Ford restraint control modules (RCMs) in low-severity collisions. They conducted 84 frontal barrier collisions with two RCM-equipped vehicles with actual speed changes as high as 13.5 km/h (8.4 mph). These authors reported that the accuracy of the EDR-reported ΔVs ranged from an underestimate of 1.8 km/h (1.1 mph) to and overestimate of 0.3 km/h (0.2 mph). They attributed these errors both to the portion of the ΔV that occurred prior to the threshold acceleration necessary to wake up the module and, in many instances, to a recording window that was too short to capture the full collision pulse. In a follow-up study, Lawrence and Wilkinson [34] reanalyzed the Ford RCM data with an updated version of the CDR software. They found differences between the ΔVs reported by the two versions of the software. When analyzed with the updated version, more accurate speed changes resulted for one of the vehicles, but less accurate for the other vehicle. Overall, the accuracy of the EDR-reported ΔVs was between −1.3 km/h (−0.8 mph) and 0.4 km/h (0.2 mph). These

---

[2] GM's name for the ACM.

authors did not report tabular data for their tests, and so these tests are not included in Figure 9.1. However, the results of these tests are consistent with other low-severity tests that are included in the figure.

Wilkinson et al. [35] examined the accuracy of EDR-reported ΔVs from the SDMs on MY 2003 and 2004 General Motors vehicles for low-severity collisions. Three MY 2004 SDM-equipped vehicles (Chevrolet Cavalier, Impala, and Trailblazer) were subjected to 136 vehicle-to-barrier and vehicle-to-vehicle frontal collisions with speed changes up to 8 km/h (5 mph). Of these tests, 65 of them were vehicle-to-vehicle and 71 were vehicle-to-barrier. The SDMs were also tested on a linear sled that allowed for replicating the crash pulses from the vehicle tests and for applying similar pulses of greater severity to the SDMs. The authors reported that "in all of the tests, the speed change reported by the SDM underestimated the actual speed change. The speed change underestimates ranged from 0.2 to 2.9 km/h [0.1 to 1.8 mph] except for several anomalous tests in which the underestimate was as high as 12.3 km/h [7.6 mph]." In comparing their results to their 2002 publication with earlier model year SDMs, the authors state that "the newer model GM SDMs, and in particular the 2004 Cavalier SDM, appear to be more accurate in reporting speed change than their predecessors. The improved accuracy is at least partially explained by the lower threshold acceleration in the new Cavalier SDM (1.1 g versus 1.3 g)." These authors did not report tabular data for their tests, and so these tests are not included in Figure 9.1. However, the results of these tests are consistent with other low-severity tests that are included in the figure.

Wilkinson et al. [5] examined the accuracy of EDR-reported ΔVs from MY 2005 to 2008 Toyota Corolla EDRs for low-severity collisions. They utilized vehicle-to-barrier tests and ACM sled tests for their evaluation. The frontal impact barrier tests ranged in speed from 2.0 to 5.2 km/h (1.2 to 3.2 mph) with ΔVs between 3.5 and 9.0 km/h (2.2 and 5.6 mph). The lowest ΔV that produced a recorded event was 4.5 km/h (2.8 mph) with a peak acceleration of 2.1 g. The authors reported that "in all in-vehicle tests, the speed change reported by the ACM underestimated the actual speed change for frontal collisions and overestimated the actual speed change for rear-end collisions. The speed change underestimates ranged from 1.3 to 2.6 km/h [0.8 to 1.6 mph] and the speed change overestimates ranged from 0.6 to 2.2 km/h [0.4 to 1.4 mph] … Threshold accelerations required to initiate the recording of an event were found to be between 2.0 and 2.1 g for all of the ACMs tested." The authors also reported that "integrating the sled acceleration pulses after a 2 g threshold was achieved and adding a constant bias of 0.4 g reproduced the temporal shift in the cumulative speed change values reported by the ACM relative to the reference speed change … it also generated maximum speed change estimates (ΔV_{model}) that were close to the ACM speed changes (ΔV_{ACM}) especially for frontal crash pulses." These authors did not report tabular data for their tests, and so these tests are not included in Figure 9.1. However, the results of these tests are consistent with other low-severity tests that are included in the figure.

Xing et al. [7] compared the response of 19 Generation 1, 2, and 3 Toyota EDRs from Toyota Corollas, Camrys, and Priuses for low-severity collisions. They used a sled to subject the EDRs to frontal and rear haversine crash pulses of varying duration (80, 120, 160, and 200 ms), peak acceleration (0.17 to 4.59 g), and ΔV (0.9 to 13.0 km/h). They found that acceleration necessary to trigger an event was approximately 2 g for Generation 1 and 2 EDRs. The acceleration necessary to trigger an event for a Generation 3 EDR appeared to vary, but events were consistently triggered for ΔVs around 8 km/h. Xing et al. theorized that "this new and different behavior for the Gen3 ACMs could be the results of software changes that simply do not report events that fall below this 8 km/h threshold despite an internal trigger that remains at 2 g." These authors further reported that "in most front impacts, the ACMs underestimated the reference speed change …." Xing et al. presented regression equations for each EDR generation/vehicle combination that would related the EDR-reported ΔV to the actual ΔV. For individual cases involving Toyotas, a reconstructionist can apply these equations. For the discussion here, it is sufficient to observe that for frontal impacts, the EDR-reported ΔV errors were between a 0.72 km/h (0.45 mph) overestimate and a 3.81 km/h (2.37 mph) underestimate. Most of the EDR-reported ΔVs underestimated the actual ΔVs. Xing et al. also noted that the magnitude of errors in the EDR-reported ΔVs could be reduced to ±1 km/h with the use of their regression equations. This suggests that testing of individual ACMs could be useful for achieving high levels of accuracy on a specific case. Some cases will warrant such testing and others will not.

In a follow-up study, Lee et al. [27] expanded their examination of Toyota ACMs to include mid-severity crashes and to confirm that sled testing of ACMs could be applied to the analysis of vehicle-to-barrier and vehicle-to-vehicle crashes. These authors reported that "in the frontal vehicle-to-barrier tests, the ACM-reported speed changes consistently underestimated the reference speed change … the ACM-reported speed change consistently overestimated the reference speed change in rear-end vehicle-to-barrier collisions. In vehicle-to-vehicle collisions, the same pattern of underestimating in frontal collisions and overestimating in rear-end collisions was observed … Overall, the ACMs underestimated frontal speed change by 1.25 km/h and overestimated rear-end speed change by 0.79 km/h." For the frontal impact vehicle tests, the range in the under-reporting for the EDR-reported ΔVs varied between 0.8 and 2.5 km/h (0.5 to 1.6 mph). These authors presented additional regression modeling with which the EDR-reported ΔV could be corrected and the accuracy improved. The vehicle tests from this study are included in Figure 9.1.

Potential Magnitude of the Error—Lateral ΔV

Haight et al. [36] reported analysis of a side impact crash test of a 2013 Kia Rio. In this test, the stationary Kia was struck on the driver's side at a 90° angle by a deformable barrier traveling 50 km/h (31.1 mph). Lateral accelerations were measured on the test vehicle, near A and B pillars on the nonstruck side of the vehicle. The impact did not induce significant yaw rotation of the Kia. The authors reported that "the lateral delta-V reported by the EDR Tool in this side impact test reasonably compares to the delta-V calculated using the IIHS accelerometer. The maximum delta-V calculated from the IIHS accelerometer is about 26.7 km/h (16.6 mph) while the maximum reported delta-V reported by the EDR tool is 25 km/h (15.5 mph)." This is an under-reporting by the EDR of 1.1 mph.

Tsoi et al. [37] evaluated the accuracy of 75 EDRs from MY 2010 to 2012 Chrysler, Ford, General Motors, Honda, Mazda, and Toyota vehicles in side impacts. These vehicles were each subjected to side impacts with a moving deformable barrier (MDB) as a part of the NHTSA Side-Impact New Car Assessment Program (SINCAP). This test procedure involved the MDB impacting the stationary test vehicle at 62 km/h (39 mph). The heading angle of the MDB is 90° relative to the test vehicle, but the wheels are crabbed, such that the velocity of the MDB is angled 27° relative to its heading. The reference ΔVs were calculated at the vehicle CGs from the laboratory accelerometers on the vehicle. The authors noted that the vehicles in these tests experienced yaw rotation and they describe the equations they used to account for this rotation during calculation of the reference ΔVs. These authors reported that the "EDRs underreported the reference lateral delta-v in the vast majority of cases, mimicking the errors and conclusions found in some longitudinal EDR accuracy studies. For maximum lateral delta-v, the average arithmetic error was −3.59 kph [2.2 mph] (−13.8%) and the average absolute error was 4.05 kph [2.5 mph] (15.9%)." Unfortunately, the authors did not calculate the ΔVs at the ACMs, and so their reference ΔVs are not directly comparable to the EDR-reported ΔVs. The authors acknowledged this as a possible error source in their study, but they believe the error would be small. Unfortunately, small errors sources are what we are trying to disentangle here. Without such analysis being completed, not a lot of stock can be put on the differences between the EDR-reported and "actual" ΔVs reported in this study.

Carr et al. [30] tested passenger car EDRs on a HYGE crash simulation sled in various orientations designed to represent different principal directions of force (PDOF). This is a useful study because it eliminated any error arising from yaw rotation, ACM location, deformation to or around the ACM, acceleration clipping, or an inadequate recording window. These authors performed direct comparison of the EDR-reported and actual longitudinal and lateral ΔVs and also examined the possibility of accurately reconstructing the PDOF orientation from the combined longitudinal and lateral EDR-report ΔVs. They reported that the maximum percentage error in the ΔV was less than 10%, with the average error magnitude for the various EDRs ranging between 0.3% and 4.3%. The magnitude of the errors was between 0 and 1 mph. They reported that the maximum PDOF angle error magnitude was 2.0°.

The specific EDRs tested by Carr et al. were from a 2012 Chevrolet Malibu, a 2012 Dodge Durango SXT, and a 2012 RAM 1500. Their test rig allowed the orientation of the EDRs to be swept through a range of yaw angles between −90° (driver side leading) to +90° (passenger side leading). The EDRs were tested throughout this range in angle increments of 22.5°. They also tested the EDRs with pitch angles between 5° and 20° in each direction. Each module was subjected to 13 simulated crash pulses with an actual ΔV of approximately 25 mph (40 km/h). This study shows that the errors in the EDR-reported longitudinal ΔVs were mostly in the direction of under-reporting, whereas the errors in the lateral ΔVs were more distributed about zero.

Accounting for Tire Forces When Incorporating EDR Data

The accelerations measured by an ACM, and thus, the ΔV that is calculated from those accelerations, include the accelerations not only because of the collision forces, but also because of tire forces present during the collision. This conflicts with some of the physical models that accident reconstructionists employ. Conservation of momentum impact models, for example, commonly assume that the collision occurs instantaneously, and thus, that there would not be any movement of the vehicle or influence of tire forces during the collision. To reconcile what the EDR measures with the assumptions of a conservation of momentum impact model, the reconstructionist may need to adjust the EDR-reported ΔV before incorporating it into the physics model. The importance of accounting for the tire forces will vary from case to case.

One point worth mentioning here: The calculations procedures used in EDSMAC4 consider the collision forces through time and account for the presence of tire forces during the collision. PC-Crash and V-Crash assume the collision forces are transferred instantaneously—there is no motion of the vehicles, and thus, no tire forces that develop during the collision. Any of these simulation packages can produce acceptable reconstruction results. However, in instances where the reconstructionist is comparing the simulation-calculated ΔV to an EDR-reported ΔV, EDSMAC4 can be a useful tool because the EDR-reported ΔV will include the effects of tire forces. PC-Crash and V-Crash can also be used (and PC-Crash is used in the illustrations that follow), but the EDR-reported ΔV may first need to be adjusted for the effects of the tire forces before an apples-to-apples comparison can be made between the simulation and the EDR data.

WREX 2016 Test #3

Consider Test #3 from WREX 2016. This test involved a 2013 H-D FXSB 103B Softail Breakout motorcycle (667 lb) impacting the passenger's side front door of a stationary 2006 Nissan Maxima (3449 lb), at a 90° angle and a speed of 43.0 mph. Figure 9.2 contains a series of frames from the video of this test. From these frames, it is apparent that lateral movement of the Nissan occurred during the collision. Figure 9.3 shows the damage to these vehicles and their rest positions, tire marks, and chalk marks identifying the passenger's side wheel positions both pre- and postcollisions. The Nissan moved 2.75 ft laterally during and following the collision. The line of action of the resultant collision force passed through the CG of the Nissan and this collision produced an insignificant magnitude of postimpact rotation of the Nissan.

In the previous chapter, conservation of energy was used to calculate a ΔV for the Nissan in this test of approximately 7.59 mph. This ΔV was then related to the impact speed of the motorcycle using conservation of momentum. The calculation of this ΔV assumed that the entire 2.75 ft of lateral movement by the Nissan occurred postcollision and that the ΔV was occurring only because of the collision forces. The impact was assumed to occur instantaneously without any movement of the Nissan or the motorcycle, and all the tire forces were lumped into the postimpact phase. However, part of the lateral movement of the Nissan did occur during the collision.

The Nissan in this test was not equipped with an EDR. If it had been, the EDR would have sensed the accelerations from the collision forces simultaneously with the tire forces present during the collision. In this instance, the tire forces would have acted in the opposite direction to the collision forces, and thus, the ΔV sensed by the EDR would have been lower than that calculated with the assumption of an instantaneous impact. If this lower

FIGURE 9.2 Video frames from WREX 2016 test #3.

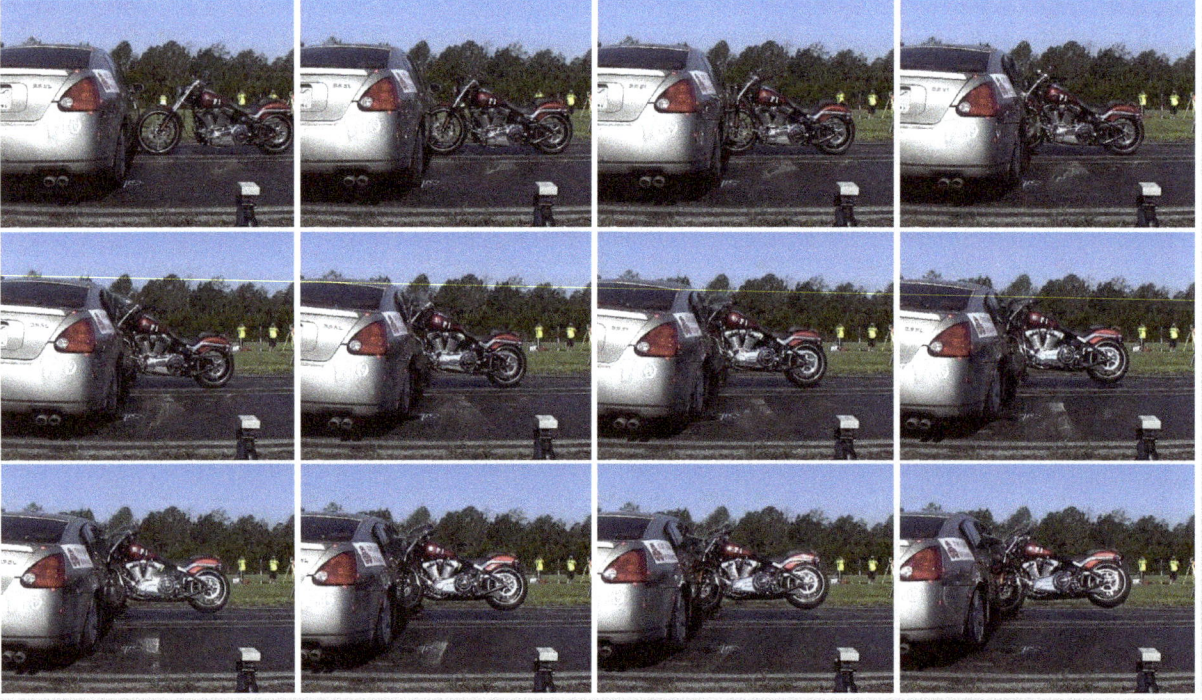

FIGURE 9.3 Impact damage and rest positions for WREX 2016 test #3.

Photo courtesy of Louis Peck

EDR-reported ΔV was going to be used for conservation of momentum analysis, an adjustment would need to be made for the effects of the tire forces, because these would be dealt with separately in the postimpact analysis.

Both vehicles were instrumented with a 3.2 kHz, ±200 g Slam Stick accelerometer, mounted close to the CG. The lateral acceleration sensed by the slam stick on the Nissan would be similar to the accelerations that an ACM would sense and that would have been used to calculate and report the ΔV if this vehicle had been EDR-equipped. The longitudinal accelerations for the H-D are depicted in Figure 9.4 and the lateral accelerations for the Nissan in Figure 9.5. In these graphs, time zero coincides with the approximate time of first contact between the H-D and the Nissan. The accelerations for both vehicles can be integrated to obtain velocities. These velocities are depicted in Figure 9.6.

Based on the accelerations and velocities, the impact duration for the motorcycle was approximately 105 ms. The accelerations for the Nissan are more ambiguous for determining impact duration. Most of the Nissan's ΔV was achieved approximately 85 ms into the collision but the maximum lateral ΔV was achieved around 105 ms. After approximately 105 ms, the motorcycle speed has been reduced to 1.8 mph. Thus, during the collision, the motorcycle experienced a ΔV of approximately 41.2 mph. After approximately 105 ms, the Nissan had been accelerated up to a lateral speed of 6.56 mph, which is the ΔV that an EDR would likely sense for this collision. These velocities yield a coefficient of restitution of 0.11, higher than the value of 0.09 obtained from the conservation of momentum modeling in Chapter 8. In addition to calculating the velocities, the displacement of the Nissan over the 105 ms collision duration can also be calculated by integrating the velocities. This led to the conclusion that the Nissan moved approximately 0.67 ft laterally during the 105 ms of the collision.

Applying the principle of conservation of energy to calculate the postimpact speed of the Nissan with this distance yields a postimpact speed or ΔV for the Nissan of 6.6 mph, approximately 1 mph lower than the ΔV calculated from the collision force alone and consistent with the ΔV calculated with the measured accelerations. This calculation is shown below. Again, the variable d_{post} is the postimpact travel distance of the Nissan. The variable η_{trans} specifies the percentage of the roadway friction that is recruited in decelerating the vehicle. In this instance, the vehicle is simply pushed sideways, and so all the available friction will be recruited. Thus, in this instance, η_{trans} can be set equal to 1. The roadway friction coefficient has been set to 0.7 for this analysis.

FIGURE 9.4 H-D impact accelerations during WREX 2016 test #3.

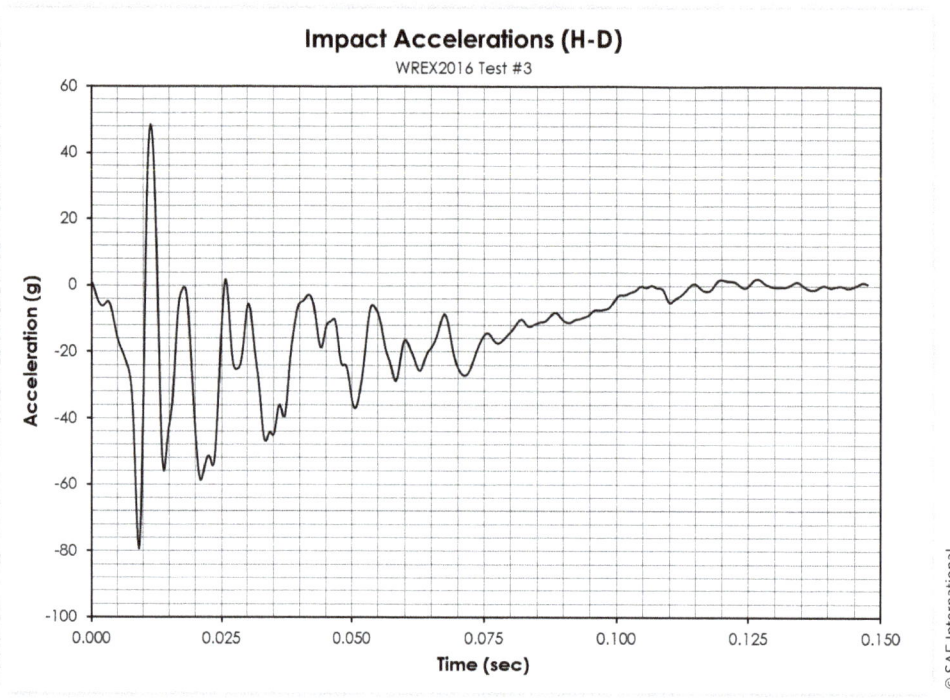

FIGURE 9.5 Nissan impact accelerations during WREX 2016 test #3.

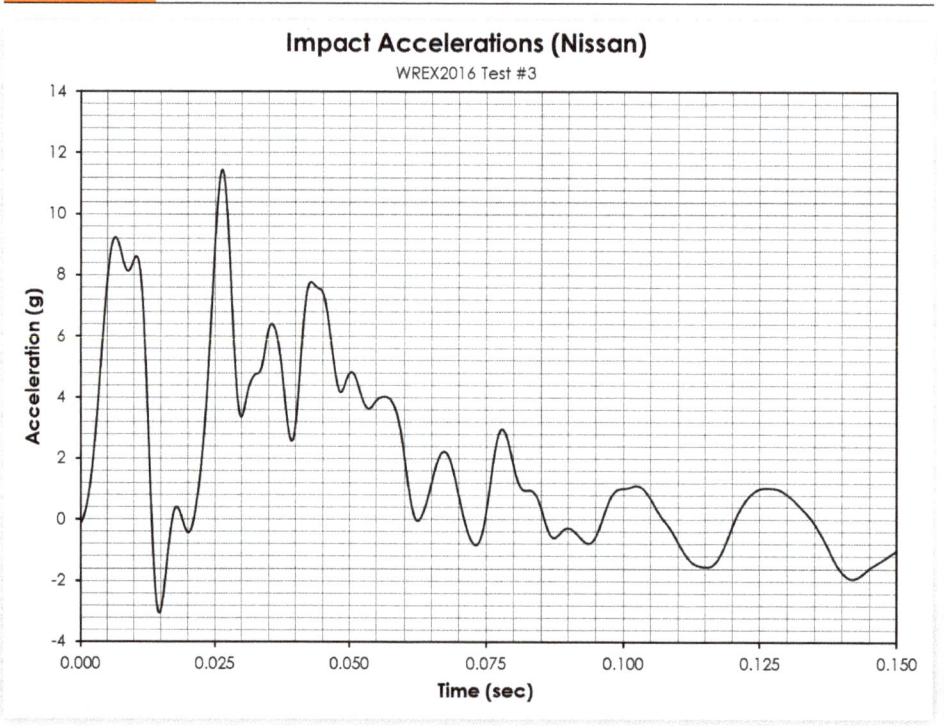

FIGURE 9.6 Velocity time histories for WREX 2016 test #3.

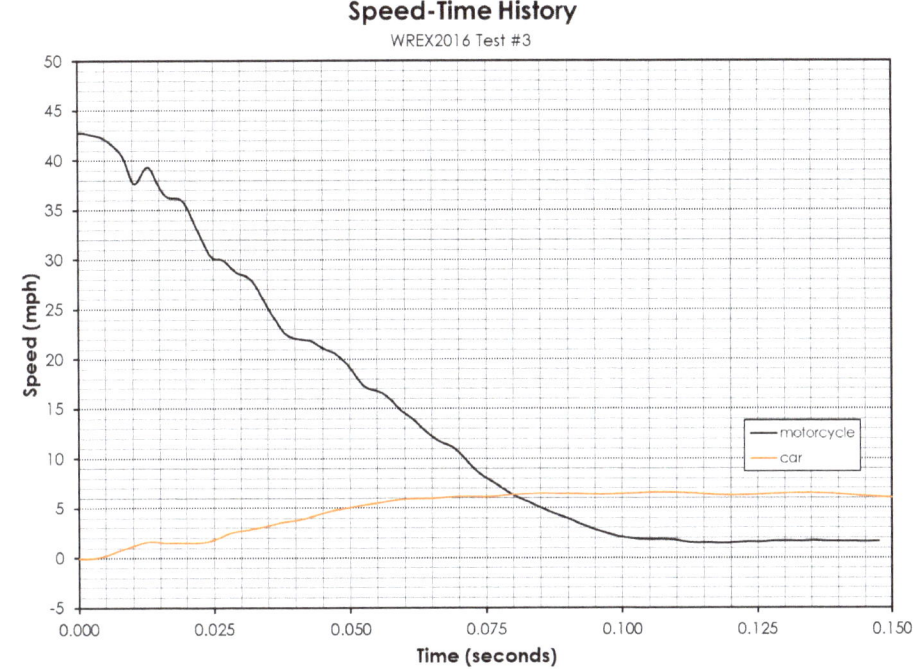

Conservation of Energy:

$$v_{post,\,nissan} = \sqrt{2\eta_{trans}\mu g d_{post}} \quad (9.2)$$

$$v_{post,nissan} = \sqrt{2 \times 1 \times 0.7 \times 32.2 \times 2.08} = 9.68\ fps = 6.6\ mph$$

Equation (9.3) is the conservation of momentum equation developed for this test in Chapter 8. In this equation, $v_{pre,mc}$ is the speed of the motorcycle immediately preceding the collision, W_{nissan} is the weight of the Nissan, W_{mc} is the mass of the motorcycle, $v_{post,nissan}$ is the speed of the Nissan immediately following the collision, and e is the coefficient of restitution.

Conservation of Momentum:

$$v_{pre,mc} = \frac{1}{1+e}\left(1 + \frac{W_{nissan}}{W_{mc}}\right) v_{post,nissan} \quad (9.3)$$

Using the coefficient of restitution of 0.09 that was obtained in Chapter 8, but using the 6.6 mph postimpact speed calculated in this chapter for the Nissan, yields the following:

$$v_{pre,mc} = \frac{1}{1+0.09}\left(1 + \frac{3449}{667}\right) 6.6 = 37.4\ mph$$

This calculated speed is 5.6 mph lower than the actual motorcycle impact speed. Using a coefficient of restitution of 0.11 would yield an impact speed of 36.7 mph, 6.3 mph lower than the actual motorcycle impact speed. Thus, in this instance, if an EDR had reported a lateral ΔV of 6.6 mph, then the ΔV because of those tire forces would need to be added onto the EDR-reported ΔV to obtain an accurate impact speed when using a conservation of momentum impact model. In this case, the discrepancy was approximately 1 mph on the lateral ΔV of the Nissan (with the 0.09 coefficient of restitution).

In this example, the accelerometer data enabled calculation of this 1 mph. For a real-world case this would typically not be possible. As an alternative, if an EDR had reported a lateral velocity change of 6.6 mph for the Nissan in this case, an estimate of the distance traveled during this change in velocity could have been calculated as follows:

$$d_{impact} = 0.636 \times \Delta V \times \Delta t \qquad (9.4)$$

$$d_{impact} = 0.636 \times 6.6 \times 1.4667 \times 0.105 = 0.65 \text{ feet}$$

Equation (9.4) is set up assuming the passenger vehicle is initially stopped and then gets accelerated up to the EDR-reported ΔV over the duration of the collision Δt. The coefficient 0.636 is a multiplier that yields the average lateral speed of the Nissan during the collision, assuming the collision pulse is sine-wave shaped. This calculation yields a distance within 0.02 ft of the distance that was calculated from the accelerometer data. Thus, this distance could be used along with the principle of conservation of energy to add the energy loss from the tire forces back into the EDR-reported ΔV, as follows:

$$\Delta V_{corrected} = \sqrt{\Delta V_{edr}^2 + 2\eta_{trans}\mu g d_{impact}} \qquad (9.5)$$

$$\Delta V_{corrected} = \frac{\sqrt{(6.6 \times 1.4667)^2 + 2 \times 1 \times 0.7 \times 32.2 \times 0.65}}{1.4667} = 7.6 \; mph$$

The EDR-reported ΔV could also be corrected iteratively using simulation. A simulation could first be completed to match the EDR-reported ΔV without any correction for tire forces. Then, the effects of the tire forces could be examined in the simulation for the first 0.65 ft after contact. A correction could then be applied to the ΔV and a new simulation run to match the ΔV corrected for tire forces. It is important to state that the speed loss in the simulation should be assessed over the distance of 0.65 ft, not over the time of 105 ms. Because the speeds in the initial simulation are not yet corrected for the effects of the tire forces, this distance will be traversed over a different time than during the actual collision.

WREX 2016 Test #5

As another example, similar analysis can be carried out for WREX 2016 Test #5. This test involved a 2013 H-D Dyna Street Bob FXDBA motorcycle (633 lb) impacting the driver's

FIGURE 9.7 Impact damage and rest positions for WREX 2016 test #5.

Photo courtesy of Louis Peck

side rear door of stationary 2006 Nissan Maxima (3449 lb) at a 90° angle and a speed of 36.9 mph. The Nissan weight was 5.45 times that of the motorcycle. Figure 9.7 shows the damage to these vehicles, their rest positions, tire marks, and chalk marks identifying the driver's side wheel positions both pre- and postcollision. The Nissan had approximately 61% of its weight on the front axle, and therefore, the CG was approximately 43.4 in. behind the front axle. The motorcycle impacted the Nissan approximately 3.25 ft behind its CG and the collision resulted in approximately 26.7° of counterclockwise rotation of the Nissan. The rear wheels of the Nissan were displaced approximately 4.8 ft from their initial positions. The front wheels were displaced approximately 1 ft from their initial positions. The right front wheel was initially displaced laterally, but as the rotation of the Nissan developed, this wheel became the approximate center of rotation.

In the PC-Crash simulation for this test discussed in Chapter 8, the Nissan experienced a 5.9 mph translational ΔV and a change in yaw angular rotation rate of approximately 80 deg/s during the impact. These velocity changes were assumed to occur instantaneously without any movement of the Nissan or the motorcycle. In reality, a portion of the Nissan's movement did occur during the collision. As with the previous example, the Nissan in this test was not equipped with an EDR. If it had been, the EDR would have sensed the accelerations from the collision forces simultaneously with the tire forces from the lateral movement occurring during the collision. The resultant tire forces in this instance would have acted in the opposite direction to the collision forces, and thus, the ΔV sensed by the EDR would have been lower than that calculated with the assumption of an instantaneous impact.

Consider the motorcycle and car accelerations that were sensed by the instrumentation in this test. The longitudinal and lateral acceleration sensed by the slam stick on the Nissan would be similar to the accelerations that the ACM sensed and that would have been used to calculate and report the ΔV if this vehicle had been EDR-equipped. The longitudinal impact accelerations for the H-D are depicted in Figure 9.8 and the resultant impact accelerations for the Nissan in Figure 9.9. In these graphs, time zero coincides approximately with the time of first contact between the H-D and the Nissan. The accelerations for both vehicles can be integrated to obtain velocities. These velocities are depicted in Figure 9.10.

FIGURE 9.8 H-D impact accelerations during WREX 2016 test #5.

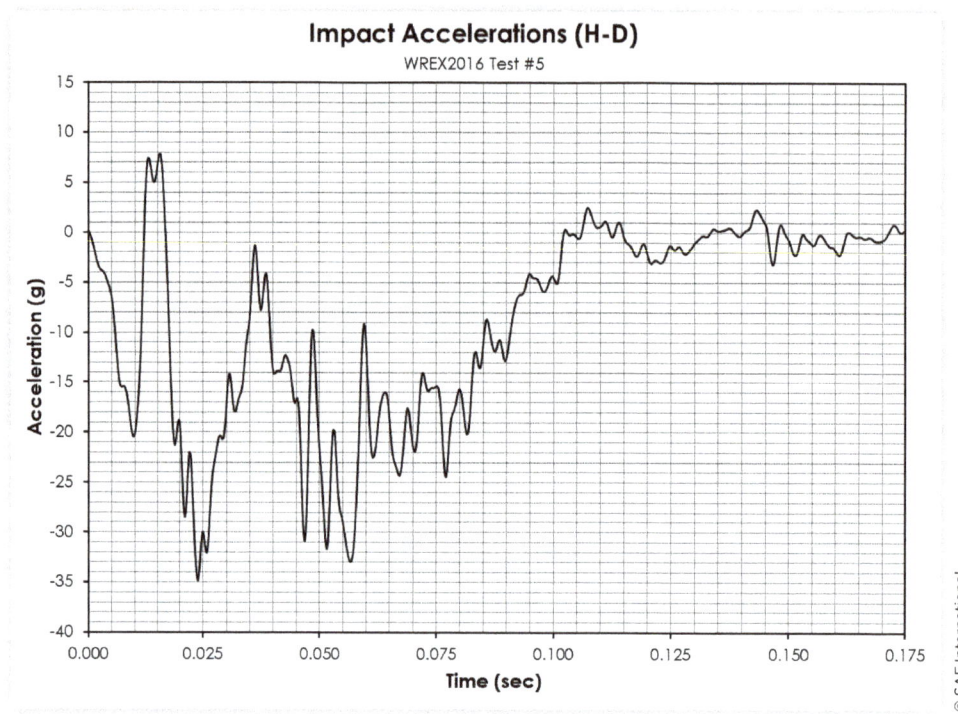

FIGURE 9.9 Nissan impact accelerations during WREX 2016 test #5.

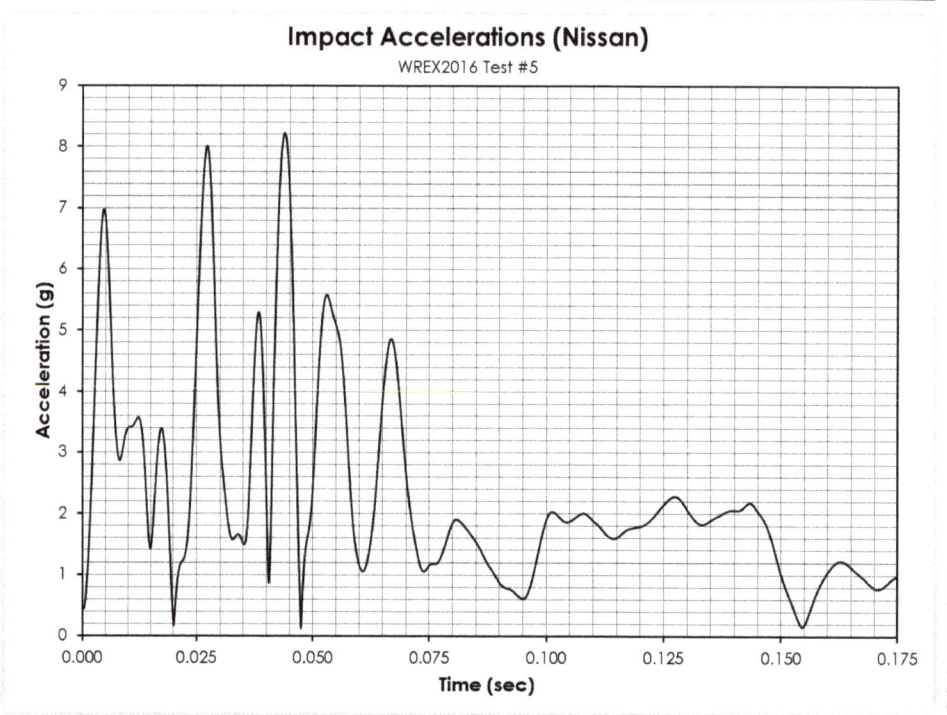

FIGURE 9.10 Velocity time histories for WREX 2016 test #5.

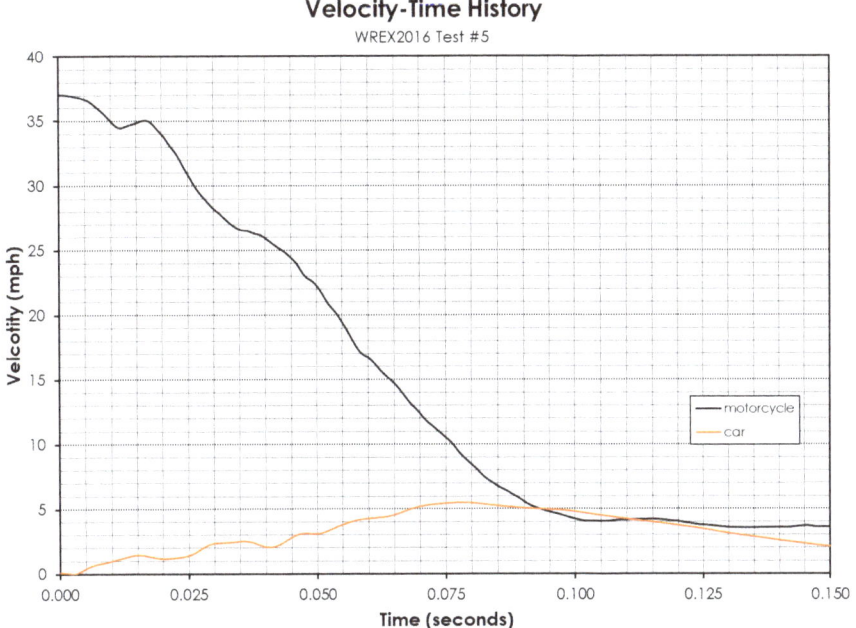

Based on the acceleration and velocity curves for the Nissan, the impact duration was approximately 76 ms. Based on the accelerations from the motorcycle, the impact duration was approximately 105 ms. The discrepancy between these impact durations can, in part, be explained by the action of the Nissan's tire forces acting during the collision. The Nissan does not begin to move laterally until the collision force applied by the motorcycle overcomes the static friction forces applied at the Nissan's tires. Not only that, once the collision force applied to the Nissan again drops below the level of the tire forces, the maximum change in velocity for the Nissan has been achieved, even though the motorcycle continues to decelerate. The tire forces applied to the motorcycle during the collision are negligible, first because the tires are initially rolling along the direction of the velocity of the motorcycle, and second, because the motorcycle tires lift off the ground as a result of the collision. So, the deceleration of the motorcycle from the collision begins before and ends after the acceleration that occurs to the Nissan.

After approximately 105 ms, the motorcycle speed has been reduced to 4 mph. Thus, during the collision, the motorcycle experienced a ΔV of approximately 32.9 mph. The Nissan reached a maximum resultant velocity of 5.5 mph at approximately 76 ms, 0.4 mph lower than the ΔV calculated with PC-Crash. The Nissan reached a resultant velocity of 4.5 mph at approximately 105 ms. Had this vehicle been EDR-equipped, the maximum lateral and longitudinal ΔVs would likely have been reported, from which the resultant 5.5 mph ΔV could have been calculated. Figure 9.11 shows the results of an updated PC-Crash simulation for Test #5, with a match to this lower ΔV. The coefficient of restitution was held constant, and the impact speed of the motorcycle was changed to achieve the match with the 5.5 mph ΔV. This resulted in an impact speed of 34.3 mph, 2.6 mph lower than the actual collision speed. Importantly, though, the Nissan in this simulation experiences too little rotation, which would be an indication to the analyst that the impact speed is lower than actual.

FIGURE 9.11 WREX 2016, test #5, PC-Crash simulation with 5.5 mph ΔV.

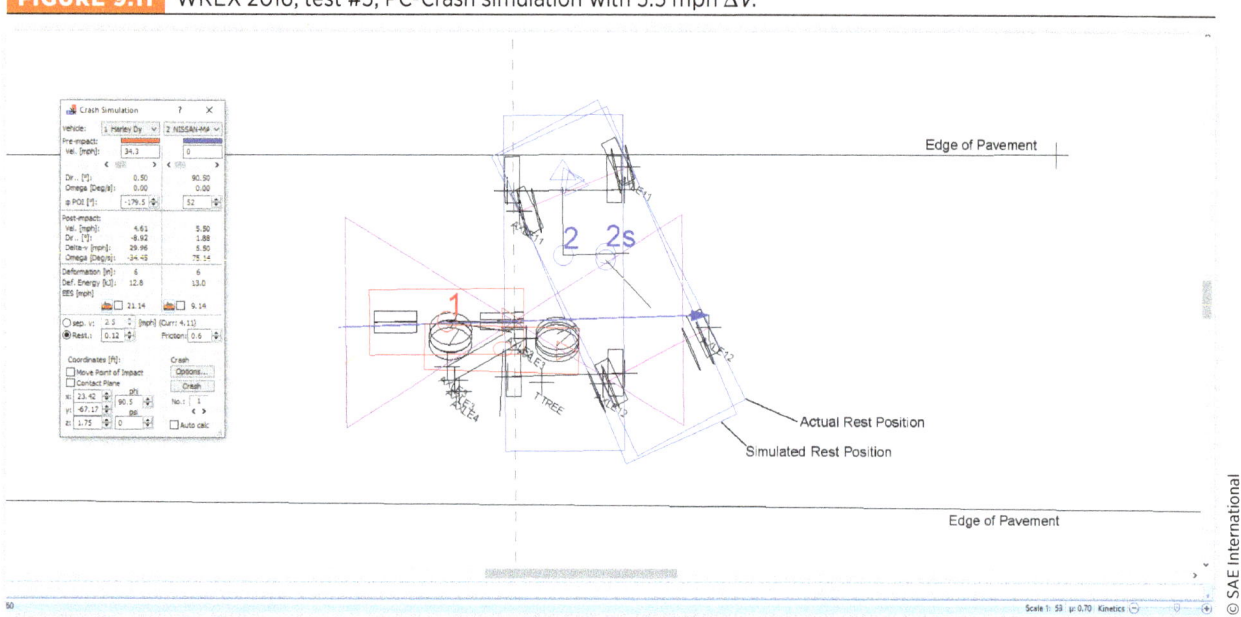

As with the previous example, an approximate travel distance for the Nissan CG can be calculated for the collision:

$$d_{impact} = 0.636 \times 5.5 \times 1.4667 \times 0.076 = 0.39 \text{ ft}$$

In the PC-Crash simulation, approximately 0.5 mph of speed loss occurred in the first 0.4 ft of movement following the instantaneous collision. If this were used to adjust the ΔV from the instrumentation, the corrected value would be 6.0 mph, approximately 0.1 mph higher than that calculated in simulation discussed in Chapter 8. Figure 9.12 shows the results of an updated PC-Crash simulation for Test #5, with a match to this corrected ΔV. The coefficient of restitution was again held constant at 0.12, and the impact speed of the motorcycle was changed to achieve the match with the 6.0 mph ΔV. This resulted in an impact speed of 37.3 mph, 0.4 mph higher than the actual collision speed. Interestingly, though, the Nissan in this simulation experiences too much rotation, which would be an indication to the analyst that the impact speed is higher than actual. Thus, the original simulation of this test presented in Chapter 8 still likely represents the best balance between accounting for the Nissan tire forces and the need to match the rest position of the Nissan. This makes sense because the tire forces were always considered in the simulation; it is just that within a PC-Crash simulation, they are lumped into the postimpact phase.

FIGURE 9.12 WREX 2016, test #5, PC-Crash simulation with 6.0 mph ΔV.

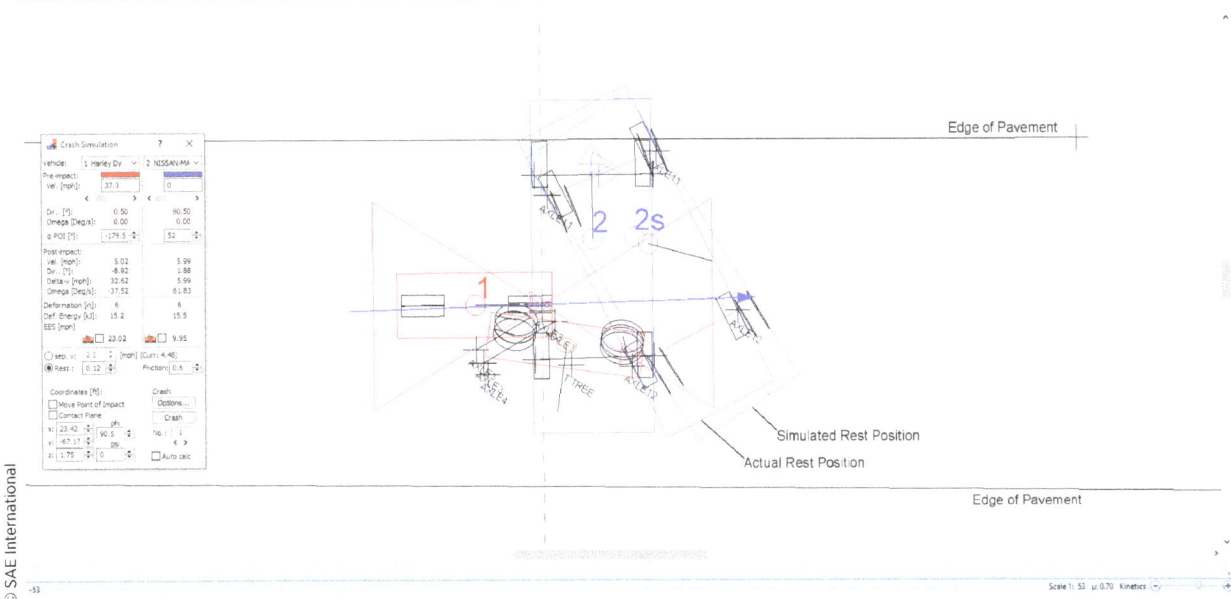

Case Study #1: Utilizing EDR Data from the Passenger Vehicle

This section describes the reconstruction of an intersection collision involving a Honda CBR 1000RR motorcycle and a BMW X3 xDrive35i sport utility vehicle (SUV). EDR data from the BMW were imaged and is incorporated into the reconstruction. This collision occurred on a rural two-lane asphalt roadway. The motorcyclist was riding southbound, and the BMW was traveling northbound and attempting to turn left onto a dirt road that intersected the north-south roadway. The BMW traveled across the path of the motorcycle and the motorcycle struck the front right corner of the BMW, engaging the bumper beam and the right front wheel. There was a tire mark on the west shoulder from the BMW being diverted by the collision. The driver of the BMW stated that he saw the motorcyclist approaching, but thought he had time to make the turn. The speed limit for north-south traffic was 60 mph.

The photographs of Figure 9.13 show the damage to the BMW and the Honda motorcycle along with their rest positions. The bumper beam and the right front wheel of the BMW were directly contacted by the motorcycle. The photograph of Figure 9.14 shows the deformation to the bumper beam from direct contact with the motorcycle. The right front wheel broke off the BMW. The motorcycle and rider also damaged the right front fender and the hood. As the collision progressed, the rider traveled across the hood of the vehicle and impacted the windshield. The rider's body came to rest well beyond the rest position of the BMW.

FIGURE 9.13 BMW and Honda motorcycle at rest following the crash.

FIGURE 9.14 Deformation to the bumper beam of the BMW from direct contact with the motorcycle.

Figure 9.15 depicts the intersection where the collision occurred. The BMW was initially traveling toward the vantage point of the camera, in the lane at the far left of the photograph. The motorcycle was traveling away from the vantage point of the camera, in the lane nearest the center of the photograph. The BMW driver was attempting to turn left into the dirt driveway depicted at the right side of this photograph.

FIGURE 9.15 Intersection where the crash occurred.

Manufacturer specifications were obtained for the BMW and the Honda. In addition, an exemplar BMW X3 xDrive35i was inspected, weighed, and tested. Based on the exemplar inspection and manufacturer specifications, the subject BMW weighed approximately 4370 lb, including the weight of the driver. This vehicle had a length of 185.5 in. and the CG was approximately 90 in. from the front of the vehicle. The ACM was under the center console and approximately 13.1 in. forward of the CG (Figure 9.16).

The Honda motorcycle had a wet weight of approximately 441 lb. The weight of the rider was approximately 170 lb, yielding a total combined weight of 611 lb. Thus, including the weight of the rider, the vehicles in this collision had a weight ratio of 7.15. Not including the weight of the rider, the weight ratio was 9.91. The Honda had a standard motorcycle braking system and was not equipped with antilock brakes (ABS).

The Bosch CDR system was used to image data from the ACM on the BMW (referred to as the Advanced Crash Safety Module (ACSM) by BMW). Precrash and crash data from one deployment event were recovered. There was 5 sec of precrash data, reported at ½-sec intervals (11 total readings). These data are included in tabular format in Figure 9.17. At a time of −5.0 sec, the vehicle speed was reported as 56 km/h (34.8 mph), the brakes were applied, and the vehicle was decelerating. Over the course of 5 sec leading up to time zero, the vehicle decelerated at an average rate of approximately −0.14 g and the reported speed at time zero was approximately 19 mph (31 km/h). At time zero, the data indicated that the driver had released the brake and applied the accelerator pedal to 100%.

FIGURE 9.16 Exemplar BMW scan with center of gravity (CG) and airbag control module (ACM) locations identified.

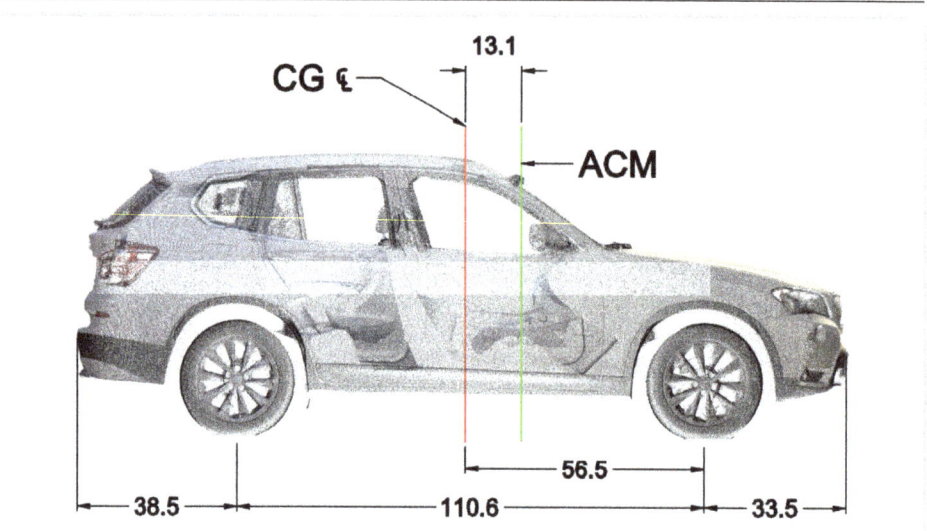

FIGURE 9.17 Precrash data from the BMW's advanced crash safety module (ACSM).

Pre-Crash Data -5 to 0 sec (Record 1, Most Recent)

Time (sec)	Speed, Vehicle Indicated (MPH [km/h])	Accelerator Pedal, % Full (%)	Engine RPM	Steering Input (deg)	Service Brake, On/Off	ABS Activity (Engaged, Non-engaged)	Stability Control (On Engaged, Non-engaged)
-5.0	35 [56]	0	1200	0	On	No ABS Activity	Non-engaged
-4.5	34 [55]	0	1200	0	On	No ABS Activity	Non-engaged
-4.0	33 [53]	0	1300	0	On	No ABS Activity	Non-engaged
-3.5	31 [50]	0	1300	0	On	No ABS Activity	Non-engaged
-3.0	29 [47]	0	1300	0	On	No ABS Activity	Non-engaged
-2.5	27 [43]	0	1400	0	On	No ABS Activity	Non-engaged
-2.0	25 [40]	0	1300	0	On	No ABS Activity	Non-engaged
-1.5	24 [38]	0	1300	-20	On	No ABS Activity	Non-engaged
-1.0	22 [36]	0	1500	-45	On	No ABS Activity	Non-engaged
-0.5	20 [32]	0	1400	-80	On	No ABS Activity	Non-engaged
0.0	19 [31]	100	1500	-270	Off	No ABS Activity	Non-engaged

The driver's steering input was reported as 0° throughout the time from −5.0 to −2.0 sec. From that point forward, the driver began steering to the left, with the steering input increasing until it reached a maximum reported value of −270°. According to the CDR report, leftward steering inputs were indicated with a negative sign and rightward steering inputs were indicated with a positive sign. The data limitations indicated that the steering angle was reported in 5° increments and was limited to a range of −250° to 250°. This is contradicted by the −270° reading at time zero. Nonetheless, it is likely that the steering input reporting does have a maximum value and the signal may have been clipped. Thus, the actual leftward steering input may have exceeded 270°. The BMW had a steering ratio of 17.8:1 and the steering wheel could be turned approximately 1½ rotations in each direction (540°).

PC-Crash was used to simulate the motion of the BMW in the moments leading up to subject collision. Within this first simulation, the roadway coefficient of friction was set at 0.76 and the TM-Easy tire model was used with default values. The integration time step was set at 1 ms. The initial speed of the BMW was set at 34.8 mph (56 km/h), consistent with the EDR-reported precrash speeds. The steering inputs from the EDR data were used in setting up the sequences within PC-Crash, with the exception that the steering input at time zero was not limited to −270°. This ending steering input was allowed to exceed a magnitude of 270° as dictated by the geometry of the turn and the location of physical evidence from the collision. The steering input was optimized so that the BMW arrived at its impact position as dictated by this physical evidence.

The time zero speed reading was the last sampled speed prior to the collision. This speed could have been sampled anytime between 0 and 0.4999 sec prior to the collision, and so, time zero in the precrash data could be the time of the collision or it could be as much as a ½ sec prior to the collision. The EDR data also indicate that the driver was inputting 100% throttle at time zero. To evaluate the effect of the uncertainty around the exact timing of 31 km/h (19.3 mph) precrash speed reading, an exemplar BMW X3 was tested. The maximum acceleration in this speed range was around 0.5 g. At this acceleration, the speed of the BMW could increase by approximately 5.5 mph (8.9 km/h) over ½ sec. Thus, the speed of the BMW at the time of the collision could be anywhere between 19.3 and 24.8 mph (31 and 40 km/h).

Sequences in PC-Crash were used to apply deceleration or acceleration to the vehicle to match the EDR-reported speeds throughout 5 sec of reported data. It took the BMW approximately 5¼ to 5½ sec in the simulation to reach the area of the collision. This implies that time zero in the precrash data was not coincident with the time of the collision. In the simulation, the BMW arrived at the point of impact at a speed near 22 mph. Figure 9.18 is a screen capture from PC-Crash showing the precrash motion of the BMW. The depicted positions are at ½-sec intervals. Figure 9.19 is a graph that compares the EDR-reported precrash speeds to the BMW speeds in the simulation. In this graph, time is plotted on the horizontal axis and speed on the vertical axis. The EDR-reported speeds are depicted with blue circles and the speeds from the PC-Crash simulation are plotted with a black line.

FIGURE 9.18 Screen capture from PC-Crash showing precollision motion of BMW (depicted positions are at ½-sec intervals).

FIGURE 9.19 PC-Crash precollision speeds plotted with the event data recorder (EDR)-reported precrash speeds.

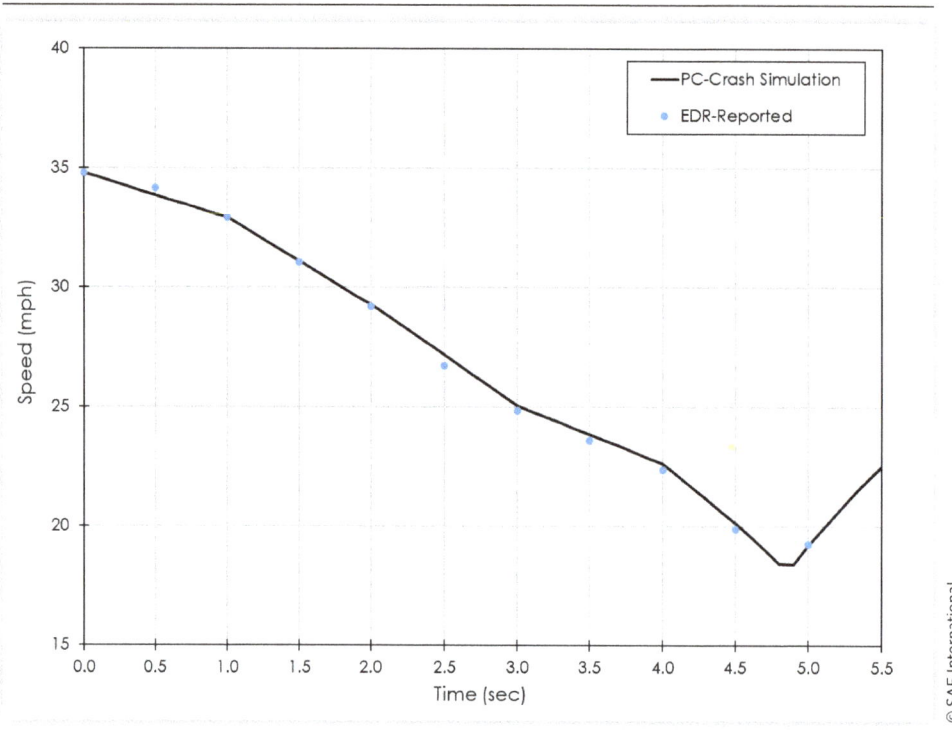

Once the left turn was simulated, a separate PC-Crash file was set up to simulate the collision. The speed of the BMW at impact was initially set at 22 mph and the vehicles were aligned consistent with the damage. A brake factor of 100% was used for the right front wheel of the BMW, 10% for the right front, and 5% for the rear wheels. The goal of this simulation was to determine the impact speed of the motorcycle that would produce a match with the tire mark deposited by the BMW, the rest position of the BMW, and the ΔV of the BMW.

The CDR report indicated that the BMW experienced a maximum longitudinal ΔV of 9.9 mph (16.0 km/h), oriented from front to rear, and a maximum lateral ΔV of 6.8 mph (11.0 km/h), oriented from right to left. The maximum recorded longitudinal ΔV was achieved at 80 ms, the maximum recorded lateral ΔV was achieved at 48 ms, and the maximum resultant ΔV was achieved at 82 ms. The longitudinal cumulative ΔV graph is included in Figure 9.20. The lateral cumulative ΔV graph is included in Figure 9.21. The longitudinal and lateral acceleration graphs from the CDR report are included in Figure 9.22 and Figure 9.23. The longitudinal acceleration peaks at −11 g, at a time between 30 and 40 ms. That lateral acceleration also peaks at −11 g, at a time between 40 and 50 ms. The reported ΔV components were analyzed through time and we determined that the maximum resultant ΔV was approximately 12.1 mph (19.5 km/h) with an approximate PDOF of 34.5°, measured clockwise relative to the front of the vehicle.

The cumulative ΔV and acceleration curves all report readings of 0 for the first 10 ms. This seems to be an indication that the ACSM is not missing the early portions

FIGURE 9.20 Longitudinal ΔV data from advanced crash safety module (ACSM).

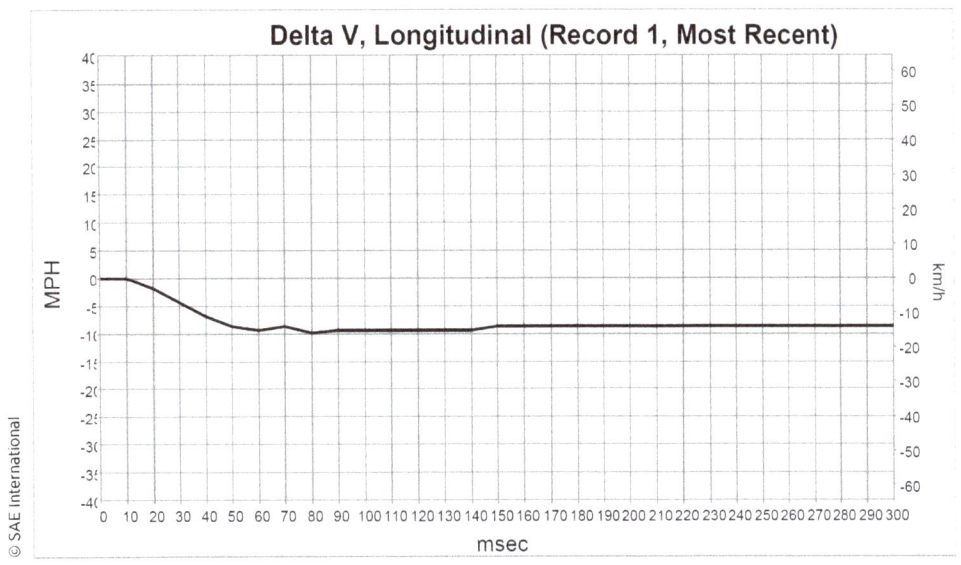

FIGURE 9.21 Lateral ΔV data from advanced crash safety module (ACSM).

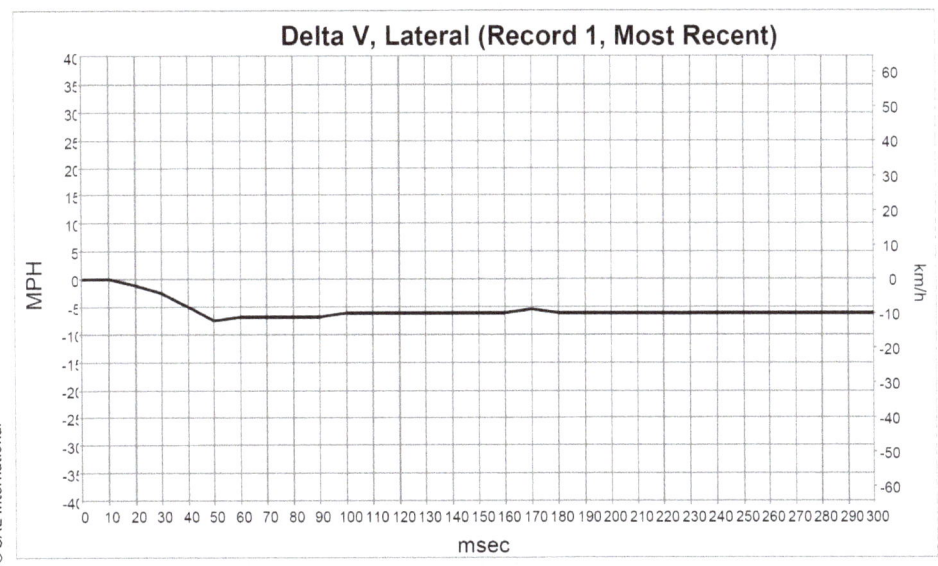

of the collision. This interpretation is consistent with the fact that the CDR report indicates a "time from time zero to algorithm wake-up start" for the frontal event of 0 ms and for the side event of 4 ms. The ACSM's ability to report this is evidence that the system is storing some of the accelerations just prior to the collision. The recording window in this instance was 300 ms, more than adequate to capture the peak accelerations and ΔVs.

FIGURE 9.22 Longitudinal accelerations from the advanced crash safety module (ACSM).

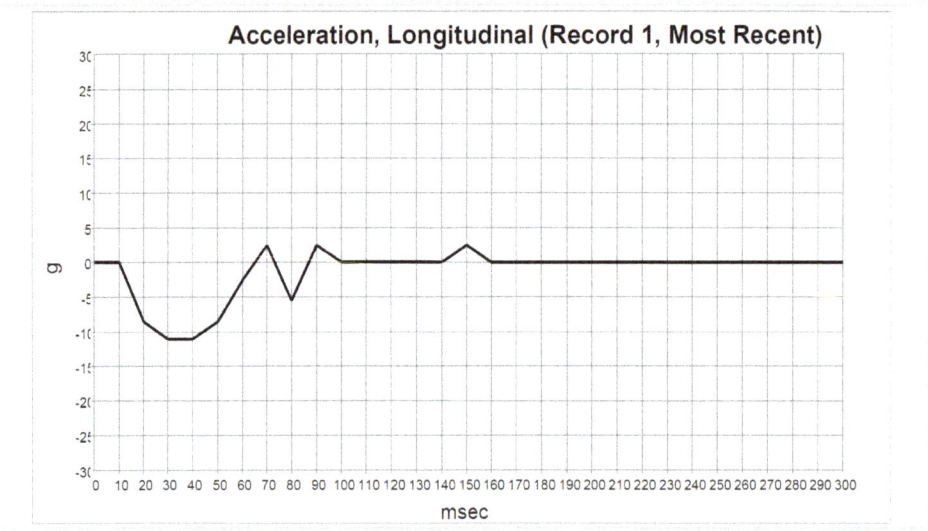

FIGURE 9.23 Lateral accelerations from the advanced crash safety module (ACSM).

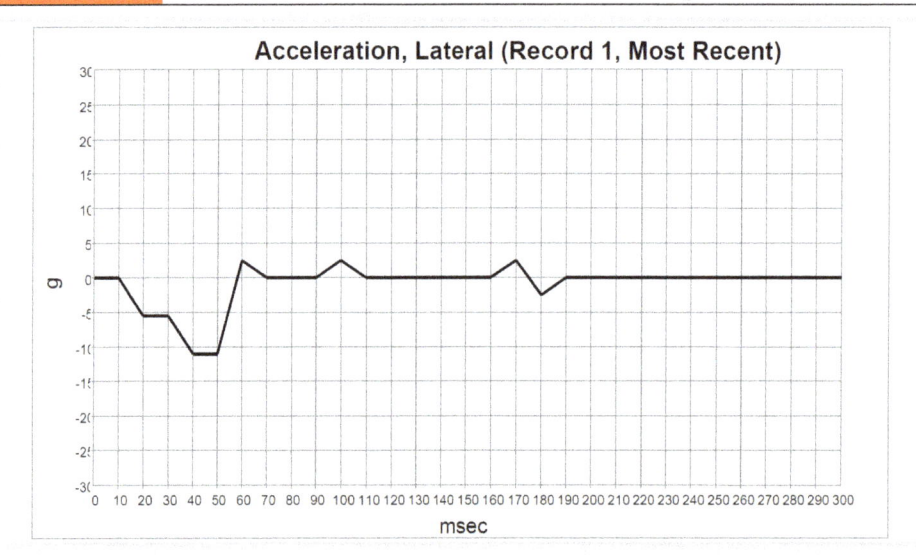

Based on the acceleration curves, the impact duration was approximately 80 ms. With a 12-mph ΔV and an 80 ms impact duration, the average acceleration would be approximately 6.8 g. Peak accelerations of 11 g make good sense with this average acceleration. Beyond that, the reported peak accelerations are well below those that would typically result in clipping (40 to 50 g). The CDR report indicated that recording the event was completed. We were able to communicate with the ACSM through the DLC and no deformation to the center console was evident during our inspection of the subject vehicle. The recorded accelerations appear to have captured the entire crash event, and the EDR reported the complete file was recorded.

CHAPTER 9 Event Data Recorders and Speedometers in Motorcycle Accidents

Now, consider error source #5. The ΔVs from the CDR report were all measured at the location of the ACM, which was under the center console on this vehicle, approximately 13.1 in. forward of the CG. During the subject crash, the BMW did experience approximately 90° of yaw rotation. The ΔV experienced at the CG of the BMW can be related to the ΔV experienced at the ACM with the following equation:

$$\Delta \vec{V}_{ACM} = \Delta \vec{V}_{CG} + \Delta \omega_{yaw} \times \vec{r}_{ACM/CG} \tag{9.6}$$

In this equation, $\Delta \vec{V}_{ACM}$ is the vector ΔV at the ACM, $\Delta \vec{V}_{CG}$ is the vector ΔV at the CG, $\Delta \omega_{yaw}$ is the change in rotational velocity about the vertical axis of the vehicle, and $\vec{r}_{ACM/CG}$ is the vector connecting the ACM position to the CG. This equation neglects changes in pitch and roll velocity experienced by the vehicle and discrepancies between the vertical position of the CG and the ACM. Evaluating the cross product in Equation (9.6) and breaking it down into longitudinal and lateral components yield the following equations:

$$\Delta V_{CG,long} = \Delta V_{ACM,long} + r_y \Delta \omega_{yaw} \tag{9.7}$$

$$\Delta V_{CG,lat} = \Delta V_{ACM,lat} - r_x \Delta \omega_{yaw} \tag{9.8}$$

In these equations, r_x and r_y are the coordinates of the ACM within the vehicle coordinate system with its origin at the CG. Within this coordinate system, a positive value of r_x implies that the ACM is located forward of the CG. A positive value of r_y implies that the ACM is situated to the right of the CG. Often, both the CG and the ACM will be on the centerline of the vehicle, and r_y will be equal to 0. If this is the case, then Equations (9.7) and (9.8) can be simplified to the following:

$$\Delta V_{CG,long} = \Delta V_{ACM,long} \tag{9.9}$$

$$\Delta V_{CG,lat} = \Delta V_{ACM,lat} - r_x \Delta \omega_{yaw} \tag{9.10}$$

A positive value for $\Delta \omega_{yaw}$ implies a clockwise rotation and a negative value implies a counterclockwise rotation. In this specific case, r_x = 13.1 in. or 1.09 ft. The change in angular velocity that the BMW experienced had a counterclockwise direction (a negative sign). This means that, in this instance, the lateral ΔV experienced at the CG of the BMW will be of lower magnitude than what the ACM experienced and reported. Based on initial simulation efforts, it was clear that the change in yaw velocity of the BMW would have to be around 100 deg/s to match the rest position. Using this Δω to correct the EDR-reported ΔVs results in the lateral ΔV at the CG being −5.5 mph, instead of −6.8 mph reported by the EDR. The resultant ΔV at the CG would be 11.3 mph, instead of 12.1-mph ΔV at the ACM. These values can be substituted into Equation (9.1) to illustrate the error that would result if this correction was not made:

©2022, SAE International

$$\Delta V_2 = \frac{W_{bmw}}{W_{honda}} \Delta V_{bmw} = 7.15 \times 12.1 \; mph = 86.5 \; mph$$

$$\Delta V_2 = \frac{W_{bmw}}{W_{honda}} \Delta V_{bmw} = 7.15 \times 11.3 \; mph = 80.8 \; mph$$

Thus, not accounting for the spin of the vehicle and the discrepancy in position between the ACM in the CG could result in a 5.7 mph error in the ΔV of the motorcycle, and thus, in its impact speed. In these calculations, the full weight of the rider has been included. This is a conservative assumption, resulting in the lowest speed for the motorcycle because a lower weight for the Honda would result in a higher weight ratio between the two vehicles.

The calculations of the previous paragraphs corrected for the spin of the vehicle using the as-reported ΔV components. However, the uncertainty because of the potential error in the EDR-reported ΔV components has not yet been considered. To carry out this uncertainty analysis, Monte Carlo simulation was used. The software Crystal Ball by Oracle was utilized to implement this Monte Carlo analysis. The longitudinal ΔV was varied between −9.9 and −11.4 mph using a uniform distribution. The lateral ΔV was varied between −5.3 to −8.3 mph using a uniform distribution. When combined with analysis using Equations (9.3) and (9.4), this led to a range on the corrected lateral ΔV between −4.0 and −7.0 mph. This in turn led to a range on the corrected resultant ΔV of 12.0 ± 0.6 mph and the corrected PDOF of 27.2 ± 3.8°. This means that, in this specific instance, the effects of the potential error in the EDR-reported ΔV offset the effects of the spin and the discrepancy in position between the ACM and the CG. This will not always be the case and these factors will need to be considered on a case-by-case basis. Table 9.1 summarizes the as-reported and corrected ΔVs for this case.

TABLE 9.1 Summary of uncorrected and corrected ΔVs

	As-reported	Corrected for airbag control module (ACM) location	Low end in uncertainty analysis	High end in uncertainty analysis
Longitudinal ΔV (mph)	−9.9	−9.9	−9.9	−11.4
Lateral ΔV (mph)	−6.8	−5.5	−4.0	−7.0
Resultant ΔV (mph)	12.1	11.3	11.4	12.6

The simulation was first optimized to match the corrected resultant ΔV obtained from the as-reported ΔV components, without consideration of the uncertainty in these components. This will result in a low-end estimate of the motorcycle impact speed. This optimization was carried out using the impact speed of the motorcycle, along with the coefficient of restitution and the impact center height. Small changes in the impact speed of the BMW were also utilized for the final step in the optimization, and it was found that an impact speed of 21 mph for the BMW resulted in an improved match with the evidence. Ultimately, an impact speed for the motorcycle of 76 mph produced a match with the corrected EDR-reported ΔV (11.3 mph), the tire mark, and the rest position of the BMW. The screen capture in Figure 9.24 shows the impact configuration, the rest positions, and the "Crash Simulation" dialog box within PC-Crash that reports the impact speeds and the calculated ΔVs. The green box in the screen capture represents the actual rest position of

FIGURE 9.24 PC-Crash screen capture of low-end speed optimized impact simulation.

the BMW. This screen capture shows that a reasonable match with the BMW's actual rest position as achieved. In this simulation, the BMW had a PDOF of 31.8°.

The high-end motorcycle speed estimate can be obtained by using a BMW resultant ΔV of 12.6 mph. A second simulation was optimized to match this value. In this instance, it was found that an impact speed of 23 mph for the BMW and an impact speed for the motorcycle of 84 mph resulted a quality match with the BMW rest position and 12.6 mph ΔV. The screen capture in Figure 9.25 shows the impact configuration, the rest positions, and the "Crash Simulation" dialog box within PC-Crash that reports the impact speeds and the calculated ΔVs. In this simulation, the BMW again had a PDOF of 31.8°. The rest position of the simulated BMW agreed very closely with the actual rest position of the BMW (again represented with the green box).

Thus, these two simulations produced a speed range for the motorcycle of 76 to 84 mph. Because the full weight of the rider was included in the simulation, this speed range is conservative. The velocity direction of the motorcycle in this crash was redirected by approximately 90°, whereas the rider continued essentially along his initial direction of travel. The rider did interact with the BMW, but not to the extent that he was redirected along with the motorcycle. In an instance like this, it would be reasonable to use ¼ to ½ of the rider's weight in the simulation. Doing so would increase the calculated speed of the motorcycle significantly. We carried out these calculations and found that using half of the rider's weight increased the range on the impact speed of the motorcycle to between 88 and 96 mph, an increase of approximately 12 mph in each end of the range. The analysis discussed here is also conservative in that no adjustment has been made for tire forces that would have been acting during the collision.

FIGURE 9.25 PC-Crash screen capture of high-end speed optimized impact simulation.

Case Study #2: Utilizing EDR Data from the Passenger Vehicle

This section describes the reconstruction of another intersection collision involving a motorcycle. EDR data from the SUV that turned across the path of and collided with the motorcycle are incorporated into the reconstruction. This collision involved a motorcyclist riding eastbound. The driver of a westbound SUV attempted to turn left to go southbound and the motorcycle struck the passenger's side rear of the SUV. The driver of the SUV stated that she did not see the motorcycle prior to initiating her turn on a yellow traffic signal. A witness who observed the motorcycle prior to the collision stated that the motorcyclist accelerated into the intersection to make it through on the yellow light. The speed limit for east-west traffic through the intersection was 60 mph. An investigating officer noted that sun glare could have been a factor contributing to the SUV driver not seeing the motorcyclist.

The photograph in Figure 9.26 depicts the motorcycle and the SUV at rest in the intersection following the collision. The photograph in Figure 9.27 also depicts the SUV at rest and shows the damage to the vehicle from the collision. The motorcycle struck and damaged the passenger's side rear of the SUV, directly contacting the rear wheel behind the axle. This caused the wheel to deform and rotate clockwise relative to the vehicle and the tire to go flat. The collision caused significant clockwise yaw rotation of the SUV, on the order of 180°, and the SUV deposited prominent tire marks between impact and rest. Figure 9.27 also shows that the front tires of the Ford are steered to the left following the collision.

FIGURE 9.26 Motorcycle and sport utility vehicle (SUV) at rest in the intersection.

FIGURE 9.27 Event data recorder (EDR)-equipped sport utility vehicle (SUV) involved in the subject collision.

The subject intersection was inspected, and its geometry mapped with a Faro Focus3D X 330 scanner. The SUV was also inspected, and its postcollision geometry documented with the same scanner. The clockwise rotation of the passenger's side rear wheel because of deformation was measured from this scan data at 17°. In the photographs of the SUV at rest in the intersection, the leftward steer angle of the right front wheel appears to be of greater magnitude than 17° the right rear wheel was angled.

The Bosch CDR system was used to image data from both the RCM and the powertrain control module (PCM) on the SUV. Data from one event were recovered from the RCM—a side deployment event. The data indicated that the driver pretensioner and both the driver and passenger curtain airbags deployed 28 ms after the system began sensing the collision. The data also indicated that the driver, who was the sole occupant of the vehicle, was buckled at the time of the collision. The RCM data included precrash data at five indicated time intervals (−4 s, −3 s, −2 s, −1 s, and 0 s). These data are included in Figure 9.28.

FIGURE 9.28 Precrash data from the sport utility vehicle (SUV)'s restraint control module (RCM).

Pre-Crash Data (First Record)

Time (sec)	-4	-3	-2	-1	0
Accelerator Pedal Position (%)	0	0	22	40	99
Vehicle Speed (MPH [km/h])	6.0 [9.7]	3.2 [5.1]	4.1 [6.6]	10.8 [17.4]	17.3 [27.9]
ABS Event in Progress	No	No	No	No	No
ESP Event in Progress	No	No	No	No	No
TCS Event in Progress	No	No	No	No	No
Brake Lamp Switch Depressed (from PCM)	Yes	Yes	No	No	No
RCM Serial Number Received by OCS	No	No	No	No	No
OCS Sensor Status	Empty	Empty	Empty	Empty	Empty
OCS System Level 1 Fault	No	No	No	No	No
OCS System Level 2 Fault	No	No	No	No	No
Vehicle Calibration ID	56	56	56	56	56
Vehicle Model Year Calibration ID	07	07	07	07	07

© SAE International

The RCM data reported a lateral change in velocity of 2.49 mph. However, the reporting window for the lateral change in velocity was only 50 ms, which is too short to capture the entire collision between the vehicles, let alone the collision between the rider and the SUV. The graphical data for this lateral change in velocity are depicted in Figure 9.29. As this figure shows, the horizontal axis of the graph starts at −49 ms and goes to 0 ms. The first nonzero acceleration reading is at −28 ms. Given that the system reports that the curtain airbags deployed 28 ms after the collision began being sensed, it is likely that, in this case, the EDR only reported the time frame from the start of the collision until the deployment of these airbags. Given this, the ΔV reported by the EDR will significantly under-report the actual ΔV, and this is an instance where the EDR-reported ΔV cannot be used in the reconstruction.

Data were also recovered from the PCM on the SUV. The graphical data, which are depicted in Figure 9.30, included the vehicle indicated speed, the accelerator pedal percentage, brake switch status, ABS status, and engine rpm for 20 sec prior to and 5 sec after time zero. Tabular data were also reported for each of the variables included in the graph. In addition, the tabular data added transmission status (reverse or not reverse), speed control status (on or off), traction control status (active or inactive), and stability control status (active or inactive). According to the data limitations in the CDR report, time zero corresponds to the time at which the PCM received a restraint deployment signal

CHAPTER 9 Event Data Recorders and Speedometers in Motorcycle Accidents **285**

FIGURE 9.29 Lateral velocity change data from the sport utility vehicle (SUV)'s restraint control module (RCM).

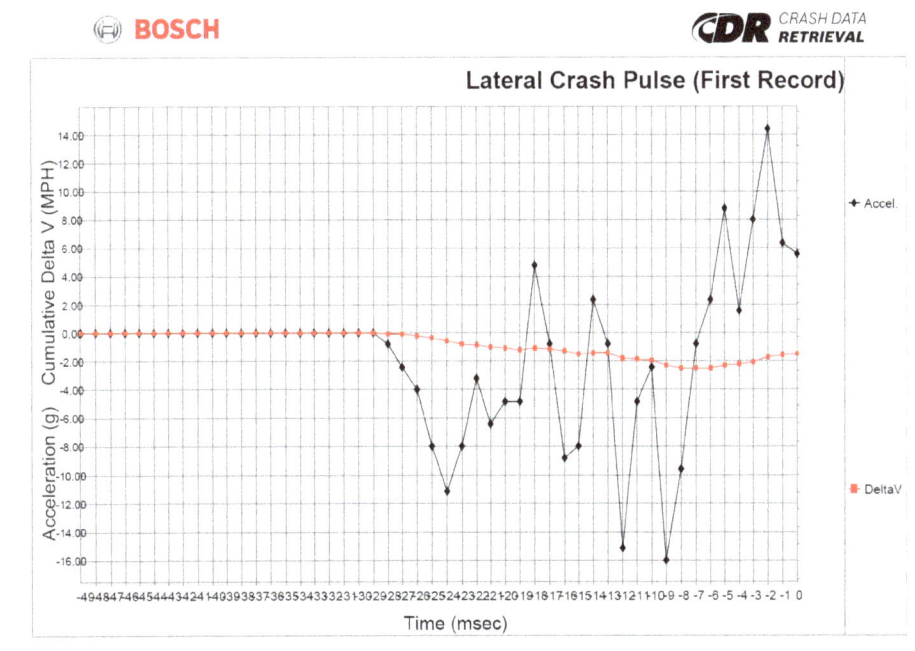

FIGURE 9.30 Powertrain control module (PCM) data from the sport utility vehicle (SUV).

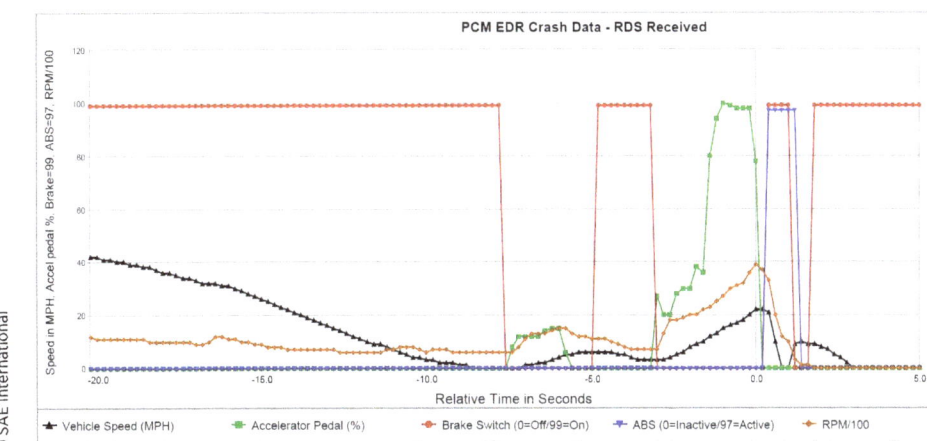

from the RCM. However, "the Restraint Deployment Signal (RDS) may not be recorded on the PCM immediately at impact. Time lags within the system may result in the Restraint Deployment Signal being recorded a few data samples after impact has occurred." The SUV was equipped with electronic stability control and the tabular data from the PCM indicate that it was activated from +0.2 sec through +1.4 s. After that, it was inactive. The AdvanceTrac system utilizes the ABS system, and the data also report the ABS system was active during the same time frame.

©2022, SAE International

Arndt et al. [38] reported a study of the accuracy of the speeds reported by the PCM on a 2005 Ford Explorer during high slip angle maneuvers and during acceleration and braking. This test vehicle was equipped with electronic stability control system (ESC). Tests were conducted with and without the ESC active. Arndt et al. noted that "in testing with ESC enabled, speed error was associated with ESC intervention ... A Ford PCM download which provides recorded speed every 0.2 seconds appears to provide enough data that an accurate speed trend can be discriminated." For the subject collision, ESC intervention *did* occur, but not until after the collision. Based on Arndt et al.'s results, there would likely be intervention-related error in the reported speeds during this time frame. In addition to that error source, the significant yaw rate of the SUV following the collision would be expected to result in errors in the reported speeds because of the discrepancy between the heading and velocity directions of the vehicle.

On the other hand, the speeds reported by the PCM prior to the collision are likely to be accurate. These speeds were recorded when the vehicle would have a low slip angle and when there was no ESC-intervention. Ruth et al. [39] reported testing to evaluate the accuracy of PCM-reported speeds under steady-state conditions. They compared PCM-reported speeds to speeds measured with a 100-Hz VBOX and with a speed trap and found a maximum difference of 0.61 km/h (0.38 mph).

The tabular data indicated that the speed of the SUV at time zero (the time of the restraint deployment signal) was 35 km/h (21.7 mph). This data also indicated that the vehicle was accelerating in the seconds leading up to the collision. The data from the RCM, on the other hand, reported a speed at time zero of 27.9 km/h (17.3 mph). This is 4.4 mph lower than the speed reported by the PCM for time zero. The RCM and the PCM are obtaining their speeds from the same sensor—a speed sensor on the transmission output shaft. However, they are sampling from this sensor at different rates. The PCM obtains a speed reading every 0.2 sec, whereas the RCM obtains a reading every 1 sec. The speed reported by the RCM for time zero would simply be the last speed reading obtained by the RCM prior to the collision. Thus, the RCM did not capture the actual speed at the time of the collision, rather it captured a speed sometime in 1-sec time window preceding the collision. Looking at the tabular data from the PCM, the 17.3 mph speed reported by the RCM would have been measured sometime between −0.6 and −0.4 sec in the PCM data. For the analysis reported here, the collision speed was estimated from the PCM data.

Camera-matching photogrammetry was utilized to locate the vehicle rest positions, tire marks from the SUV, and scrapes and gouges on the asphalt from the motorcycle. As was described in Chapter 5, camera matching involves reconstructing the location and characteristics of the camera that took the photograph being analyzed. Once the camera location and characteristics are obtained, physical evidence within the photograph can be located. This process was carried out for several police photographs that depicted evidence related to the subject collision. A sample of these photogrammetry results is included in the following figures. Figure 9.31 is one of the police photographs that depicts the tire marks deposited by the SUV following the collision. Figure 9.32 shows the scan data from the accident site overlaid on this photograph, indicating that the camera location and characteristics have been reconstructed. Figure 9.33 shows the tire marks located and traced with dark blue outlines. A set of scrapes and gouges on the asphalt are also shown in this image (with light blue lines) that were traced from another police photograph that was also analyzed.

One indicator of the impact speed of a motorcycle with a passenger vehicle is the magnitude of the translation and rotation experienced by the passenger vehicle following

FIGURE 9.31 Police photograph selected for analysis.

FIGURE 9.32 Scan data overlaid onto the police photograph via camera matching.

the impact. PC-Crash simulation software was used to simulate the subject collision, specifically to determine the motorcycle impact speed necessary to cause the documented post-collision rotation of the SUV. In setting up this simulation, manufacturer specifications were obtained for each of the vehicles to obtain the geometric dimensions and weights. The SUV was equipped with a 3.5L, V6 gasoline engine, an automatic transmission, all-wheel drive, ABS, and ESC. Based on the manufacturer specifications for this vehicle, the weight at the time of the collision was estimated at 4385 lb, including the weight of the driver. The motorcycle was equipped with a 1584cc V-Twin engine, a manual transmission, and a conventional braking system. The weight of the motorcycle at the time of the collision was approximately 714 lb and the weight of the rider was approximately 170 lb. The rider's weight was included in the weight of the motorcycle within the simulation.

FIGURE 9.33 Physical evidence located in the photograph.

In the simulation of this collision, the roadway coefficient of friction was set at 0.76 and the TM-Easy tire model was used with default values for both vehicles. The integration time step was set at 5 ms. The coefficient of restitution was set at 0.25. Based on the EDR data, the impact speed of the SUV was initially set at 21.7 mph. Sequences in PC-Crash were used to steer the passenger's side rear wheel of the Ford 17° to the right immediately following the collision. The simulation was optimized using the impact speed of the motorcycle, the precise location of the collision on the roadway, the inter-vehicular friction, the impact center height, the steering angles of the front wheels, and the brake factors for each wheel. Small changes in the impact speed of the Ford were also utilized for the final step in the optimization. Consistent with the orientation of the front wheels of the Ford when it was at rest in the intersection, leftward steering inputs of 23° at the left front wheel and 19° at the right front wheel were ultimately utilized, developing over 2.3 sec. The SUV was an all-wheel drive vehicle, and brake factors of 5% were used for the front wheels and the left rear wheel. A brake factor of 100% was used for the wheel that was struck and deformed by the motorcycle—the right rear wheel. The exception to these brake factors was during the interval of ESC-intervention, during which a 100% brake factor was assigned to the left front wheel, and when the system indicated that the driver applied the brakes near the rest position. Brake factors were not applied to the wheels of the motorcycle because it fell on its side shortly after the collision. The coefficient of friction between the body of the motorcycle and the ground was ultimately set at 0.4, though this parameter was also iterated in optimizing the simulation.

Figure 9.34 is a screen capture from PC-Crash showing the final impact parameters and the motion of the vehicles following the collision. A high-quality match was obtained with the rest positions for both vehicles and the postcollision motion of the SUV matched very well with the documented tire marks. The optimization process led to the conclusion that the motorcycle was traveling 69 mph at the time of the collision. During the optimization process, it was recognized that the overall rotation of the Ford could be matched with a wide range of motorcycle impact speeds. However, the rate of rotation, as represented by the tire marks that were deposited, could only be matched with a narrow band of motorcycle impact speeds. In this case, a high-quality match was obtained with the motorcycle rest position, though the trajectory of the motorcycle was slightly below the documented

FIGURE 9.34 Motion from optimized simulation.

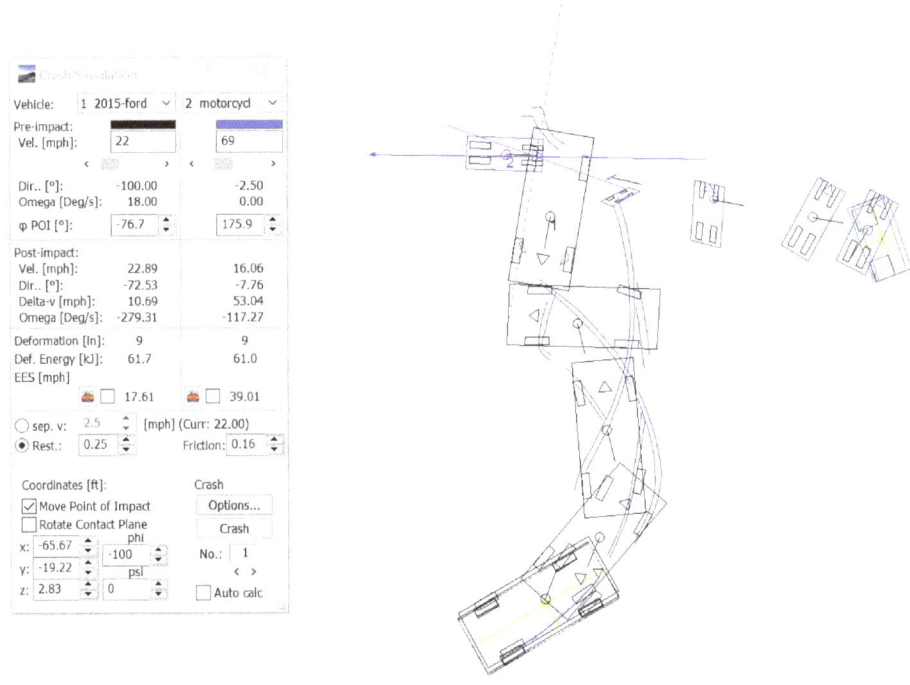

scrapes and gouges. Such a quality match of the rest position of the motorcycle would not be expected in every case, but often matching at least the postimpact velocity direction of the motorcycle will help with the simulation optimization.

The change in velocity calculated for the SUV in this simulation was 10.7 mph with a principal direction of force of 81°. Thus, the total lateral velocity change for the SUV was approximately 10.6 mph, 8.1 mph more than what was reported by the RCM. The simulation calculated a change in yaw rotational speed of the SUV of approximately 280 deg/s.

Event Data from the Motorcycle

For now, it is less common to have EDR data from the motorcycle because most motorcycles do not have EDRs. Fatzinger and Landerville [40] categorized the motorcycle EDRs that do exist into three groups: (1) those that give diagnostic codes; (2) those that log data for measuring performance; and (3) those that record crash-related data. The second and third categories have the greatest potential for use in crash reconstruction. Performance data loggers have the limitation that they are typically only present and turned on in instances where the motorcycle operator has chosen this. EDRs that record crash-related data are now installed on few motorcycles. Many Kawasaki motorcycles now include crash-sensing EDRs on their motorcycles and (at times) they will make the data available to crash reconstructionists. The owner's manual will typically specify whether a motorcycle is equipped with an EDR. Some Can-Am motorcycles and ATVs also include EDRs that report "last instant" data, including in some instances 60 sec of vehicle speed, engine speed, throttle position, and much more.

Fatzinger and Landerville examined the EDR data on 2013-2016 Kawasaki Ninja 300 motorcycles. They noted that "a thorough research effort was conducted in order to establish methods for downloading EDR data, determining which data was recorded in collision events, validating its accuracy, and determining the triggering conditions." For the physical testing that Fatzinger and Landerville conducted, they utilized a 2013 Ninja EX300A equipped with a Denso ECU. They were able to trigger EDR events by positioning the motorcycle in a rear wheel stand, idling the engine in sixth gear so that the rear wheel was spinning at approximately 16 mph, then "applying a hard brake application to stop the rear wheel and motor [and] then manually tilting the tip-over sensor … After learning how to trigger events, it was quickly discovered that the three EDR event locations in the ECU memory were all 'locked' events, or events that could not be overwritten. Once all three EDR event memory locations were full, no new EDR events could be recorded. The events were populated in chronological order 1-3, with Event 3 being the most recent." Fatzinger and Landerville also discovered that "if the rear wheel speed gradually came to a stop, followed by an immediate tip-over, an EDR event was not recorded. It was apparent that a rear wheel deceleration rate threshold had to be exceeded in order to trigger an EDR event if the rear wheel stopped rotating before time zero (emergency engine shut-down)." They discovered that the braking deceleration threshold to trigger an event was approximately 0.6 g. Finally, "if the rear wheel was still moving at the time of emergency shut-down, an EDR event would always be triggered … at an indicated speed of approximately 2 mph or higher … Anything below this value would not trigger an EDR event." After generating events, Fatzinger and Landerville accessed the following data for EDR events at a rate of 2 Hz: vehicle speed, gear position, inlet air temperature, coolant temperature, battery voltage, and diagnostic trouble codes. They accessed the following data at a rate of 10 Hz: throttle position, engine RPM, clutch in/out, fuel injector pulse, timing BTDC, and fuel cutout.

In another study, Fatzinger [41] tested the EDR capabilities and behavior on Kawasaki Ninja ZX-6R and ZX-10R motorcycles. Like the prior study, he reported that EDR recording would be triggered by activation of the tip-over sensor. Activation of this sensor also triggered the engine to shut-off. Fatzinger reported that "an EDR event was only recorded if the motorcycle was commanded to shut-down by the tip-over sensor, and either had rear wheel movement at the time of shut-down or the rear wheel experienced a certain amount of deceleration in the several seconds prior to shut-down. The 'time zero' data element was synchronous with the tip-over commanded shut-down signal."

Peck [42] described several data loggers that fall in the category of those that log data for measuring performance. For instance, the Woolrich Racing Log Box, installation of which requires some modification of the wiring system of the motorcycle. This data logger can access a plethora of data from the ECU of the motorcycle. The data available depend on the specific motorcycle but could include wheel speed, throttle position, brake switch status, transmission gear position, engine speed, and other potentially useful parameters. Captured data are stored on a micro-SD card, allowing for straightforward retrieval. The HP Race Datalogger is a similar device manufactured by BMW. This data logger can also record GPS data. This system can be plugged directly into the original wiring harness of certain BMWs, but the owner must choose to install the device. Some more technologically advanced motorcycles, such as Yamaha's 2015+ YZF-R1M, are equipped with GPS-enabled onboard data acquisition systems and can pair with smartphones and tablets. Peck recommended that reconstructionists be aware of the potential presence of such data and preserve any devices where it might be stored.

Peck reported testing of the Ducati Data Analyzer (DDA). Introduced in 2007, this system consisted of a single component resembling a USB drive that which had the ability to log vehicle speed, engine speed (RPM), engine temperature, throttle aperture, gear position (calculated value), engine temperature, and total distance traveled. Ducati introduced a GPS-enabled version of the device in June 2012, which added a GPS sensor to record the vehicle position. Peck reported testing of a 2008 Ducati 848 equipped with an early version of the DDA. He rode this motorcycle while collecting data with the DDA and with a VBOX Sport, a GPS data acquisition device with a sampling rate of 20 Hz (Racelogic, Farmington Hills, MI). He compared the data from these two devices. Peck concluded that the vehicle speeds reported by the DDA were accurate and aligned well to those reported by the VBOX Sport.

Video as a Source of Data from the Motorcycle

Video from a camera on a motorcyclist's helmet or chest, or surveillance footage from a security camera on a nearby business, can be another source of data (an EDR of sorts) about the events leading to a collision. When conducting analysis of video, keep the following in mind:

- The view of a camera has a limit. There will always be what the camera shows and what it does not. A complete reconstruction may require knowledge of more than what is captured within the viewing area of the camera, and so, additional information may need to be obtained from physical evidence, principles of physics, or electronic data.

- Video is composed of a series of still images. Understanding how these images were captured and how they have been processed can sometimes be important to understanding the motion depicted, particularly when quantifying the uncertainty in the analysis. Questions relevant here are: Is each of the images independent? Did the camera capture and save the entire image or only part of it? Has any kind of compression been applied?

- A camera captures a finite number of frames in any given second. Just as there is motion that occurs outside the view of the camera, there is also motion that can occur between frames and during the time an image is being captured. There can be uncertainty about the precise time between frames. Beauchamp et al. [43] reported that "for cameras with a constant frame rate, uncertainty in time … can be ignored. However, uncertainty in time can arise when the frame rate is variable. Cameras often including an embedded time stamp for each frame of the video. In some cases, the camera/software company can be contacted to verify the accuracy of the time stamp. This mean that although a camera's frame rate is variable, its time stamp can be considered accurate eliminating this type of uncertainty …."

- Each of the images contained within a video is likely to have lens distortion that may need to be removed for analysis. This distortion will be the same across a series of frames captured with a single camera, lens, and focal length. Chapter 5 discusses the process of removing distortion from a photograph or video image.

- Video analysis will often need to determine the position and motion (translation, rotation, and zooming) of the camera before the motion of objects within the video can be quantified. For a static camera, this process is simplified.

- There will be uncertainty in the reconstructed translation, rotation, and zooming of a camera. There will also be uncertainty in the reconstructed position and orientation of objects within a video frame.

- Some objects depicted by a video frame may be out of focus (blurry). Blurriness can lead to uncertainty in determining the position of the camera, determining the position and orientation of objects depicted in a frame, and tracking the motion of objects across a series of frames.

- Postprocessing can alter the brightness, contrast, and color balance of a video image. That may make the position and orientation of objects within the image more distinct.

- Video analysis results should be consistent with the physical limitations on the motion of objects in the real world. For example, there are limits on the rate at which vehicles can accelerate or decelerate. If video analysis produces unrealistic results, the source will be an inaccuracy in the calculated position of the tracked object, the frame rate of the video, or both.

A Residual Speedometer Reading as a Source of Data from the Motorcycle

Sometimes after a collision, the speedometer or tachometer of a motorcycle will exhibit a residual reading other than zero. In some instances, other physical evidence will make it evident that the residual speed reading is not a reliable indicator of the impact speed of the motorcycle. However, if the motorcycle is equipped with a speedometer with a stepper motor, then there is the potential for the residual speedometer reading to be accurate and useful in the reconstruction [44–47]. As Montalbano et al. [44] explain: "A stepper motor converts electrical current into mechanical movements. The feature of these motors that serves as the critical factor … is that they require continuous electrical power to adjust the speedometer indications either upward or downward—this includes returning to a 0-mph indication. In the case of instantaneous loss of electrical power to the stepper motor, there is no way to generate the stepper motor movement needed to return the needle to its zero position. In the absence of any other applied force, a needle attached to an unpowered stepper motor will remain in its position indefinitely. As a result of this feature, speedometers equipped with electrical stepper motor driven needles will maintain the speed indication that was displayed at the moment of the power loss, colloquially described as a 'frozen' speedometer."

Based on the research by these authors, the process of establishing the reliability of a residual speedometer or tachometer reading on a motorcycle could involve the following steps:

- Determine if the speedometer or tachometer on the subject motorcycle employed a stepper motor.

- If a stepper motor was present, determine the stepper motor manufacturer, model, and/or type.

- Quantify the torque resistance of the speedometer or tachometer needle to rule out that the inertial forces of the collision or postcollision handling did not change the position of the speedometer needle. The authors noted that, "for motorcycle speedometers with stepper-motors of high torque resistance, the frozen needle indication should be a reliable indication of the motorcycle's impact speed in situations where there is no pre-impact lock-up of the speed-sensing wheel." On the other hand, Anderson [48] observed that "residual readings on speedometers with low resistance to needle motion are not reliable indicators of that vehicle's speed at impact, even with a sudden power loss coincident with the time of the collision."

These authors noted that the presence of a stepper motor can be confirmed with an exemplar motorcycle by instantaneously removing "power to the exemplar vehicle's gauge cluster while the indicator needle(s) is displaying a non-zero value … The simplest method of achieving this instantaneous power loss is to remove the motorcycle's primary fuse, or if it can be located, pulling the motorcycles gauge cluster fuse, while the gauge needle is displaying a non-zero value. However, due to the complexity of some motorcycle electronic systems, identifying the correct fuse for ensuring instantaneous loss of power to the vehicle's gauge cluster can be challenging. If the analyst were to be unsuccessful at locating a fuse directly linked to the instrument cluster, the next step would be to remove the wiring harness directly from the speedometer body. This typically requires the removal of some body panels and the removal of the speedometer from its mount. Once the gauge cluster has been removed, and while the drive-wheel is rotating, the primary harness into the gauge cluster should be quickly disconnected. If the needle freezes at the pre-power loss value, then it can be concluded that the subject vehicle's gauge is driven by a stepper motor … The subject speedometer, or a replacement part from the original manufacturer with the same part number as the subject gauge, can be cut open in the back for visual inspection of the component parts and structure. Because manufacturers parallel source some components such as stepper motors, it is important to identify, when possible, the specific make and model of stepper motor inside the subject instrument gauge. This process renders the speedometer unusable, but it is also the most conclusive procedure to determine the stepper motor type."

Through their testing, Montalbano et al. identified the following motorcycles that had stepper motor speedometers and tachometers:

- 2008 Harley Davidson Heritage Softail Classic FLSTCI
- 2014 Harley Davidson Heritage Softail FLSTC103
- 2015 Harley Davidson Street Glide Special FLHXS
- 2014 Triumph Bonneville
- 2015 BMW R1200RT
- 2015 Indian Chieftain
- 2014 Triumph Tiger Explorer (tachometer only)

Fatzinger et al. [49] tested six electronic needle-display speedometers. They elevated the rear wheels of the motorcycles and accelerated them to predetermined speeds. They then disconnected the speedometer wiring harnesses. They found that the "dial indicator would move slightly up, down, or remain in place depending on the model of the speedometer. The observed change of indicated speed was within ±10 mph upon power loss." In addition, these authors subjected speedometers to impact testing on a linear drop rail and quantified the minimum acceleration to cause needle movement. This type of testing constitutes an alternate method to establishing high torque resistance in the stepper motor and has the advantage of being more directly related to the type of loading that would be expected during a collision. The results obtained by these authors are summarized below.

1999 Harley Davidson FXDX [gear driven (spur)]—"During the power interruption testing, the FXDX model speedometer stayed within ±1 mph of the indicated speed prior to the power interruption … During the drop testing, the FXDX speedometer dial … indicator rotated approximately 2-3 mph due to the impact, in the direction of pre-impact momentum … the deceleration reached nearly 350 g, over an impact duration of roughly 7 ms."

2013 Harley Davidson FLTRU [gear driven (worm)]—"During the power interruption testing, the maximum drop in speed of the FLTRU model speedometer was approximately 8 mph … At no point during the power interruption testing did the dial indicator observably increase in speed … During the drop testing, the FLTRU was very resistant to rotation at relatively high deceleration rates. After a six foot drop test, the dial indicator on the FLTRU remained within a 1 mph deviation. The six foot drop test again produced deceleration values at nearly 350 g over 7 ms."

2015 Honda GL 1800 Gold Wing [direct drive]—"During the power interruption testing, the GL 1800 model speedometer returned to zero after every test, regardless of dial indicator position."

2015 BMW R1200GS [gear driven (worm)]—"During the power interruption testing, the R1200GS speedometer stayed at the indicated speed. There was no visible increase or decrease in speed at any dial indicator position. During the drop testing, the R1200GS speedometer dial was very resistant to motion at relatively high deceleration rates. The dial indicator remained at the pre-test position below roughly 200 g. At impact levels reaching 230 g, the dial indicator rotated in the direction of pre-impact momentum roughly 1-2 mph. At impact levels of 310 g, the dial indicator rotated approximately 2-3 mph."

2014 Triumph T100 Bonneville [gear driven (worm)]—"During the power interruption testing, the T100 speedometer stayed at the indicated speed, very similar to the R1200GS speedometer. There was no visible increase or decrease in speed at any dial indicator position. During the drop testing, the T100 was similar to the FLTRU speedometers and very resistant to moving at relatively high deceleration rates. After a six foot drop test, the dial indicator on the T100 rotated in the direction of pre-impact momentum approximately 1–2 mph. The six foot drop test again produced deceleration values at nearly 350 g over 7 ms."

2012 Kawasaki ZX-14R [direct drive]—"During the power interruption testing, the ZX-14R speedometer moved ±10 mph. The dial indicator would either increase or decrease in indicated speed, or not move at all depending on the position of the stepper motor before power loss … During the drop testing, the ZX-14R speedometer was more likely to rotate than the other speedometers tested." The speedometer rotated approximately 7 mph because of an impact that reached 50 g over 13 ms.

The drop testing by Fatzinger et al. subjected the speedometers to significant impact forces. The forces applied to the speedometer during a crash would need to be evaluated on a case-by-case basis. These forces would depend on several factors, including the relative speed of impact between motorcycle and the passenger vehicle, the distance from the front of the motorcycle to the speedometer, the stiffness of the motorcycle forks, the stiffness of the passenger vehicle structure, and whether the speedometer itself became a part of the crushing region of the motorcycle during the collision. As an illustration, consider a scenario in which an upright motorcycle strikes a nonmoving, nondeforming barrier at a speed of 40 mph. Assume an impact duration of 55 ms and a haversine collision pulse. This pulse has the following functional form for the acceleration [50]:

$$a = a_{peak} sin^2\left(\pi \frac{t}{\Delta t}\right) \tag{9.11}$$

In this equation, a_{peak} is the peak acceleration achieved during the pulse, t is time, and Δt is the total duration of the collision. With this collision pulse shape, a peak acceleration of approximately 72.9 g gives the motorcycle a ΔV of 44 mph, and thus, yields a collision with 10% restitution. Under this scenario, the motorcycle will experience a maximum dynamic crush of approximately 1.5 ft at approximately 40 ms into the collision. If the speedometer is 1 ft back from the front of the motorcycle, then it will be traveling approximately 31.8 mph when it reaches the barrier. Now, assume that portion of the motorcycle comes to a stop against the barrier in another 10 ms that will produce an average acceleration on the speedometer of approximately 145 g, and perhaps a peak acceleration of 290 g. This would be likely to be on the high-end of the accelerations that would be experienced by a speedometer that remains attached to the motorcycle, because this illustration assumes a nonmoving, nondeforming barrier. On the other hand, consider a scenario in which the motorcycle impacts a soft region of a passenger car, the speedometer of the motorcycle is outside of the crushing region of the motorcycle, and the impact duration is longer. Assume an impact duration of 155 ms and a haversine pulse. In this scenario, the speedometer would experience a peak acceleration of around 26 g, significantly below the acceleration level tested by Fatzinger et al.

Based on these studies, a frozen speedometer reading would reliably report the reading that was on the motorcycle speedometer at impact when:

- The speedometer is driven by a stepper motor.
- A power loss to the speedometer occurs during the collision.
- The speedometer needle does not move significantly when a power loss occurs.
- The speedometer needle has high torque resistance and does not move significantly under impact loading.

In instances where a motorcyclist employs emergency level braking prior to impact, there may be some discrepancy between the actual speed of the motorcycle and the reading on the speedometer. One reason for this could be latency in the speedometer (i.e., how quickly the speedometer responds to actual changes in speed). Another reason could be wheel-slip because of the heavy braking. This is similar to the issue that comes up with the precrash speeds reported by a passenger car EDR, where when heavy braking is present, there can be a discrepancy between the vehicle-indicated speed and the actual speed of the car. Neither of these issues has been addressed in prior literature related to motorcycle speedometers, and these topics could benefit from additional testing and research.

To examine the issue of wheel slip, consider a heavy braking scenario for a motorcycle that remains upright with insignificant lean, continues straight, and does not yaw substantially. Equation (9.12) defines the longitudinal slip ratio (s_{long}) for a single tire [51].

$$s_{long} = \frac{V - r\omega}{V} = \frac{V - V_{rolling}}{V} \tag{9.12}$$

In this equation, V is the travel speed of the wheel center which, for a heavy braking scenario in which the motorcycle continues traveling in a straight line, would coincide with the over-the-ground travel speed of the motorcycle. The variable r is the effective rolling radius of the tire and ω is the angular velocity of the wheel. The quantity $r\omega$ is the rolling speed of the tire ($V_{rolling}$). Equation (9.13) can be rearranged as follows:

$$V = V_{rolling} \frac{1}{1 - s_{long}} \tag{9.13}$$

©2022, SAE International

For a straight-line maximum effort braking with an ABS-equipped motorcycle, the variable $V_{rolling}$ can be treated as the speed reported by the speedometer and V as the actual over-the-ground speed of the vehicle. Assume an ABS system that keeps the longitudinal tire slip at 20% or below. If this slip percentage is correct, then during maximum braking, the actual speed of the motorcycle could be as much as 25% greater than the rolling speed of the drive wheel.

Another issue that could cause a discrepancy between the speed on the speedometer and the actual over-the-ground speed of the motorcycle is lean of the motorcycle. To see this, first consider a motorcycle traveling straight down the road without any lean. If this motorcycle uses a rear wheel speed sensor to feed the speedometer speed, then the motorcycle would need to have a rolling radius for the rear tire listed within its programming. The effective circumference of the rear tire would be given by the following equation:

$$C_{tire} = 2\pi r_{rolling} \qquad (9.14)$$

In this equation, C_{tire} is the effective circumference of the tire and $r_{rolling}$ is the rolling radius of the tire in feet. This circumference is the distance covered during one rotation of the tire. The speed of the motorcycle in feet per second is then given by the following equation:

$$S = 2\pi r_{rolling} \omega_{rolling} \qquad (9.15)$$

In this equation, $\omega_{rolling}$ is the angular velocity of the tire in radians per second. Now, if the motorcycle is traveling and continues to travel at a steady speed as the rider leans the motorcycle, the motorcycle will roll onto a portion of the rear tire that has a lower effective rolling radius. For the actual over-the-ground speed of the motorcycle to stay constant, the wheel will have to roll at a higher rate. However, the rolling radius value within the programming will not have changed, and so, the motorcycle will calculate and report a speed that is higher than actual because it will use a higher rolling radius than actual. Thus, if a motorcyclist leans to swerve prior to a collision, the reading on the speedometer at the time of impact could be slightly higher than the actual speed.

Case Study: Speedometer Testing on a 2019 Harley-Davidson Heritage Softail Classic

Testing was conducted on the speedometer on a 2019 Harley-Davidson Heritage Softail Classic. This motorcycle is show in **Figure 9.35**. The front and rear wheels of the motorcycle were elevated on wheel stands. GoPro cameras were positioned to capture the speedometer and the fuse controlling power to the speedometer, which was located under the front seat of the motorcycle. The clutch lever was depressed with the motorcycle in first gear, the power was turned on, and the motorcycle was started. The speedometer needle remained at zero and did not perform a calibration procedure. When the clutch lever was released with the motorcycle in first gear, the rear wheel would begin rolling and the speedometer would read at a speed of approximately 10 mph. The motorcycle was shifted up to fourth gear and the throttle was used to accelerate the rear wheel until the speedometer read a steady speed of 65 mph. The fuse for the speedometer was then pulled. This procedure was then repeated for speeds of 55, 45, 35, 25, and 15 mph. In each case, the speedometer immediately became fixed in its current position with no movement. When the fuse was replaced, the speedometer would return to zero.

CHAPTER 9 Event Data Recorders and Speedometers in Motorcycle Accidents   297

FIGURE 9.35 Exemplar 2019 Harley-Davidson heritage softail classic.

In addition to the exemplar motorcycle, an exemplar speedometer was purchased that could be used for additional testing. The speedometer on the exemplar motorcycle was compared to the exemplar speedometer to verify a match. As Figure 9.36 shows, these speedometers were both manufactured by Continental Automotive, and they had the same part numbers. According to Goddard [46], "Another common type of stepper motor utilizes a motor and gearbox manufactured by Continental Automotive (formerly Siemens VDO). This arrangement has a worm gear connecting the rotor shaft to the needle. Employing a worm gear results in a high static torque and so provides a high degree of resistance to

FIGURE 9.36 Comparing the speedometer on the exemplar motorcycle to the exemplar speedometer.

©2022, SAE International

free movement of the needle when the instrument is not powered." The speedometer on the exemplar motorcycle was unplugged and the wiring harness plugged into the exemplar speedometer. The rear wheel of the motorcycle was accelerated until this speedometer read approximately 45 mph, and then the fuse was pulled so that the exemplar speedometer would remain at 45 mph for the drop testing (Figure 9.37). The speedometer for the exemplar motorcycle was then reinstalled.

FIGURE 9.37 Exemplar speedometer with the needle positioned at approximately 45 mph.

FIGURE 9.38 Drop test setup.

This exemplar speedometer was next drop tested to determine if the needle would move under impact loading. To carry out this drop testing, the speedometer was attached to a wood block. The wood block was attached to wire rope rails on either side via two eyebolts on each side. These rails would guide the wood block during the drop. The wood block and speedometer were dropped from a both a 1-ft height and a 5-ft height onto another wood block placed on the floor. A piece of soft foam was placed on top of the wood block on the floor to attenuate the collision. This drop test rig is depicted in Figure 9.38. The drop fixture was not instrumented. Based on the drop heights, the speed of the drop fixture at impact in the 1-ft drop would have been around 5½ mph, and in the 5-ft drop, it would have been approximately 12.2 mph. The speedometer needle did not move in either 1-ft or 5-ft drop tests.

References

[1] Bortles, W., Biever, W., Carter, N., and Smith, C., "A Compendium of Passenger Vehicle Event Data Recorder Literature and Analysis of Validation Studies," SAE Technical Paper 2016-01-1497, 2016, doi:10.4271/2016-01-1497.

[2] Ruth, R. and Wright, B., "Using E.D.R. Pre-Crash Data to Calculate a Range for Speed at Impact," *Accident Reconstruction Journal*, 27(3): 31–37, 2017.

[3] Rose, N. and Carter, N., "Motorcycle Accident Reconstruction: Applicable Error Rates for Struck Vehicle EDR-Reported ΔV," *Collision: The International Compendium for Crash Research*, 13(1): 88–109, 2019.

[4] Ruth, R., "Applying Automotive EDR Data to Traffic Crash Reconstruction," SAE Course #C1210, Course Slides, November 2018.

[5] Wilkinson, C., Lawrence, J., Nelson, T., and Bowler, J., "The Accuracy and Sensitivity of 2005 to 2008 Toyota Corolla Event Data Recorders in Low-Speed Collisions," *SAE International Journal of Transportation Safety*, 1(2): 420–429, 2013, doi:10.4271/2013-01-1268.

[6] Ruth, R. and Muir, B., "Longitudinal Delta V Offset between Front and Rear Crashes in 2007 Toyota Yaris Generation 04 EDR," SAE Technical Paper 2016-01-1496, 2016, doi:10.4271/2016-01-1496.

[7] Xing, P., Lee, F., Flynn, T., Wilkinson, C. et al, "Comparison of the Accuracy and Sensitivity of Generation 1, 2 and 3 Toyota Event Data Recorders in Low-Speed Collisions," *SAE International Journal of Transportation Safety*, 4(1): 172–186, 2016, doi:10.4271/2016-01-1494.

[8] Code of Federal Regulations, 49 CFR 563 – Event Data Recorders, in effect as of September 1, 2012.

[9] Bundorf, R.T., "Analysis and Calculation of Delta-V from Crash Test Data," SAE Technical Paper 960899, 1996, doi:10.4271/960899.

[10] Marine, M.C. and Werner, S.M., "Delta-V Analysis from Crash Test Data for Vehicles with Post-Impact Yaw Motion," SAE Technical Paper 980219, 1998, doi:10.4271/980219.

[11] Marine, M. and Werner, S., "Crash-Induced Yaw Motion Effects on Airbag Control Module Delta-V," *Collision: The International Compendium for Crash Research*, 15(1): 82–92.

[12] Rose, N.A., "Quantifying the Uncertainty in the Coefficient of Restitution Obtained with Accelerometer Data from a Crash Test," SAE Technical Paper 2007-01-0730, 2007, doi:10.4271/2007-01-0730.

[13] Haight, S. and Haight, R., "Analysis of Event Data Recorder Delta-V Reporting in the IIHS Small Overlap Crash Test," *Collision* 8(2): 8–23, 2013.

[14] Emori, R., "Analytical Approach to Automobile Collisions," SAE Technical Paper 680016, 1968, doi:10.4271/680016.

[15] Varat, M. and Husher, S., "Vehicle Impact Response Through the Use of Accelerometer Data," SAE Technical Paper 2000-01-0850, 2000, doi:10.4271/2000-01-0850.

[16] Comeau, J.-L., German, A., and Floyd, D., "Comparison of Crash Pulse Data from Motor Vehicle Event Data Recorders and Laboratory Instrumentation," *Proceedings of the Canadian Multidisciplinary Road Safety Conference XIV*, Ottawa, Ontario, Canada, June 27–30, 2004.

[17] Niehoff, P., Gabler, H.C., Brophy, J., Chidester, C., et al., "Evaluation of Event Data Recorders in Full System Crash Tests," Paper No. 05-0271, *19th ESV Conference*, Washington, D.C., June 2005.

[18] Wilkinson, C.C., Lawrence, J.M., and King, D.J., "The Accuracy of General Motors Event Data Recorders in NHTSA Crash Tests," *Collision: The International Compendium for Crash Research*, 2(1):70–76, 2007.

[19] Gabler, H.C., Thor, C.P., and Hinch, J., "Preliminary Evaluation of Advanced Air Bag Field Performance Using Event Data Recorders," DOT HS 811 015, August 2008.

[20] Comeau, J.-L., Dalmotas, D., and German, A., "Evaluation of the Accuracy of Event Data Recorders in Chrysler Vehicle in Frontal Crash Tests," *Proceedings of the 21st Canadian Multidisciplinary Road Safety Conference*, Halifax, Nova Scotia, May 2011.

[21] Comeau, J.-L., Dalmotas, D., and German, A., "Event Data Recorders in Toyota Vehicles," *Proceedings of the 21st Canadian Multidisciplinary Road Safety Conference*, Halifax, Nova Scotia, May 2011.

[22] Exponent Failure Analysis Associates, "Testing and Analysis of Toyota Event Data Recorders," October 2011, https://pressroom.toyota.com/exponent-report-testing-and-analysis-of-toyota-event-data-recorders/, accessed on November 8, 2021.

[23] German, A., Dalmotas, D., and Comeau, J.-L., "Crash Pulse Data from Event Data Recorders in Rigid Barrier Tests," Paper No. 11-0395, *22nd ESV Conference*, Washington, D.C., June 2011.

[24] Tsoi, A., Hinch, J., Ruth, R., and Gabler, H., "Validation of Event Data Recorders in High Severity Full-Frontal Crash Tests," *SAE International Journal of Transportation Safety*, 1(1): 76–99, 2013, doi:10.4271/2013-01-1265.

[25] Ruth, R. and Tsoi, A., "Accuracy of Translations Obtained by 2013 GIT Tool on 2010-2012 Kia and Hyundai EDR Speed and Delta V Data in NCAP Tests," SAE Technical Paper 2014-01-0502, 2014, doi:10.4271/2014-01-0502.

[26] Vandiver, W., Anderson, R., Ikram, I., Randles, B. et al., "Analysis of Crash Data from a 2012 Kia Soul Event Data Recorder," SAE Technical Paper 2015-01-1445, 2015, doi:10.4271/2015-01-1445.

[27] Lee, F., Xing, P., Yang, M., Lee, J. et al., "Behavior of Toyota Airbag Control Modules Exposed to Low and Mid-Severity Collision Pulses," SAE Technical Paper 2017-01-1438, 2017, doi:10.4271/2017-01-1438.

[28] Correia, J., Iliadis, K., McCarron, E., and Smolej, M., "Utilizing Data from Automotive Event Data Recorders," *Presented at the Canadian Multidisciplinary Road Safety Conference XII*, London, Ontario, June 10–13, 2001.

[29] Beck, R., Casteel, D., Phillips, E., Eubanks, J. et al., "Motorcycle Collinear Collisions Involving Motor Vehicles Equipped with Event Data Recorders," *Collision: The International Compendium for Crash Research*, 1(1): 82–96, 2006.

[30] Carr, L., Rucoba, R., Barnes, D., Kent, S. et al., "EDR Pulse Component Vector Analysis," SAE Technical Paper 2015-01-1448, 2015, doi:10.4271/2015-01-1448.

[31] Chidester, A., Hinch, J., Mercer, T., and Schultz, K., "Recording Automotive Crash Event Data," *Proceedings of the International Symposium on Transportation Recorders*, Arlington, VI, 1999.

[32] Lawrence, J., Wilkinson, C., King, D., Heinrichs, B. et al., "The Accuracy and Sensitivity of Event Data Recorders in Low-Speed Collisions," SAE Technical Paper 2002-01-0679, 2002, doi:10.4271/2002-01-0679.

[33] Wilkinson, C., Lawrence, J., Heinrichs, B., and Siegmund, G., "The Accuracy of Crash Data Saved by Ford Restraint Control Modules in Low-speed Collisions," SAE Technical Paper 2004-01-1214, 2004, doi:10.4271/2004-01-1214.

[34] Lawrence, J. and Wilkinson, C., "The Accuracy of Crash Data from Ford Restraint Control Modules Interpreted with Revised Vetronix Software," SAE Technical Paper 2005-01-1206, 2005, doi:10.4271/2005-01-1206.

[35] Wilkinson, C., Lawrence, J., Heinrichs, B., and King, D., "The Accuracy and Sensitivity of 2003 and 2004 General Motors Event Data Recorders in Low-Speed Barrier and Vehicle Collisions," SAE Technical Paper 2005-01-1190, 2005, doi:10.4271/2005-01-1190.

[36] Haight, R., Gyorke, S., and Haight, S., "Hyundai and Kia Crash Data, The Indispensable Compendium: Section 2 – Crash Testing Involving Hyundai and Kia Vehicles," *Collision*, 8(2): 77–86, 2013.

[37] Tsoi, A., Johnson, N., and Gabler, H., "Validation of Event Data Recorders in Side-Impact Crash Tests," SAE Technical Paper 2014-01-0503, 2014, doi:10.4271/2014-01-0503.

[38] Arndt, M.W., Rosenfield, M., Stevens, D., and Arndt, S., "Test Results: Ford PCM Downloads Compared to Instrumented Vehicle Response in High Slip Angle Turning and Other Dynamic Maneuvers," SAE Technical Paper 2009-01-0882, 2009, doi:10.4271/2009-01-0882.

[39] Ruth, R.R., West, O., Engle, J., and Reust, T.J., "Accuracy of Powertrain Control Module (PCM) Event Data Recorders," SAE Technical Paper 2008-01-0162, 2008, doi:10.4271/2008-01-0162.

[40] Fatzinger, E. and Landerville, J., "An Analysis of EDR Data in Kawasaki Ninja 300 (EX300) Motorcycles," SAE Technical Paper 2017-01-1436, 2017, doi:10.4271/2017-01-1436.

[41] Fatzinger, E., "An Analysis of EDR Data in Kawasaki Ninja ZX-6R and ZX-10R Motorcycles Equipped with ABS (KIBS) and Traction Control (KTRC)," SAE Technical Paper 2018-01-1443, 2018, doi:10.4271/2018-01-1443.

[42] Peck, L.R., "Exploration and Validation of the Ducati Data Analyzer (DDA)," *Accident Reconstruction Journal*, 28(1): 19–22, 2018, ISSN 1057-8153.

[43] Beauchamp, G., Pentecost, D., Koch, D., Hashemian, A. et al., "Speed Analysis from Video: A Method for Determining a Range in the Calculations," SAE Technical Paper 2021-01-0887, 2021, doi:10.4271/2021-01-0887.

[44] Montalbano, P., Melcher, D., Keller, R., Rush, T. et al., "Testing Methodology to Evaluate Reliability of a "Frozen" Speedometer Reading in Motorcycle/Scooter Impacts with Pre-Impact Braking," SAE Technical Paper 2016-01-1482, 2016, doi:10.4271/2016-01-1482.

[45] Goddard, C. and Price, D., "A review of speedometers and the criteria to be considered before accepting "frozen" readings and other marks," *Proceedings of the 10th ITAI International Conference*, Leeds, UK, 2011.

[46] Goddard, C., "Crash Testing Stepper Motors and Speedometers," *Impact: The Journal of the Institute of Traffic Accident Investigators*, 21(3): 30–34, 2013.

[47] Goddard, C. and Price, D., "Speedometers and Collision Reconstruction," SAE Technical Paper 2017-01-1412, 2017, doi:10.4271/2017-01-1412.

[48] Anderson, R.D., "Post-Collision Speedometer Readings and Vehicle Impact Speeds," *Collision: The International Compendium for Crash Research*, 5(2): 33–41, 2010.

[49] Fatzinger, E., Shaw, T., and Landerville, J., "The Effects of Power Interruption on Electronic Needle-Display Motorcycle Speedometers," SAE Technical Paper 2016-01-1474, 2016, doi:10.4271/2016-01-1474.

[50] Varat, M. and Husher, S., "Vehicle Impact Response Analysis Through the Use of Accelerometer Data," SAE Technical Paper 2000-01-0850, 2000, doi:10.4271/2000-01-0850.

[51] Gillespie, T.D., *Fundamentals of Vehicle Dynamics*, Society of Automotive Engineers, Warrendale, PA, 1992.

10

Human Factors in Motorcycle Crashes

By Nathan Rose and Steve Arndt, PhD, CHFP

In discussing the human factors issues that arise when reconstructing motorcycle crashes, epidemiological studies can provide some guidance. Epidemiological studies are useful in illuminating factors that may contribute to crashes, but they cannot reveal which of these factors contributed to any particular crash. That must come out of an evaluation of the evidence and facts specific to each crash. For example, epidemiological studies related to motorcycle crashes have demonstrated that, after being involved in a crash with a motorcycle, passenger car drivers often report not having seen the motorcycle. Researchers have identified many factors that could contribute to the motorcycle not being seen—the small and narrow profile of some motorcycles, the passenger car driver not expecting to see a motorcyclist, the motorcycle being occluded by other traffic or some geometric feature of the site, a lack of lighting to make the motorcycle detectable, or a lack of contrast between the rider and surrounding environment. For any particular crash, though, these factors should be evaluated in relationship to the evidence rather than simply assuming that they contributed to the crash. It is possible that none of them contributed to a crash and that the physical evidence will show that the driver actually did see the motorcyclist. For example, there may be preimpact skid marks from the car, even though the driver reported not having seen the motorcyclist, or the EDR data could indicate a driver came off the throttle and began to brake prior to impact. A discrepancy between the driver's description and what actually occurred can often be explained by understanding the limitations of human attention, memory, or recall associated with driving and collisions and postcollision attempts to reconstruct a memory or story that explains what happened.

General categories of factors that can lead to a crash emerge from epidemiological studies. Hurt and Dupont [1] reported research they conducted as a part of the Traffic Safety Center at the University of Southern California, under the sponsorship of the National

Highway Traffic Safety Administration of the US Department of Transportation. They reported that this research was to involve "on-scene, in-depth multidisciplinary investigation of at least 900 motorcycle accidents, and the acquisition of at least 3600 police traffic accident reports for comparison." At the time of their 1977 report, 300 of the on-scene investigations had been completed, and 900 traffic crash reports had been gathered. Based on this data, Hurt and Dupont stated that "the motorcycle is particularly sensitive to environmental problems such as animals in the roadway, oil, water, and gravel contamination of the roadway, grooved freeways, railroad tracks, etc. Also, it is clear from the accident investigations that vehicle mechanical problems have far more serious consequences for the motorcycle than for the contemporary passenger automobile. A puncture flat on the freeway essentially guarantees a disaster for the motorcycle rider while the same occurrence in a passenger car would only cause anxious moments ... Of course, in the study of any system of motor vehicle accidents, the problems of inattention, alcohol, risk-taking behavior, etc., will appear and contribute to accident causation. [In addition are problems of] motorcycle conspicuity, rider skill, training and licensing, and protective equipment."

A more recent study—referred to as the Motorcycle Accidents In Depth Study (MAIDS)—examined the causes of motorcycle accidents in five European countries (France, Germany, the Netherlands, Spain, and Italy) [2]. This study was conducted by the Association of European Motorcycle Manufacturers (ACEM), and, as the final report (Version 2.0) for this study indicates, 921 accidents were investigated and reconstructed. The study concluded that the cause of most motorcycle accidents "was found to be human error. The most frequent human error was a failure to see the [motorcycle] within the traffic environment, due to lack of driver attention, temporary view obstructions or the low conspicuity of the [motorcycle] ... When the accident riders were compared to the exposure population, the data demonstrated that the use of alcohol increased the risk of being in an accident, although the percentage was lower than in other studies. Unlicensed [motorcycle] operators who were illegally riding ... were also found to be at greater risk of being involved in an accident when compared to licensed [motorcycle] riders."

Physical Factors Affecting the Visibility of Motorcycles

Motorcycles are often narrower than other vehicle types. Because of this, foliage, signage, power poles, bridge columns, or other structures can at times occlude the visibility of a motorcycle. Curves and hills can also create obstructions that limit a rider's or driver's view. In addition, a motorcycle operator's decisions about where to ride can affect visibility. A motorcyclist can occupy different positions within a lane, and these positions can influence both the motorcyclist's ability to see other traffic and the ability of other drivers to see the motorcyclist. As an example, consider the three photographs of Figure 10.1. Each of these photographs shows the view to the left from a different vantage point at an intersection. Each of these views represents a possible stopping position and vantage point for a passenger car driver intending to make a left turn at this intersection. Depending on where the driver chooses to stop, the bushes and utility pole would have varying influences on what the driver is able to see. If the driver chose to initiate a turn from the first position, they would be doing so with a much more restricted view of the approaching traffic than the view afforded them by the third position.

©2022, SAE International

FIGURE 10.1 View of the oncoming traffic lane from three different vantage points at an intersection.

FIGURE 10.2 Occlusion of a passenger car and a motorcycle by a power pole.

Another example is shown in Figure 10.2. The first three images of this figure show a white passenger vehicle traveling behind the power pole. Because of the size of the vehicle, the pole only partially occludes the view of the car as it passes the pole. The next three images show a motorcycle passing behind the same pole. Because the motorcycle is smaller, there is a time frame when the motorcycle becomes entirely obstructed by the pole. How long the motorcycle is occluded and the degree to which this may play a role in a crash will depend on several factors, including the speed of the motorcycle, the relative position between the pole, the viewer, and the motorcycle, and when and for how long the driver is looking in the direction of the approaching motorcycle.

When reconstructing a crash where a geometric obstruction could have influenced the visibility of a motorcyclist, on-site photography and testing can be performed. Scaled diagrams and interactive three-dimensional computer models can also be used to analyze and illustrate the influence that various objects have on the view available to the driver. The use of computer modeling allows for the analysis of multiple positions for both the observer and the motorcycle. Figure 10.3 shows an example of such diagrams. In these graphics, visibility cones have been created to show the influence of the utility pole and bushes on the driver's view from two different vantage points at the same intersection.

FIGURE 10.3 Influence of vantage point on the degree to which a utility pole obstructs a driver's view.

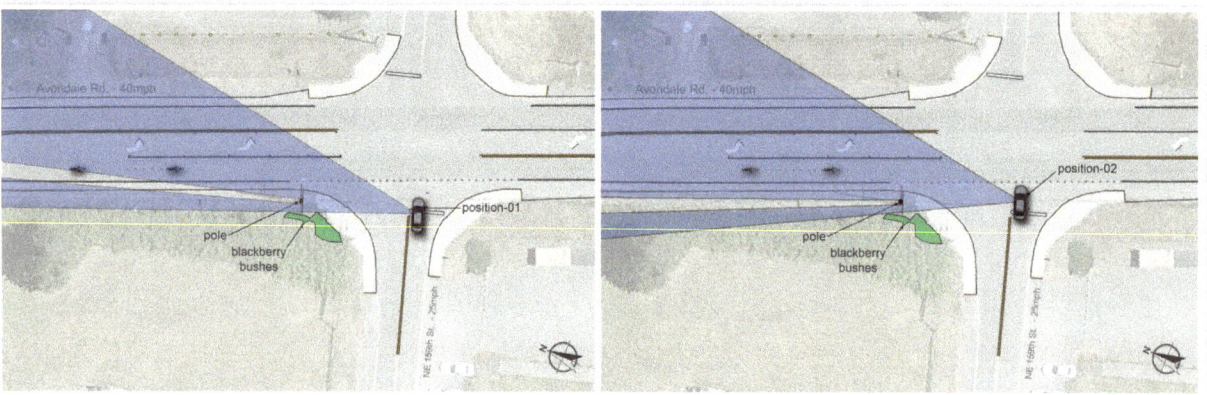

The geometry of a hillcrest can also create a visual obstruction, limiting visibility for drivers and motorcycle operators on either side of the crest. During the day, for passenger cars and trucks, it is typically their roof line and windshield that first emerges over the hill crest. Because of a lack of contrast at night, this part of the vehicle may not be visible or conspicuous, and it may not be until the headlamps emerge that the vehicle is both visible and conspicuous. For a motorcycle, it is the head or helmet of the operator that is typically the first object to emerge over the crest of the hill. This portion of the motorcyclist may not be visible at night, and in the daytime, may not be conspicuous. The headlamp of a motorcycle, which will usually be on during both the day and night, may add to the visibility and conspicuity of the motorcycle when it emerges over the hill crest. The images in Figure 10.4 demonstrate the effect that a hill crest has on visibility for both a passenger car and a motorcycle.

FIGURE 10.4 Passenger car and motorcycle cresting a hill.

Why do Drivers Sometimes Fail to "See" Motorcyclists?

Hurt and Dupont [1] noted that "the most likely comment of an automobile driver involved in a traffic collision with a motorcycle is that he, or she, did not SEE the motorcycle …" (emphasis in original). Hurt and DuPont continued: "The origin of this problem seems to be related to the element of conspicuity (or conspicuousness) of the motorcycle; in other words, how easy it is to see the motorcycle. When the motorcycle and the automobile are on collision paths, or when the vehicles are in opposing traffic, the conspicuity due to motion is very low, if it exists at all. Consequently, recognition of the motorcycle by the automobile driver will depend entirely upon the conspicuity due to contrast. If the approaching motorcycle and rider blend well with the background scene, and if the automobile driver has not developed improved visual search habits which include low-threat targets … the motorcycle will not be recognized as a vehicle and a traffic hazard exists."[1] Without discounting the factors described by Hurt and Dupont, their statements are not fully supported by later research. Additional factors identified by later researchers include physical obstructions of motorcycles from other traffic, inattention and distraction on the part of a passenger car driver, a driver conducting a visual search of inadequate duration and overlooking a motorcycle, a lack of expectation to encounter a motorcycle, and excessive speed on the part of a motorcyclist.

Olson [4] examined the literature related to why passenger car drivers sometimes fail to detect motorcyclists. Although he noted that "considered logically, it seems reasonable that motorcycles should be less conspicuous than cars because they are smaller," Olson questions motorcycle conspicuity as the likely explanation for car drivers missing motorcycles. He observed that "the strongest support for the conspicuity hypothesis may be that the offending operator often reports a failure to detect the other vehicle." However, Olson noted that "the conspicuity hypothesis has not been seriously challenged. Almost all investigators have accepted it as fact, concentrating their efforts on means to improve conspicuity rather than on asking whether the hypothesis is correct. This is unfortunate because alternative hypotheses can be advanced. Some have research data to support them; some are speculative. All are consistent with the known facts …" Olson noted that drivers claiming to have *not* seen another vehicle is *not* unique to motorcycle-car intersection collisions.[2] He stated: "Violations of right of way are a common cause of collisions between automobiles, and afterward the errant driver often claims not to have seen the other vehicle. This should not be surprising. Of all the reasons that someone would deliberately move into the path of an oncoming vehicle, failure to detect it must be high on the list. But if the claimed failure to detect is not unique to motorcycle collisions, then it is not evidence for a special conspicuity problem with motorcycles." Olson discussed other explanations for

---

1. Helman et al. [3] cited a number of sources and noted that conspicuity can be defined as "the extent to which an object stands out from its surroundings. Conspicuity is different than visibility (although in practice the same factors affect both) which is usually defined as the ease with which an object can be detected when an observer is aware of its location. It is generally acknowledged that the most important determinant of an object's conspicuity/visibility is its contrast with its surroundings, although other features such as an object's movement relative to its background also play a role."
2. Herslund and Jørgensen [5], for instance, reported a study of crashes in which drivers reported that they failed to see approaching bicyclists. They reported that experienced drivers may be more likely to make these errors than inexperienced drivers, and they suggested that this may relate to the expectations and search patterns drivers develop over time.

©2022, SAE International

why passenger car drivers sometimes miss motorcyclists, including visual obstructions and errors in the drivers' estimates of how far away a motorcycle is and how fast it is traveling.

Similarly, Hole [6] observed that "it is unarguable that [motorcycles] have a much smaller frontal area than a four-wheeled vehicle. However, being harder to detect is not the same as being difficult to detect. Motorcycles may still be well above sensory thresholds for detection. As mentioned earlier, claims that motorcycles are hard to detect because of sensory limitations rest heavily on drivers' statements that they did not see the motorcycle. Precisely what a driver means when he or she say this is a little ambiguous. Such statements are interpreted as meaning that the motorcycle was not detectable, in sensory terms. However, an equally valid interpretation is that the motorcyclist was detectable, but that for some reason this information did not reach conscious awareness."

Along these same lines, Sager and his colleagues [7] noted that "much previous research has focused on motorcycle properties, such as size, shape, and color to explain its inconspicuousness … Much of the motorcycle safety research conducted since has focused on making motorcycles more conspicuous, generally through various lighting treatments such as headlight modulators, additional lights, and bright reflective garments … There is some debate, however, regarding the effectiveness of these measures … it has been suggested that the problem may not be one of conspicuity at all … collision statistics remain largely unchanged, suggesting that the issue may not be related solely to the motorcycle's static properties." Sager et al.'s research suggests that the motorcycle and rider's dynamic properties, such as lane position, also make a difference to the likelihood a motorcyclist will be detected.

Sager and his colleagues demonstrated this using a driving simulator to examine the motorcyclist's lane position as a factor in crashes where a passenger car driver turns left and violates the right-of-way of the motorcycle. They used a North American driving configuration with cars driving on the right (rather than left) side of the road. They described their experiment as follows: "Seventeen participants faced oncoming traffic in a high-fidelity driving simulator and indicated when gaps were safe enough for them to turn left at an intersection. We manipulated the size of the gaps and the type of oncoming vehicle over 135 trials, with gap sizes varying from 3 to 5 s, and vehicles consisting of either a car, a motorcycle in the left-of-lane position, or a motorcycle in the right-of-lane position. Our results show that drivers are more likely to turn in front of an oncoming motorcycle when it travels in the left-of-lane position than when it travels in the right-of-lane position." Sager and his colleagues had determined, based on the intersection geometry and the acceleration capabilities of the vehicle, that "a three-second gap in a stream of oncoming traffic would not allow for the safe execution of a left turn, that a four-second gap would allow for the safe execution of a left turn, but leave very little safety margin, and that a five-second or more gap in the stream of traffic would allow for the execution of a left turn and leave a reasonable safety margin."

For each of the three gap sizes—3, 4, and 5 sec—participants chose to turn more frequently when the motorcyclist was in the left-of-lane position than when the motorcycle was approaching in the right-of-lane position. Sager et al. conclude that, "these results are consistent with our hypothesis that the right-of-lane position offers more motion cues to an oncoming driver and is therefore more likely to deter oncoming drivers from crossing in front of a motorcyclist's path as they approach an intersection. However, our findings are inconsistent with some motorcycle rider training which motorcyclists generally leave with the belief that they should always ride in the left portion of the lane. Our results suggest that the right-of-lane position may be a safer riding position when entering an intersection."

Ouellet [8] also examined the optimal lane positioning for motorcyclists in terms of the time they had available for collision avoidance, noting that "lane positioning as the rider approaches a potentially threatening situation is a simpler, more reliable and more effective means of reducing collision risk than reliance on emergency braking." His study revealed that "the motorcycle rider can do more to avoid a collision by moving laterally away from a threatening vehicle, putting at least one lane-width between them, before a vehicle begins to violate his right-of-way, than he can be effective braking after the other vehicle has begun to violate his right-of-way" (emphasis in original). Depending on the intersection geometry and what other vehicles are present, these statements could dictate a left-of-lane, right-of-lane, or center-of-lane positioning.

Hole et al. [9] reported three experiments related to motorcycle conspicuity. These experiments involved showing the test subjects a series of images containing traffic. Less than half of the images contained motorcyclists, so that the test subjects could not assume there would be a motorcyclist in each image. Hole and his colleagues recorded the time it took the subjects to determine if a motorcyclist was present in each image. They varied if the motorcycle headlight was on or not, the type of clothing worn by the motorcyclists (plain dark, plain bright, patterned dark, and patterned bright), the distance of the motorcycles from the viewer, and the driving situation (urban or semirural). They also examined the influence of background clutter on the conspicuity of the motorcyclists. These researchers reported that "the effectiveness of the conspicuity aids used, especially clothing, may depend on the situation in which the motorcyclist was located: bright clothing and headlight use may not be infallible aids to conspicuity. Brightness contrast between the motorcyclist and the surroundings may be more important as a determinant of conspicuity than the motorcyclist's brightness *per se*. Motorcyclists' conspicuity is a more complex issue than has hitherto been acknowledged."

Specific findings by these researchers included the fact that "motorcyclists were detected more quickly the nearer they were to the viewer, and in both locations the biggest difference between the headlight-off and headlight-on conditions was at the furthest viewing distance"; "the effectiveness of the headlight as a conspicuity aid was much less clear-cut in the urban setting than in the semi-rural environment … headlight use in the urban location enhanced conspicuity only when the motorcyclist was wearing plain bright or patterned dark clothing: when patterned-bright or plain dark clothing were worn, subjects responded faster when the headlight was off than when it was on. In the urban setting, a consistent advantage for headlight use was demonstrated only when the motorcyclist was wearing patterned-dark clothing"; "in both locations, many more motorcyclists were undetected at the furthest distance from the viewer than when the motorcyclist was nearby … for the semi-rural location, at all three distances, the error-rate for the slides in which the motorcyclist's headlight was lit was half that for the slides in which the headlight was unlit … For the urban location, at all three distances, the error-rate for the slides in which the motorcyclist's headlight was lit was lower than that for the slides in which the headlight was unlit, but not markedly so … in both locations, there was little effect of clothing type except possibly at the furthest distance."

These authors noted several limitations of their study. Among these was their observation that "problems are also caused by the fact that instructing subjects to look for motorcyclists may cause them to process a traffic scene in ways that are different to those used in normal driving … Cole and Hughes distinguish between two types of conspicuity. 'Attention conspicuity' refers to the capacity of a stimulus to be noticed when the observer is not actively looking for it. 'Search conspicuity' refers to the capacity of a stimulus to be noticed when the observer is specifically looking for it. The experiments reported here have examined factors affecting motorcyclists' search conspicuity, but in real life, attention conspicuity may also be important." Helman et al. [3] further clarified the relationship among visibility, search conspicuity, and attention conspicuity with the following three statements:

- If the observers are directed to look *at* the location of the motorcycle to see if they can detect it, we are measuring visibility.

- If the observers are directed to look *for* the motorcycle in the scene but are not told where it is, we are measuring search conspicuity.

- If the observers are simply asked to report the things in the road scene that grab their attention, we are measuring attention conspicuity.

Finally, Hole et al. observed that "the fact that there were few differences between conditions when the motorcyclist was nearby implies that motorcyclists' conspicuity *at the close range within which accidents often occur* might be relatively unaffected by such factors: within this range, it is possible that the psychological state of the driver may play a more important role than the physical characteristics of the motorcyclist … inappropriate expectancies may be more important in accident causation than the motorcyclist's physical properties."

Gershon et al. [10] reported two experiments related to motorcycle conspicuity. The first experiment "evaluated the influence of [the motorcycle and rider's] attention conspicuity on the ability of un-alerted viewers to detect it." The second experiment "evaluated the [motorcycle and rider's] search conspicuity to alerted viewers." Gershon and his colleagues varied the driving scenario (urban and interurban), the motorcycle rider's outfit (black, white, and reflective), and the distance of the motorcycle from the viewer. In the first experiment, 66 students were individually presented with a series of pictures. They viewed each picture for 0.6 sec and then were asked to report all the vehicle types they observed in each picture. In the second experiment, 64 participants viewed the same pictures utilized in the previous experiment. In the second experiment, though, the participants were instructed to look for motorcycles and to report if each photograph showed a motorcycle.

For the first experiment, Gershon et al. reported that the detection of the motorcycles "depended on the interaction between its distance from the viewer, the driving scenario and [the] rider's outfit … when the [motorcycle] was distant the different outfit conditions affected its' attention conspicuity. In urban roads, where the background surrounding the [motorcycle and rider] was more complex and multi-colored, the reflective and white outfits increased its attention conspicuity compared to the black outfit condition. In contrast, in inter-urban roads, where the background was solely a bright sky, the black outfit provided an advantage for the [motorcycle's] detectability."

For the second experiment, Gershon et al. reported that the "detection rate of the alerted viewers was very high and the average reaction time to identify the presence of a [motorcycle] was the shortest in the inter-urban environment. Like the results of experiment 1, in urban environments the reflective and white clothing provided an advantage to the detection of the [motorcycle and rider], while in the inter-urban environment the black outfit presented an advantage. Comparing the results of the two experiments revealed that at the farthest distance, the increased awareness in the search conspicuity detection rates were three times higher than in the attention conspicuity." In other words, the rider's clothing made a difference, but the driver's awareness or expectation that there would be motorcyclists in some of the pictures made a bigger difference. As Gershon et al. noted, "unfortunately, detectability—especially attention conspicuity—is compromised by the perceptual characteristics of the environment that change continuously along a route. Thus, to increase detectability, [motorcycle] riders need to be aware of the perceptual aspects of their riding environment. In parallel, the results of the second experiment with alerted viewers demonstrate that other road users (e.g., car drivers) can improve their detection performance when they increase their level of expectancy and awareness concerning a possible existence of a [motorcycle] on the road (as drivers with high expectation obtained nearly 100% detection rates)."

Lenné and Mitsopoulos-Rubens [11] reported a study in which they subjected 43 experienced drivers to a series of trials in a driving simulator. The task given to these subjects was to "turn ahead of an oncoming vehicle if they felt that they had sufficient room to do so safely." In some trials, subjects had to turn in front of a motorcycle with its headlight on, and, in other trials, the headlight was off. The gap available for the turn was also varied (short, medium, and long). Lenné and Mitsopoulos-Rubens reported that "at short time gaps low-beam headlights may confer some benefit in gap acceptance by encouraging drivers to accept fewer gaps ahead of a motorcycle with headlights on than ahead of a motorcycle with headlights off. No statistically significant differences in gap acceptance between the headlight conditions were found at either the medium or long time gaps."

Crundall et al. [12] noted that the most common cause of motorcycle collisions in the United Kingdom "was that of another vehicle pulling into the path of a motorcycle when exiting from a side road onto the main carriageway." These authors observed that the statistics related to the number of look-but-fail-to-see collisions with motorcyclists may be inflated "by self-report biases. One could imagine alternative causes: a failure to look in the appropriate direction; or having looked and perceived the approaching motorcycle, the car driver might fail to judge the level of risk that the conflicting motorcycle presents." To further examine these issues, they developed a test in which subjects viewed video of an intersection on multiple screens simultaneously. The video was from the vantage point of a driver wanting to pull out at a T-junction, and the screens were set up such that the subjects could turn their heads to the left and right to look for conflicting traffic. Crundall et al. noted that "Mirror information was edited into the forward-facing video footage, providing a left-side mirror in the bottom-right of the left screen, a right-side mirror in the bottom-left of the right screen, and a rear-view mirror at the top of the central screen. The three televisions were angled from each other at 120° providing an immersive video, wherein participants could look to the left and right, as if looking through the side windows of their car, to check for conflicting vehicles on the main carriageway." Both novice and experienced drivers were tested, as was a group of drivers with considerable experience driving both cars and motorcycles. "Specifically, we were interested in when drivers first fixate the conflicting vehicles approaching the T-junction (when they look), how long they looked for (a measure of whether they perceive) and when they press a button to pull out from the junction (which, given that the necessary—but not sufficient—preconditions of looking and perceiving are met, can be considered a measure of appraisal)."

Crundall et al.'s study included 74 test subjects—25 novice car drivers with a mean age of 20.6 years and a mean experience level of 1.6 years, 25 experienced car drivers with a mean age of 33.4 years and a mean experience level of 14.8 years, and 24 drivers with significant experience with both cars and motorcycles (dual drivers) with a mean age of 44.9 years and a mean experience level of 25.7 years with cars and 20.0 years with a motorcycle. The videos used in the study include ten scenarios with conflicting motorcycles, ten with conflicting cars, and ten with no conflicting vehicles. Conflicting vehicles could appear from either the right or the left. The clips included the approach phase to the T-junction, the stop, and then the time for the participants to decide about when they would pull out. Crundall et al. also noted that "a further 42 clips (not analysed in the current paper) were randomly interspersed which required a different response; either a lane-change decision … or a hazard perception response. Participants could not predict when a hazard might appear, and thus had to remain vigilant to hazards even during the T-junction scenarios. Response times reflecting when the participants thought it was safe to pull out were recorded, along with the participants' eye movements."

Crundall et al. reported that "the most immediate finding from the analyses was the greater caution given to conflicting motorcycles than to conflicting cars. Both the percentage of safe responses and the [reaction times] reflect a greater safety margin in responding to motorcycles ... In regard to group differences, dual drivers were more cautious than the novice drivers, with the experienced group falling in between. This pattern held regardless of whether or not there was conflicting traffic. While the overall means improved with experience, the differentiation between motorcycle clips and car clips seemed greatest for the dual drivers followed by the novice drivers...dual drivers were the most sensitive to the presence of a conflicting motorcycle, while experienced drivers appeared the least sensitive."

Crundall et al.'s research suggests that car drivers who also ride motorcycles are more aware of approaching motorcycles and less likely to violate their right-of-way. This is consistent with the findings of other researchers. Magazzù et al. [13] for instance, found that "having gained experience in riding any motorcycle...results in drivers being less prone to cause crashes with motorcycles with respect to drivers with no motorcycle license. It is reasonable to assume that car drivers who hold a motorcycle license have acquired more ability in riding and controlling two wheeled vehicles than drivers without a license. Therefore, it is possible to infer that some riding ability and knowledge of the risk annexed to riding, could protect drivers, maybe by helping them in the detection of oncoming motorcycles and the prediction of their manoeuvres."

Rogé et al. [14] reported a study aimed at determining "whether the low visibility of motorcycles is the result of their low cognitive conspicuity and/or their low sensory conspicuity for car drivers." These authors defined sensory conspicuity as "the extent to which an object can be distinguished from its environment because of its physical characteristics: angular size, eccentricity in relation to the point of gaze, brightness against the background, color, and so on ... in other words, sensory conspicuity reflects an object's ability to attract visual attention and to be precisely located as a result of its physical properties." Rogé et al. relate cognitive conspicuity to driver expectations, noting that "an observer's focus of attention is strongly influenced by his or her expectations, objectives, and knowledge ... in many cases, inappropriate expectations may be more important in accident causation than the motorcyclist's physical properties." These authors tested a sample of 42 car drivers in a simulator. Half of the drivers were motorcyclists, and the other half were not. These subjects were subjected to three test sessions lasting 12 min each, with a break in between sessions. During each session, the subjects drove on roads with a speed limit of 90 km/h and on a highway with a speed limit of 130 km/h. They also passed through junctions and roundabouts where the speed limit was 50 and 30 km/h. The traffic encountered by the test subjects in the simulator included 49 vehicles—small cars, vans, buses, tractor-trailers, and motorcycles. The authors noted that "the participants could not anticipate when and from where a motorcycle might appear because they never came back to the same section of the circuit and had to detect a motorcycle in several different situations." The authors of this study concluded that both sensory and cognitive conspicuity had an influence on drivers' detection of motorcyclists. Specifically, "a high level of color contrast between the motorcycles and the road surface enhanced the visibility of motorcycles," and car drivers who were also motorcyclists detected the motorcyclists sooner than car drivers who were not motorcyclists.

Helman et al. [3] identified and reviewed 27 studies (including some of those reviewed above) that sought to improve motorcyclists' visibility or conspicuity or to improve the accuracy of judgments of motorcyclists' speed or time to contact by other road users. These authors reported that "both [bright clothing and DRLs] seem to be capable of improving conspicuity, when this is measured in terms of detection (under search and attention conspicuity conditions), and when measured in terms of a behavioural response (such as size of gap accepted in front of a given motorcycle). The majority of studies covered in this review support this conclusion, although there are limitations … due to the number of different visual contexts in which motorcyclists find themselves when riding. For example, coloured clothing is more effective when viewed against a contrasting background. In terms of lighting, although it appears that dedicated daytime lighting on motorcycles is effective in increasing conspicuity, this effect may be smaller when other vehicles have their lights on … When lighting is arranged in such a way as to accentuate the form of the motorcycle (and to provide greater information for judging approach speed), this aids the observer in determining the time to arrival of the approaching bike (especially at night) … Across all treatments there is evidence that colour can play a role in effectiveness; this may be especially true in settings where coloured motorcycle lights aid in the motorcycle standing out from surrounding vehicles which have white lights. Although most studies reviewed show benefits of bright clothing, dark clothing may be better if the background is also brightly coloured. In line with the underlying mechanisms proposed, higher contrast with background surroundings to enable better visibility, search conspicuity, and attention conspicuity would be beneficial. Given that environments may differ over even fairly small changes in time or location, there is not likely to be a one-size-fits-all solution, meaning that motorcyclists need to be aware of the limitations of whichever interventions they use."

In addition to these factors, Brenac et al. [15] observed that "the hypothesis of a link between motorcycle speed and low conspicuity may indeed be advanced: for a given time to potential collision, the higher the motorcyclist's speed, the greater is the distance from the other vehicle. And therefore, for a given time to potential collision, the higher the motorcyclist's speed, the smaller is the motorcyclist's apparent size in the field of vision of the other driver." The other implication, of course, is that the slower the motorcyclist is traveling, the greater the time available for the intruding driver to clear the intersection before the motorcyclist arrives. Brenac and his colleagues conducted in-depth investigations of 22 collisions occurring in urban areas between motorcycles and other vehicles, many of which were situations where drivers pulled out into the path of an approaching motorcycle. Based on these reconstructions, Brenac and his colleagues concluded that there was "a significant relation between problems of conspicuity and the motorcyclist's high level of speed in accident cases occurring in urban areas."

A study by Labbett and Langham [16] examined the visual search patterns of drivers at two intersections using a hidden video camera. One of the intersections had a visibility obstruction on the corner, and the other did not and allowed for an unobstructed view of several hundred meters. The intersections were on the campus of the University of Sussex, and the video footage enabled these researchers to determine which vehicles had a university parking pass and which did not. This was used to determine which drivers were likely familiar with the intersections and which were not. Labbett and Langham concluded that, on average, "the drivers observed spent less than 0.5 seconds searching for hazards." They also found that the drivers tended to only search in one direction. Labbett and Langham also did an experiment in which participants were shown video clips of approaching traffic. They found different search patterns between novice and experienced drivers. "The experienced drivers tended to fixate on only small areas of the screen whilst novice drivers tended to search many parts of the scene."

Review of these studies leads to the following observations related to collisions where a passenger car driver violates a motorcyclist's right-of-way and then states that they did not see the motorcyclists. First, for a driver to avoid making an unsafe turn in front of a motorcyclist, they need to *detect* the motorcyclists. If they do detect the motorcyclist, they will then need to make a reasonable judgment about the time available to complete their turn before the motorcyclist arrives at the intersection. The following factors may contribute to drivers failing to detect approaching motorcyclists: (a) inattention and distraction on the part of the driver; (b) occlusion of the motorcycle caused by the geometry of the intersection, by other traffic, by the geometry of the driver's vehicle, or by the small size of the motorcycle relative to a passenger car; (c) drivers not expecting to see motorcyclists on the road; and (d) a lack of conspicuity of the motorcycle and rider. The influence of the motorcycle headlight and the color and characteristics of the rider's clothing on the likelihood the motorcycle will be detected depend on the specific environment in which the accident unfolds and on how far away the motorcycle is when it needed to be detected.

The Size-Arrival Effect

Sometimes, a driver who is wanting to pull out at an intersection detects an approaching motorcyclist, but still pulls across the motorcyclist's path when there is inadequate time to do so. In these instances, the driver is misjudging the time gap between themselves and the approaching motorcyclist. This time gap is created by the combination of the distance between the driver and the motorcycle and speed of the approaching motorcycle. Thus, to accurately assess this time gap, a driver will need to perceive the distance between themselves and the approaching motorcycle and make a reasonable assessment of the rate at which the motorcycle is covering that distance. Of course, these need not be explicitly numerical assessments.

Olson and Farber [17] observed that "drivers can use several cues to judge distance to another vehicle. The most important of these is its angular size. Thus, if a car or truck appears tiny, we know it is far away. Other cues that a driver may use are the length of the intervening roadway, the interposition of other traffic, and intervening scene or background elements such as buildings or regularly spaced telephone poles."[3] In relationship to speed, Olson and Farber stated that "driver judgement of the speed of other vehicles is generally less reliable than judgements of distance, especially when the other vehicle is moving directly toward or away from the observer. The cues drivers use to judge speed include the motion of the other vehicle relative to the background, the apparent lateral motion of vehicles traversing a curve and the rate of increase or decrease in the angular size of the vehicle as it comes closer or moves farther away ... When

---

[3] Olson and Farber also observes: "It is important to make a distinction between a person's ability to judge distance in relative terms and the ability to express that judgement in feet, yards, car lengths, fractions of a mile or, as is sometimes asked by lawyers, in multiples or fraction of a football field. When we say drivers can judge distance to within 15% it means they can tell the difference between two distances when they differ by 15%. It does not mean that if a driver is asked to estimate a distance in feet or yards, he or she will be able to do so within 15%. In fact, witnesses' judgements of distance in the usual measuring units are likely to be unreliable. What drivers can do reasonably well, based on experience, is judge a given distance as adequate or inadequate relative to the distance needed for a maneuver."

the path of the other vehicle is close to being directly toward or away from the observer, the primary closing rate cue, and perhaps the sole cue, is *rate* of change of image size."

Hancock et al. [18] reported an experiment examining the decisions that drivers made when turning left across a line of traffic. They presented oncoming traffic to these drivers at an intersection via a closed-loop, interactive, fixed-base driving simulator. The setup and signage of the intersection was permissive, such that the drivers could turn left at any time they chose. The intersecting roadways were each two lanes wide. The lanes were separated by double yellow lines. The drivers were instructed to turn left across a stream of oncoming traffic when they judged it safe to proceed. The approach speed of the oncoming traffic was varied (10 to 70 mph), as was the gap between vehicles (3 to 9 sec), and the vehicle type (motorcycle, compact car, large car, and delivery truck). Forty participants were tested in this study, which ranged in age from 19 to 44 years old. There were 16 males and 24 females. These authors concluded that "left turn decisions appeared to result from the complex interplay of rate-of-change perceptual variables such as 'time-to-contact' and the perceived characteristics of the vehicles themselves." Further, there was a "tendency of participants to accept more turns at any particular gap size as the velocity increases … The second major cross comparison finding is that the frequency of turns decreases as the size of the vehicle increases" (emphasis in original). Or, said differently, the frequency of the turns increased as the size of the vehicle decreased. The authors' data showed that the left-turning drivers more frequently accepted a 3-sec gap when motorcycles were approaching than when any other vehicle type was approaching. At many intersections, a 3-sec gap would be insufficient for completing a safe left turn.

In describing the limitations of their results, these authors noted that "it is important to caution against simplistic interpretations that might be drawn. For example, the presentation of a line of traffic of the same vehicle type all equally spaced and all travelling at a consistent velocity is a highly unusual driving condition." Also, "while the verbal instructions given to participants emphasized making turns when they thought that it was safe to do so, some may not have heeded these suggestions and adopted other risk-taking sets given the novelty of simulator driving. For example, a shift in risk taking behavior tended to accompany the collisions between participants and on-coming vehicles for a variable number of trials before returning to 'normal' turning. In addition, perceptual learning as a function of the number of trials may have influenced the pattern of results. As with any set of results, caution in interpretation is conceded, however, randomization of trial conditions and randomization of participants to conditions assisted in controlling for parts of these concerns."

Hancock and Caird [19] reported a follow-up study that utilized older drivers, ranging in age from 51 to 84 years old. They concluded that "older drivers were uniformly more conservative in their turn decisions than their younger contemporaries. For the former group, as the frontal surface area of the on-coming vehicle increased, larger vehicles tended to elicit a greater hesitance to turn." The authors' data showed a significantly lower percentage of the older drivers accepting 3-sec and 4-sec gaps when a motorcycle was approaching than did the younger drivers. Still, there were differences in the gap-acceptance decisions of the older drivers related to vehicle type. The older drivers were most conservative with their gap selection when the delivery truck was approaching and not as conservative when motorcycles were approaching (see also, the literature review by Pai [20]).

Horswill et al. [21] noted that "drivers adopt smaller safety margins when pulling out in front of motorcycles compared with cars. This could partly account for why the most common motorcycle/car accident involves a car violating a motorcyclist's right of way. One possible explanation is the size-arrival effect in which smaller objects are perceived to arrive later than larger objects. That is, drivers may estimate the time to arrival of motorcycles to be later than cars because motorcycles are smaller." These authors conducted two experiments to test this hypothesis. In the first experiment, test subjects (28 drivers who had never ridden a motorcycle) were shown video footage of traffic approaching the viewing position of the camera. Four vehicles were used to create the video footage—a small motorcycle, a large motorcycle, a car, and a van—and these vehicles were driven toward the scene at speeds of either 30 or 40 mph. The scene blacked out 4 sec before the vehicle reached the camera's position. Subjects were asked to press a response button when they estimated the vehicle would have reached the viewing position of the camera. This experiment resulted in the conclusion, consistent with the authors' hypothesis, that "time-to-arrival estimations were significantly longer for motorcycles than for the larger vehicles …."

In the second experiment, Horswill et al. varied the time at which the video was blacked out (1, 2, 4, and 7 sec prior to arrival). For these scenarios, each subject viewed the approaching vehicle for 4 sec prior to the screen going black, but the starting position of the vehicle varied. From this experiment, the authors concluded that the "motorcycles were estimated to arrive significantly later than cars, and this was significant when vehicles disappeared 1 s before they arrived (both at 30 and 40 mph). This indicated that vehicle differences were unlikely to be a result of threshold differences in detecting object expansion as all vehicles in the 1-s condition would be well above threshold before occlusion." From both experiments, these authors concluded that "one reason that motorcyclists could be at greater risk of being hit at road junctions is because of an unfortunate optical illusion. People estimated that motorcycles reached them later than cars when time-to-arrival was actually the same … This effect is consistent with the size–arrival effect … in which participants judge, incorrectly, that approaching smaller objects will arrive later than larger objects."

In 2003, Horswill and Helman [22] had examined motorcyclists' behavior in comparison to that of car drivers and reported that "motorcyclists chose faster speeds than the car drivers, overtook more, and pulled into smaller gaps in traffic, though they did not travel any closer to the vehicle in front." If a motorcyclist does choose to approach an intersection at a high speed, this will exacerbate the size-arrival effect and make it harder for a left-turning driver to judge the gap available for to complete their turn.

Effectiveness of Daytime Running Lights on Motorcycles

Extensive research has been conducted related to the front lighting of motorcycles and its influence on the likelihood drivers will detect motorcycles during the day. In a 1985 study, for example, Zador [23] examined the effectiveness—in terms of reducing fatal motorcycle accidents—of laws requiring the daytime and nighttime use of motorcycle headlights and taillights. At the time of his study, 14 states in the United States had these laws. Zador used daytime motorcycle crash data from the Fatal Analysis Reporting System (FARS) from 1975 to 1983. This study concluded that "the risk of daytime crashes was

13 percent lower in states with motorcycle daytime headlight laws than in states without such laws."

Hole and Tyrrell [24] reported a study to examine if motorcyclists not utilizing a headlight in the day would be more at risk if most motorcyclists were utilizing a headlight. They observed that "it has been suggested that drivers might scan for lights rather than for motorcyclists *per se*." To explore this possibility, they conducted two experiments. In both experiments, test subjects viewed a series of slides depicting traffic, and they had to decide as quickly as possible if each image depicted a motorcyclist. In the first experiment, approximately half of the slides contained a motorcyclist. The final slide in the series shown to each subject always contained a motorcyclist, with the headlight either on or off. For the preceding slides that showed motorcyclists, the "headlight use was either consistent or inconsistent with that of the motorcyclist in the last slide." In the second experiment, Hole and Tyrrell "manipulated the probability with which motorcyclists with and without headlights were presented to the subjects." These researchers reported that the first experiment "showed that headlight-using motorcyclists were more quickly detected than unlit motorcyclists, especially when they were far away ... [and] repeated exposure to headlight-using motorcyclists significantly delayed detection of an unlit motorcyclist." They further reported that the second experiment showed "that this delayed-detection effect occurred [even] when only 60% of the motorcyclists shown were using their headlight."

Yuan [25] reported a study of the effectiveness of the "ride-bright" legislation implemented in Singapore in November 1995. This legislation required motorcyclists to ride with their headlights on during the daytime. Yuan examined crash data for the years 1992 through 1996 and found that there was "a sharp decline in daytime fatalities compared with nighttime fatalities in 1996." Yuan observed that, "one possible reason why daytime headlights have successfully reduced fatal and serious injury accidents may be that the improved conspicuity gives time to the other motorists to react. It is reasonable to assume that if other motorists can be alerted earlier, they can stop their vehicles earlier and have a longer distance to travel before hitting the motorcycle or motorcyclist. The longer braking time available with daytime headlights may result in lower impact speeds, which in turn may lower the probability of fatal or serious injury accidents. If this is the case, then one way to make headlights more effective is to make them more conspicuous, say, by having multiple, more powerful headlights to send even earlier warning signals."

Jenness et al. [26] reported a daytime field experiment to determine if the gap acceptance behavior of drivers turning left in front of a motorcycle changed with forward lighting added to a motorcycle above and beyond the standard low beam headlamp that turns on automatically. The intent of daytime running lights (DRL) on motorcycles is to increase conspicuity, and thus, to overcome passenger car drivers' lack of expectancy and increase the likelihood passenger car drivers will recognize the presence of motorcycles. In his 1977 study, Hurt and DuPont [27] had noted that "one important countermeasure is the use of a lighted motorcycle headlamp during daylight ... The data collected by the USC-DOT Motorcycle Accident Research Teams shows that the motorcycles NOT using the headlamp-on during daylight are overrepresented in the accident population. The bouncing, flickering headlamp of the moving motorcycle is a powerful attention-getting mechanism, which greatly improves motorcycle conspicuity in traffic." A study by Olson et al. [28] confirmed this finding, noting that "the most effective means of improving daytime conspicuity ... is to require motorcyclists to drive during the day with their low-beam headlamp turned on."

The study by Jenness et al. [26] utilized 32 drivers (19 to 67 years old) and five experimental lighting systems with various configurations of auxiliary lighting. No experienced motorcycle riders or people with motorcycle riders in their immediate family were included as test subjects. Subjects viewed the approaching traffic on an active roadway (including a motorcycle) and indicated when it would and would not be safe to initiate their left turn across the approaching traffic. Jenness and his colleagues added a distracting element to the study by giving the subjects a secondary visual task that would, on occasion, occupy their attention. Jenness et al. concluded that, *on average*, the safety margin that the test subjects gave the motorcycle did not differ significantly between any of the experimental lighting systems and the baseline lighting system. "However, having either low-mounted auxiliary lamps or modulated high beam lamps on the motorcycle significantly reduced the probability of obtaining a potentially unsafe short safety margin as compared to the baseline lighting treatment. Overall, the results suggest that enhancing the frontal conspicuity of motorcycles with lighting treatments beyond an illuminated low beam headlamp may be an effective countermeasure for daytime crashes involving right-of-way violations."

In another study, Jenness et al. [29] examined the possibility that widespread use of DRL on passenger vehicles might make the use of DRL on motorcycles less effective at increasing the conspicuity of motorcycles. They referred to this as the Fleet DRL Hypothesis. These authors examined crash data from Canada, where DRL use was mandatory for the entire vehicle population and compared it to crash data from 24 US states where DRL use was not mandatory and "fleet penetration of DRL was modest." US crash data were drawn from the FARS, and Canadian crash data were drawn from the Canadian National Collision Data Base (NCDB). Crashes in the years 2001 through 2007 were studied and the authors stated that "crash scenarios that were plausibly relevant to frontal conspicuity of the involved vehicles were defined as DRL-relevant. The proportion of DRL-relevant crashes was modeled by country, year, and whether the crash involved a motorcycle." Jenness et al. concluded that "the Fleet DRL Hypothesis may be true for urban roadways (but may not be true for rural roadways). These results suggest that there could be negative consequences for motorcycle riders of widespread DRL use in the vehicle fleet. For urban roadways especially, the proportion of two-vehicle fatal motorcycle crashes that are relevant to frontal conspicuity of the vehicles (DRL relevant) is higher in Canada than in the USA. This result and other related predictions verified by the modeling results support the Fleet DRL Hypothesis for urban roadways, that *widespread use of DRL in the vehicle fleet increases the relative risk for certain types of multi-vehicle motorcycle crashes*" (emphasis in original).

Cavallo and Pinto [30] also examined the influence of DRL on cars on the conspicuity of motorcycles. They showed 24 adult licensed drivers color photographs representing complex urban traffic scenes with low-luminosity conditions (overcast skies). The photographs were displayed for 250 ms and the subjects had to identify vulnerable road users (motorcyclists, cyclists, and pedestrians). The vulnerable road users were located at varying distances and eccentricities from the vantage point of the camera. The motorcycles in the scenes always had their front light on, whereas the researchers varied if the other vehicles in the scene had DRLs on or not. Only half of the photographs contained a vulnerable road user. The limited viewing time given to the subjects was intended to resolve a methodological weakness Cavallo and Pinto point out in other studies. Namely, "we contend that the sensorial conspicuity of motorcycles should be assessed in complex environments, with a time-limited task where the observer has to 'notice' a motorcycle, not look for one. In short, the situation must call upon selective attention, in such a way that attentional conspicuity rather than search conspicuity is at stake." Cavallo and Pinto concluded that "the present study revealed a detrimental effect of car DRLs on the perception of vulnerable road users:

motorcyclists, cyclists, and pedestrians were less well detected when the headlights of cars in their vicinity were on … The car-DRL effect tended to occur when the motorcycle was far away, and when it was located in the center of the visual scene. The negative effect of car DRLs at greater distances suggests that the DRLs generated competing light patterns in conditions where the motorcycles were hard to see because of their small angular size. At shorter distances, where the angular size of the motorcycle and thus its inherent conspicuity were greater, car DRLs had little or no impact."

Rößger et al. [31] reported a study in which they aimed "to identify a front signal pattern created by additional light sources that would make motorcycles clearly and quickly distinguishable from other vehicles …." They conducted two experiments. In the first, they presented a set of 40 photographs, one at a time, to the test subjects. Nine of these photographs contained a motorcycle and the other 31 did not. These other photographs showed other types of traffic or no traffic at all. The frontal appearance of the motorcycle was varied in the following configurations: (a) a single headlight on the front of the motorcycle; (b) a t-shaped pattern of lights on the front of the motorcycle; and (c) a t-shaped pattern of light on the motorcycle and additional lights on the operator's helmet. These researchers found that "motorcycles with a T-shaped light configuration are more quickly identified, particularly when the motorcycles are in visual competition with other motorized road users. Furthermore, analysis of gaze behavior showed that they were faster fixated by the subjects in the experiment, and the mean duration of fixations was shorter."

The Perception-Response Process

To determine if a crash could have been avoided by a reasonably alert and prudent operator, it may be necessary to determine a range of probable perception-response times (PRTs) for a motorcyclist or driver. Olson [32] divided the perception-response process into four steps: detection, identification, decision, and response. According to Olson, "perception-response time begins when some object or condition of concern enters the driver's field of vision … [detection] concludes when the driver develops a conscious awareness that 'something' is present. The something may be within the field of view of the driver for some time before it is detected. Hence, there is the potential for a significant delay between the presentation of the stimulus and its being detected." During the identification step, "sufficient information is acquired about the object or condition to be able to reach a decision as to what action, if any, is required." During the decision step, the driver decides whether to change their speed or direction in response to the hazard, and during the response step, they enact that decision by moving their hands or feet to operate the controls of the vehicle. Consistent with Olson's definition, Ayres and Kubose [33] state that the PRT is the time that elapses from "perceptual availability" of a hazard until the "onset of a visible useful response" (such as braking or steering). Muttart [34], on the other hand, asserts that the PRT should not begin at the point of first perceptual availability or at the first point of detection but at the first point a detectable object or situation becomes an immediate hazard. He states, "perception-response should be from perception as an immediate hazard up until first *vehicle* response." In the same reference, Muttart states: "Most importantly, perception-response time is exactly that—it is not vision-response time. Therefore, the time period starts from when the hazard is easily identifiable as an immediate hazard until the vehicle first starts to respond due to driver response choice. Objects that cannot be easily identifiable as an immediate hazard should be analyzed further before PRTs are utilized, if utilized at all."

©2022, SAE International

It is sometimes assumed that 1.5 sec (or some other stock value) is a reasonable perception-response time for a driver, regardless of the situation, or that 1.5 sec is a good place to start for a PRT analysis. The idea of a single PRT for all responses has never been the scientifically accurate assumption for a PRT analysis. As early as 1868, Donders recognized that PRT was variable depending on the conditions of the testing. Given the potential complexities of a human response to the dynamic situation of collisions avoidance it is scientifically naïve to assert that the human response can be characterized by with a single PRT value. Research by Muttart over the last two decades has done much to advance the way that accident reconstructionists evaluate PRTs [34–37], and it has become less common for reconstructionists to assume a single value. Muttart's work has demonstrated that PRTs are situation dependent and that there is variability in the response times within a population of drivers. These facts were perhaps obvious, and others had pointed them out, but they were not often used by reconstructionists in practice. Muttart's work has given reconstructionists the ability to apply these ideas in practice by providing mathematical equations for predicting driver PRTs for various situations and implementing these equations in his software, IDRR (Interactive Driver Response Research).

These equations are based on data from hundreds of driver response studies, including simulator studies, closed track studies, and most recently naturalistic driving studies. The following scenarios are addressed: (1) drivers responding to lead vehicles that were either stopped or moving slowly; (2) drivers responding to being cut off; (3) drivers responding to vehicles intruding into their path; and (4) drivers responding to traffic signals. Within each of these categories, Muttart has demonstrated that apparent discrepancies between the results of various studies can be explained in terms of differences in methodology. He noted that, "in general, as the methodology of the experiment became closer to that of real life, the response times increased." Also, "response times increased from laboratory to simulator to closed course and then to road studies."

As an example, Figure 10.5 depicts sample data obtained from IDRR for a path intrusion scenario. The orange bars are average PRTs for car drivers responding to a path intrusion in the daytime at an intersection. The blue bars are average PRTs for motorcyclists responding to a path intrusion in the daytime at an intersection. These values are approximately 0.1 sec longer than those for the car drivers. In his 4-h webinar related to the path intrusion scenario, and in his 40-h advanced human factors course, Muttart [37] indicates that motorcyclists' PRTs are generally about 0.1 sec longer than passenger car drivers' PRTs. Muttart attributes this to the number and complexity of the avoidance responses that motorcyclists can have.

The gray bars in Figure 10.5 are 85th percentile PRTs for motorcyclists responding to a path intrusion in the daytime. These values vary from 0.5 sec to 0.8 sec longer than the average values. The yellow bars are 85th percentile PRTs for motorcyclists responding to a path intrusion at night. These values are 0.2 sec longer than the daytime values. This graph also shows that PRTs depend on the eccentricity, which is the angle between the heading of the through vehicle and the area from which the intruding vehicle is coming (the stop bar of an intersection roadway, for instance). As this angle increases, PRTs increase. In practice, this angle depends on the geometry of the roadway and on the relative positions of the vehicles when the intruding vehicle begins pulling out.

One question often asked when assessing a motorcyclist's ability to avoid a crash is: "Which PRT should be used—the average value or the 85th percentile (or, perhaps, some other high percentile comparable to the 85th percentile)?" The best answer is that there is no single value. If we assume that for a population of drivers, the variability in PRTs can be represented with normal distribution, then 50% of drivers will have a PRT faster than the

CHAPTER 10 Human Factors in Motorcycle Crashes   323

FIGURE 10.5 Path intrusion perception-response times (PRTs) from Interactive Driver Response Research (IDRR).

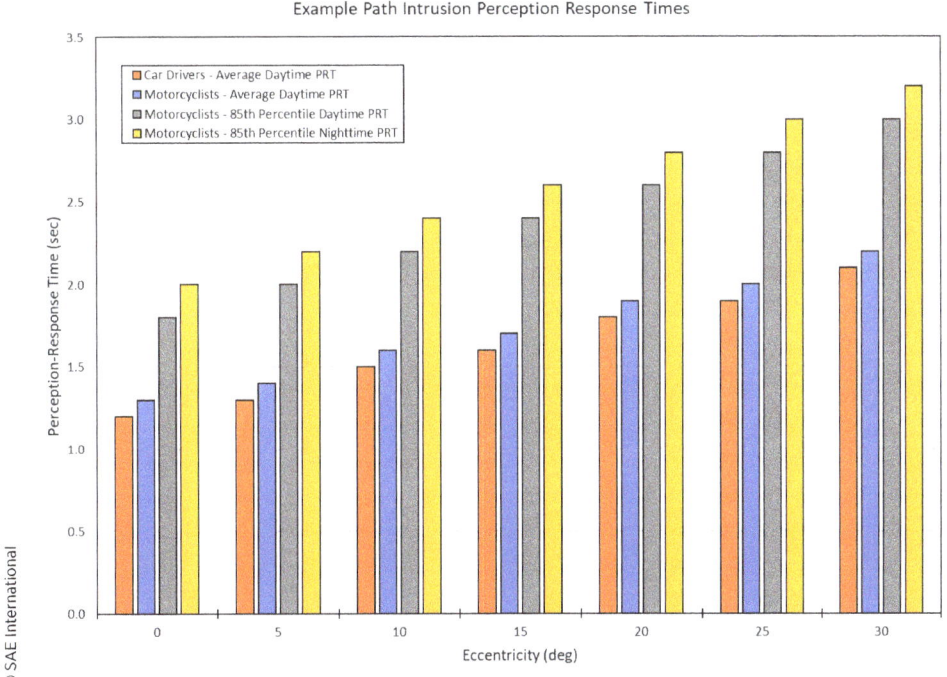

average, and 50% of drivers will have a PRT slower than the average. To use the average PRT when assessing a motorcyclist's opportunity to avoid a hazard is to impose the assumption that the motorcyclist will have a PRT faster than 50% of the population. It is not reasonable to limit the analysis to the single assumption. In fact, we cannot know where a randomly selected motorcyclist (the motorcyclist involved in our crash) will fall within the population of motorcyclists when it comes to their PRT. There is also no basis for saying that the PRT of a motorcyclist who has a 75th percentile PRT (for instance) is unreasonable, for they are still within the spread of the population for normal, attentive motorcyclists. Thus, when assessing the avoidability of a crash to a motorcyclist, it more appropriate to report a range of PRT values. Typically, one can adequately describe the distribution of PRTs by reporting the 50th percentile (average) and the 85th percentile.[4]

---

[4] This point is affirmed in the book *Forensic Aspects of Driver Perception and Response*, 4th ed., by David Krauss: "While they are undoubtedly useful and very commonly used, it is important to remember that the measures of central tendency [i.e., the average] represent but a single point in a distribution. They tell us nothing about the scatter in performance, the shape of the distribution, or how that distribution relates to other distributions that may be of interest. In addition, they are virtually no help in predicting what can be expected from one randomly selected individual … Accident investigators are generally interested in assessing the performance of a given individual, who is drawn from an unknown part of the distribution of interest. In such cases measures of central tendency are virtually useless as a guide …."

In any human motor control situation, there will always be a speed-accuracy trade-off. The less tolerant a response is to movement errors the slower the physical action will need to be in order to execute the motion successfully. This is described by Ayres and Kubose [33], who noted that "it has long been understood that faster responses tend to be less accurate … Clearly there is no single value that can be said to represent the reaction time of even one person for one task, without considering the accuracy of task performance." This is relevant to crash avoidance because "a driver would tend to respond quickly although perhaps not very accurately when a collision is imminent, but would take a more careful approach (accurate, appropriate) with more time available. There is some evidence that drivers are sensitive to the immediacy of a hazard and can adjust accordingly." Thus, "one should not expect … that drivers will efficiently use all time available to make an optimum avoidance maneuver … it is unreasonable to expect optimal response timing and maneuver performance by drivers faced with emergencies."

Another issue that comes up related to PRTs of motorcyclists is that strategy of a motorcyclist prepositioning their fingers on the front brake lever. Hurt et al. [38] examined the influence of hand position on motorcyclist's brake response times. This study utilized a Kawasaki KH100 that was equipped with "an electromechanical timer to record response times in brake actuation. A red light of approximately 25 in^2 surface was mounted on the front of the test motorcycle in the central visual space of normal motorcycle operation. The timer was coupled with controls so that the timing would start simultaneously with the illumination of the red signal light, and the timer would stop as the participant applied the front brake by lever actuation." The motorcycle was stationary during the tests, and the participants knew that they were to apply the brake lever when the light became illuminated. Each of the 100 subjects was tested in two initial hand positions—full grip on the throttle and with fingers prepositioned on the brake lever. The mean response time for all subjects with a full grip on the throttle was 0.433 sec, and with fingers prepositioned on the brake lever it was 0.219 sec. There was also less variability in the response times with the fingers prepositioned on the brake lever ($\sigma = 0.053$ sec versus $\sigma = 0.086$ sec).

Ecker et al. [39] examined the brake reaction times of Austrian motorcyclists by testing more than 300 individuals, most of which were participants in the motorcycle safety courses offered by the Austrian Automobile Association. A Honda CB500 motorcycle was utilized, which was equipped with two digital timers to measure brake reaction times on both the front and rear brakes. Ecker et al. reported that "A red signal light was mounted on the instrument panel of the motorcycle, being positioned on the peripheral of the visual field for motorcycle operation … The light could be activated at any time by the test coordinator via remote control. Thereby the trigger signal for starting the braking maneuver was rather unexpected for the riders. At the same moment the signal light was triggered, the digital timers also were started." The test subjects were instructed to drive at a speed of approximately 60 km/h (37 mph) and to make a "full stop emergency braking maneuver when the bright red flare of the signal light went on." Thus, while the riders may not have been able to predict when the light would turn on, they knew it would turn on and what the appropriate response was. The "brake reaction time" was defined as "the period of time that starts with the trigger signal and ends when the brake light switch registered a brake application … It was neither feasible nor advisable to detect the first contact with the brake lever because some riders already had minimal contact prior to the trigger signal. However, the electrical switches at the front-wheel brake and the rear-wheel brake were adjusted for minimal travel of the respective lever …."

Ecker et al. found that "hand- and foot-position during riding can have a significant effect on the brake reaction time." They also found that riding experience, measured by "total driven distance" and "number of years of motorcycle use" influenced the average BRT. "Both parameters show a negative correlation coefficient with BRT on the front wheel brake (and on the rear wheel) ranging from −8% to −16% … it can be concluded that riding experience improves the brake reaction time and (at least) compensates adverse effects of age. Since the latency period can hardly be shortened by experience (i.e. training) it can be supposed that the movement period is shorter for experienced riders. A possible explanation could be that experienced riders are more frequently in a 'ready-to-brake' position, it might also be that these riders have better trained muscular actions and therefore a faster response." Of course, it should be kept in mind that these riders were not responding to an actual emergency situation, and it is unknown how this would play out in real-world emergency situations. At any rate, the Hurt and Thom study and the Ecker et al. study show that a rider can decrease their PRT by prepositioning their fingers on the front brake lever. The overall influence is around 0.2 sec.

Motorcyclists' Responses to a Laterally Incurring Vehicle

Many motorcycle crashes occur at intersections, when a car driver attempts to cross the path of an approaching motorcyclist. Consider a scenario in which the driver of a passenger vehicle turns left across the path of an approaching motorcyclist. Initially, as depicted in Figure 10.6, the left-turning driver sits stopped in the turn pocket waiting to make the turn and assessing the oncoming traffic.[5] Two possible lane and lane positions are depicted for the approaching motorcyclist in Figure 10.6: (1) a left-of-lane position in the left lane and (2) a left-of-lane position in the right lane.

FIGURE 10.6 A left turn across path (LTAP) scenario involving a motorcycle.

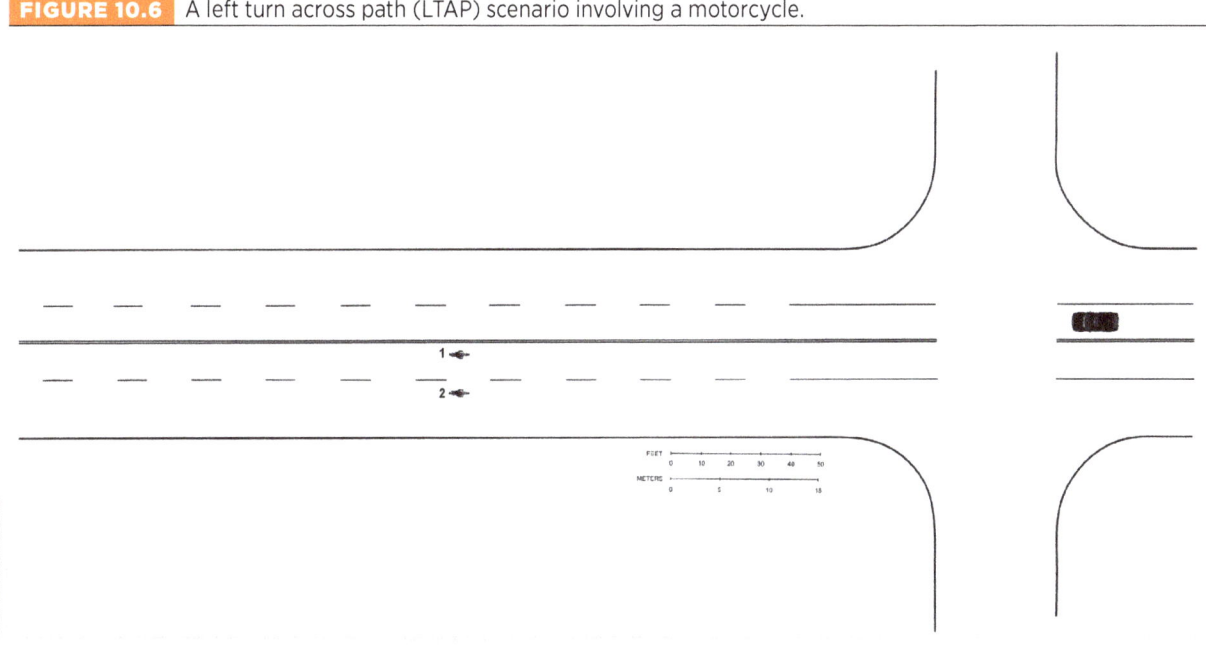

---

[5] This is the same intersection geometry used by Ouellet in his excellent study titled "Lane Positioning for Collision Avoidance: An Hypothesis." See the Proceedings of the 1990 International Motorcycle Safety Conference: The Human Element, vol. 2, Orlando, Florida, pp. 9.58–9.80.

Assume that this scenario is occurring during the day and that there are no other vehicles approaching the intersection with the motorcyclist. The motorcyclist has a green light. The car driver also has a green light and is permitted to turn left across the motorcyclist's path, if the turn can be completed safely. As the left-turning vehicle sits stopped and the motorcyclist approaches, there is no way for the motorcyclist to know which of these will occur:

- The left-turning driver will wait to turn until the motorcycle has passed through the intersection. This is what occurs most of the time.
- The left-turning driver will turn left across the path of the motorcyclist.
- The left-turning driver will begin a turn and then stop, blocking some portion of the intersection.
- The left-turning driver will begin to nudge out as the motorcyclist approaches, intending to time a turn into the intersection immediately after the motorcyclist passes. This scenario presents the motorcyclist with a different psychological experience than the first, but from the left-turning driver's perspective is equivalent to the first.

The first of these possibilities represents a scenario in which the left-turning driver detects the motorcyclist and decides the available gap is insufficient to complete the turn. The second is a scenario in which the left-turning driver either does not detect the motorcyclist or misjudges the time available to make the turn. The third possibility is a scenario in which the left-turning driver does not initially detect the motorcyclist, but eventually does; or, perhaps, a scenario where the left-turning driver's assessment of the available gap changes through the course of their maneuver. From the left-turning driver's perspective, the fourth possibility is equivalent to the first. But from the motorcyclist's perspective, the fourth highlights the uncertainty and ambiguity inherent in this intersection approach.

For each of these possibilities, the left-turning vehicle initially represents a *potential* hazard for the motorcyclist. For the first, the left-turning vehicle never becomes anything more than a potential hazard and calls for no emergency response. For the second and third, the vehicle proceeds into the intersection and becomes an actual, *immediate* hazard to which the motorcyclist must respond.[6] For the fourth possibility, the motorcyclist is left to wonder what the driver's intentions are. If the motorcyclist could know the other driver's thoughts, no response would be called for; but the opposing driver's thoughts are not known. So, presented with these four possibilities, what is the dividing line between a potential and an immediate hazard, and when should the motorcyclist conclude that an emergency response is warranted?[7]

[6.] See the following publication for further discussion of potential versus immediate hazards: Muttart, J.W., *Drivers' Responses in Emergency Situations*, Crash Safety Solutions, February 2019.

[7.] Muttart compiled studies that examined driver perception-response times (PRT) for this scenario. This compilation is contained in some of his publications and within his software IDRR (Interactive Driver Response Research). For this scenario, Muttart recommends referencing the start of the approaching driver's or motorcyclist's PRT to the first lateral movement of the left-turning vehicle. This reference for the PRT enables a consistent comparison of a crash scenario to the data contained in the relevant studies. However, it is important to state that we do not know the precise timing of the approaching rider's perceptions or thoughts. Fortunately, we do not need to know this timing for our analysis. I raise this issue simply to point out that, despite the method we use to achieve consistency in analysis, for the motorcyclist in the real-world scenario, the dividing line between a potential and immediate hazard in this scenario is ambiguous. This is particularly a problem for approaching motorcyclists because there is the possibility of the motorcycle capsizing from the heavy braking if the motorcyclist responds more severely than needed. The problem for the motorcyclist is to discern how severe of a response is needed without having access to the left-turning driver's thoughts or intentions or future actions.

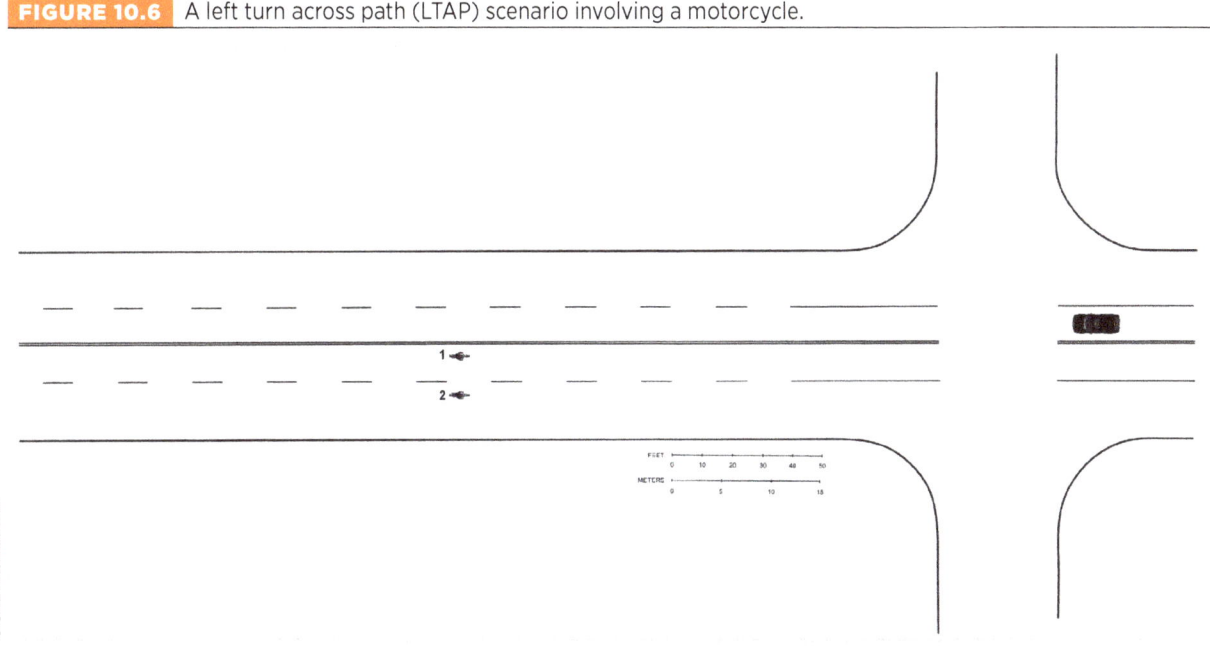

Given the uncertainty in this scenario, there are actions the motorcyclist could take to improve their ability to cope with any of the choices the left-turning driver could make. For example, they could:

- Ride at a safe speed as they approach the intersection.[8]
- Move as far to the right as practical.
- Position two fingers on the front brake lever (or hover over both brakes). Hovering over the brakes seems to achieve about a one to two-tenths of a second reduction in the motorcyclists' PRT [38–40].

Even with these actions, ambiguity remains: *the motorcyclist does not know what the left-turning driver is going to do.* Sometimes it is suggested that a motorcyclist should initiate an emergency response to a left-turning vehicle before that vehicle makes any lateral movement into the turn. This is unreasonable, impractical, and ill-advised. As expected, most of the time, the left-turning driver lets the motorcyclist pass without initiating their turn, even if they have their turn signal on and are nudging forward as the motorcyclist approaches. Initiating an emergency response to every potential hazard cannot be expected of any vehicle operator. While operators should drive defensively, there must be the allowance for the expectation that vehicle operators, in general, will operate with the laws that govern the roadways. Muttart has noted: "Even with a directional signal activated, a through driver might have no reason to believe the turning driver would violate a right of way. If drivers were expected to respond to hazards when first visible, drivers would be braking for every car they drove past … the [approaching] driver needs to first recognize actionable information that the opposite vehicle will turn."[9]

The idea that drivers can expect other roadway users to abide by the laws is supported by case law in many states. This case law generally states that approaching drivers are not obliged to assume or anticipate that opposing vehicle will fail to yield the right of way. As an example, the Supreme Court in Washington State noted: "To rule differently, would, we fear, make shambles of the right-of-way rule. Everyone driving upon an arterial highway observing vehicle at the intersections, approaching or waiting to enter, would be obliged to slow his vehicle to a near halt until he could ascertain with reasonable certainty whether the approaching vehicles intended to allow a fair margin of safety before entering upon the arterial. This, of course, defeats the very idea of arterial highways and the right of way at uncontrolled intersections, both of which are designed to allow a continue flow of traffic at safe speeds."[10]

[8.] What is a safe speed? The answer is, of course, that it depends. We could perhaps start with equating a safe speed with the speed limit. However, keep in mind that, for motorcyclists, there are sometimes legitimate reasons for not riding the speed limit while approaching an intersection. For example, a motorcyclist can gain a safety advantage by traveling through an intersection with other traffic—using that traffic as a shield. Sometimes the other traffic will be exceeding the speed limit and it will be safer for the motorcyclist to speed to keep up with that traffic than to fall back and travel through the intersection alone. Different scenarios call for different "safe speeds."

[9.] Dinakar, S. and Muttart, J., "Driver Behavior in Left Turn across Path from Opposite Direction Crash and near Crash Events from SHRP2 Naturalistic Driving," SAE Technical Paper 2019-01-0414, 2019, doi:10.4271/2019-01-0414.

[10.] *Petersavage v. Bock*, 72 Wn.2d 1 (1967), 431 P.2d 603, The Supreme Court of Washington, https://law.justia.com/cases/washington/supreme-court/1967/38286-1.html, accessed on September 2, 2021.

A Numerical Example of the Left Turn Across Path Scenario

To illustrate the effect of ambiguity in this scenario, consider a numerical example based on Figure 10.6. A Range Rover is positioned in the left turn pocket, facing to the left on the page and about to initiate a left turn across the path of an approaching motorcycle, a Triumph Street Cup, approaching from left to right on the page. The speed limit is 35 mph. Assume that the approaching motorcyclist is traveling 35 mph. There are two through lanes, and the motorcyclist will need to decide which lane and lane position to ride in while approaching the intersection. The two possible lane and lane-position choices depicted in Figure 10.6 are not the only options for the motorcyclist, but these will be adequate for illustration purposes. Assume also that this scenario is playing out on an asphalt road surface.

Assume, for now, that the left-turning Range Rover is a time-limited hazard [41]. This means that the driver of the Range Rover will make a continuous turn and be in the motorcyclist's path for a limited time. Formulated this way, the avoidance action required by the motorcyclist is to use braking to lengthen the time it takes to reach the intersection, so the Range Rover will be afforded the necessary time to pass through. This scenario can deteriorate for the motorcyclist if the driver of the Range Rover recognizes the presence of the motorcyclist partway through the turn and makes the choice to stop in the intersection, perhaps ending up within the path of the motorcyclist. The scenario then becomes a distance-limited hazard where the motorcyclist must stop short of the stopped or stopping vehicle.[11]

We make the following additional assumptions in developing this scenario:

- The driver of the Range Rover rolls straight forward for 1.5 sec with an acceleration of 0.07 g, and then initiates the turn.

- As the driver initiates the turn, the acceleration increases to 0.15 g.

- The driver of the Range Rover releases the throttle when the lateral acceleration of the vehicle reaches approximately 0.22 g [42].

- The PRT for the approaching motorcyclist begins when the heading angle of the Range Rover begins changing (i.e., the first lateral movement of the Range Rover; this occurs 2.0 sec into the scenario).

- A 1.5 sec PRT is assumed for the motorcyclist.[12] This PRT includes the time it takes for the motorcycle to respond to the operator's avoidance input (braking in this illustration). Thus, the motorcyclist will achieve the maximal braking level at 3.5 sec into the scenario.

- A 0.25 sec brake lag is assumed for the motorcycle. This lag time is included in the PRT, so the braking starts building up 3.25 sec into the scenario.

[11.] And, of course, from the motorcyclist's perspective, these scenarios may not be distinguishable at the time the motorcyclist needs to begin responding. The motorcyclist may respond as if the scenario was a distance-limited hazard regardless of what the left-turning driver ends up doing.

[12.] One factor worth mentioning: Muttart and others have shown that the angle between the approaching driver's eyes and the left-turning vehicle (the eccentricity) influences the PRT, with the PRT getting longer as the eccentricity angle increases. The time-to-collision, which would be different for the left and right lane scenarios, can also have an influence on the PRT. For this illustration, we are assuming that motorcyclists riding in the left and right lanes would have the same PRT, but in actuality, the rider in the right lane would have a higher eccentricity to the left-turning vehicle, and thus potentially a longer PRT. We will take up a more nuanced illustration in the next section. The PRT used here in this section is for illustration purposes.

©2022, SAE International

Two braking levels are considered—0.35 and 0.6 g. The first of these is equivalent to moderate braking with the front and rear brakes or emergency level braking with the rear brake only. Most riders will be able to achieve this deceleration. The second braking level is equivalent to an emergency level of braking with the front and rear brakes (or the front only). Many riders will be able to achieve this level of deceleration. Some riders will be able to exceed this level of deceleration. Some riders will exceed a 0.6 g deceleration, but will lock up their front tire doing so, and will lay the motorcycle down or reach a deceleration sufficient to cause a pitch-over. This paragraph assumes the motorcycle does not have antilock brakes (ABS). For motorcycles with ABS, motorcyclists will generally be able to achieve a higher deceleration on a more consistent and predictable basis without causing capsizing.

These assumptions were implemented in the software package PC-Crash. This analysis resulted in the following zones for the two possible lane and lane positions (see Figure 10.7):

Red Zone: An impact is unavoidable by the motorcyclist. The left extent of this zone is defined by the motorcyclist achieving a braking deceleration of 0.60 g.

Yellow Zone: An impact is avoidable by the motorcyclist assuming they can achieve a braking deceleration between 0.35 and 0.60 g.

Gray Zone: An impact is avoidable by the motorcyclist assuming they can achieve a braking deceleration between 0.00 and 0.35 g.

Brown Zone: In this zone, the motorcyclist would not need to brake to avoid a collision. However, because the motorcyclist cannot read the mind of the left-turning driver and cannot predict the exact timing of the left turn, they might choose to brake. This braking could lead to a collision that would not have otherwise occurred. The precise size of the brown zones will vary from rider-to-rider because of variations in the psychology and perceptions of riders.

The zones in Figure 10.7 were established using motorcycle positions corresponding to the beginning of the Range Rover's turn, as signified by the heading of the Range Rover beginning to change (2 sec into the simulations). Several observations follow from these zones [8]. First, the size of red zone is significantly larger for the rider who chooses the left lane over the rider who chooses the right lane. The combined size of the yellow and gray zones (the area where braking by the motorcyclist makes a difference) is also smaller for the rider in the left lane compared to the rider in the right lane. The rider who chooses the right lane benefits from increasing the lateral distance between themselves and the left-turning vehicle. This would be reversed for a scenario where the Range Rover is attempting to cross the intersection from the motorcyclist's right.

The zones in Figure 10.7 are established assuming the motorcyclist begins perceiving and responding as soon as the heading angle of the Range Rover begins to change. With a PRT of 1.5 sec, this implies the motorcyclist begins applying the brakes when the Range Rover is in the position shown in Figure 10.8. Thus, particularly for the motorcyclist in the right lane, this scenario remains ambiguous when their braking response must begin. Will the driver make a continuous turn or suddenly recognize the presence of the motorcyclist and make the choice to stop in the intersection? There is no way for the motorcyclist to know before they must respond. They could wait to respond until the situation becomes less ambiguous, but this delay would increase the likelihood of a collision for the scenarios in which the Range Rover proceeds through the path of the motorcyclist. For an accident reconstructionist, this means that the response of the motorcyclist in this scenario should be judged based on what could be known about the scenario by the motorcyclists at the time the response was required, not based on the hindsight of the accident reconstructionist who knows what the left-turning driver ultimately chose to do.

FIGURE 10.7 Avoidance zones for this scenario (red = unavoidable; yellow = avoidable with a braking deceleration between 0.35 and 0.60 g; gray = avoidable with a braking deceleration between 0.00 and 0.35 g; brown = braking not necessary but might be employed).

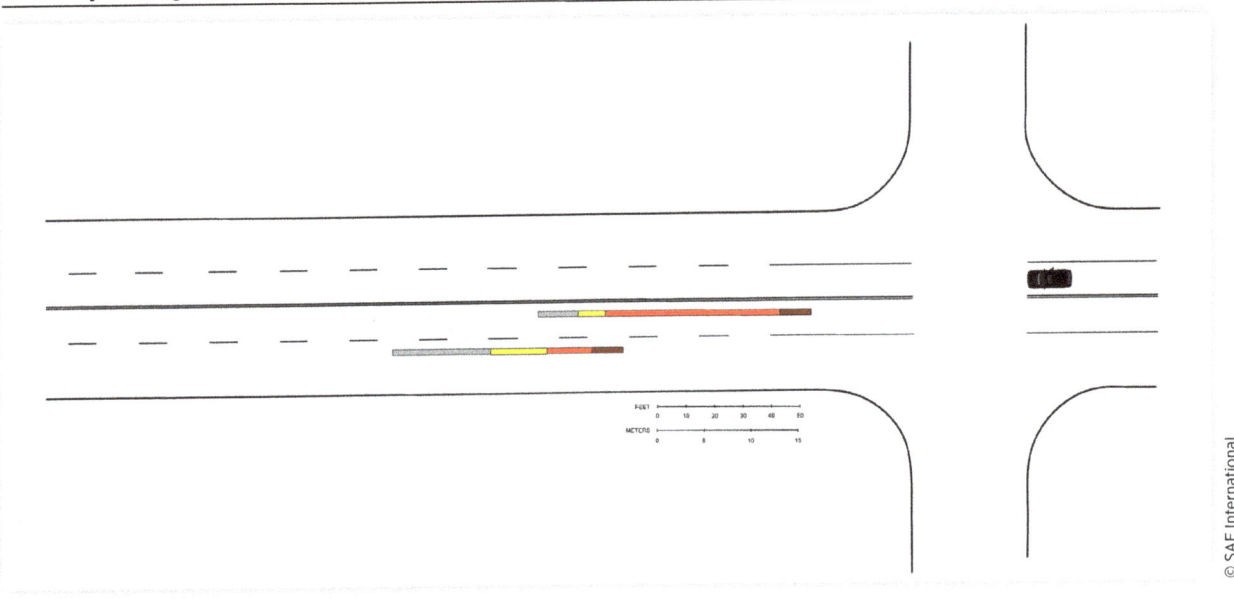

FIGURE 10.8 The position of the Range Rover when the motorcyclists begin braking.

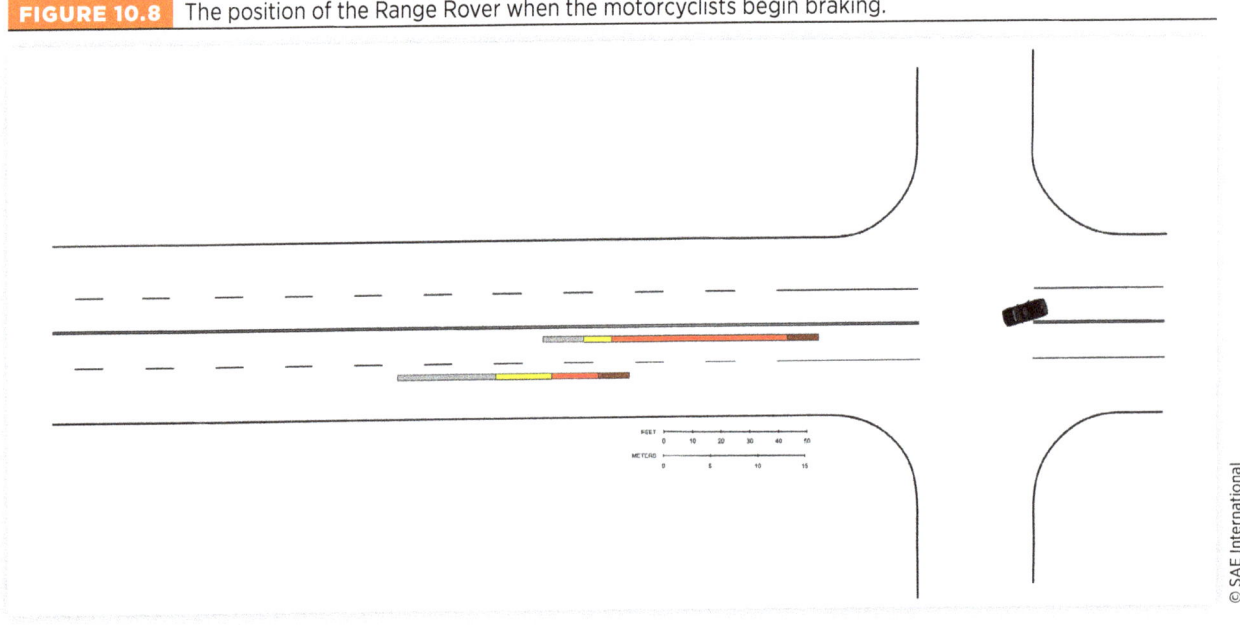

The zones in Figure 10.7 also assume that the approaching rider transitions immediately from their perception-response process to their maximum level of braking. Ising et al. [43] observed that, in accident reconstruction, "there has traditionally been an assumption that full braking is occurring upon completion of the mechanical brake lag. This assumption is challenged by a growing body of research indicating a concurrent driver-related delay between brake application and full braking. Such a delay would have upstream consequences on the calculated relative positions and speeds of the vehicles at the moment of hazard onset." In other words, in responding to a developing hazard, drivers do not typically move immediately to maximum braking. Other authors make similar observations. Writing about a different, but likely related scenario, Lee [44] observed that "a driver does not normally initiate full-power braking as soon as he sees that a lead vehicle is braking—among other things this would risk being run into from behind. Rather, the driver most likely adjusts his braking on the basis of his assessment of the urgency of the situation … Emergency braking is blind braking, its purpose being to stop in the shortest possible distance. Normal braking, however, is smooth and controlled."

Prynne and Martin [45] noted that "accident studies have shown that the full braking capability of vehicles is not often used in emergencies … Emergency braking, therefore, is often a two-stage process with drivers rapidly depressing the brake pedal to the normal limit of depression (about a third of the full range available) and then depressing the pedal further to some lower position after they have thought about the situation." Prynne and Martin also note that "as a general rule, drivers brake if the obstacle is some distance away and swerve if the obstacle is very close … these are automatic reactions."

Ising et al. [43] examined this issue through analysis of data from Mazzae et al. [46]. The Mazzae et al. study included both dry and wet road testing. The dry road testing utilized 104 male and 88 female drivers between the ages of 25 and 55 years old. The wet road testing utilized 26 males and 27 females. Test subjects were told that the study was to assess steering and speed maintenance in typical driving conditions. They drove several laps on a track, passing through a simulated intersection several times. Initially, real vehicles were situated at this intersection as if waiting to travel across the road on which the subjects were driving. Between the third and fourth laps, the real vehicles were replaced with foam replicas, and as the subjects approached the intersection on their fourth lap, the foam vehicle to the right was towed rapidly into their path, blocking half of their lane. This study is discussed in greater detail in the next section.

Ising et al. analyzed Mazzae et al.'s data and observed that "there is a driver-related braking delay occurring after initial brake application and prior to full emergency braking. In the first phase of brake application, lasting approximately 0.3 seconds, the vehicle reaches a moderate deceleration (about 0.4 g). Thereafter, the vehicle deceleration profile is dependent on [time-to-intersection (TTI)] with those drivers that did ultimately apply full braking taking approximately 0.4 to 0.8 seconds longer to do so … this interpretation of our TTI data is consistent with the suggestion of Prynne and Martin who hypothesized a two-phase braking process in which drivers monitored their approach and adjusted their braking accordingly. However, if drivers were capable of rapidly incorporating visual feedback of hazard distance into their braking behavior, it is not clear why there was no difference found between drivers whose braking attempts resulted in a crash and those who avoided impact. Relative vehicle position perceived after braking is initiated may play only a modest role in braking behavior." Ising et al. conclude their article with this statement related to crash avoidance analysis: "Finally, while the calculated braking profile can be applied when full braking is known to have been applied, this study showed that many drivers never achieved full, or even moderate, braking in this lateral incursion scenario." These studies were related to passenger car drivers, and a similar study related to motorcyclists would be useful. However, it is reasonable to think that the findings of these studies would apply to motorcyclists. In fact, for a motorcyclist riding a motorcycle with a conventional braking system, the loss of stability and capsizing that could occur by locking up one of their wheels adds additional incentive to not utilize full emergency braking.

©2022, SAE International

A More Nuanced Example

We can make this example more nuanced by incorporating the influence of eccentricity and time-to-collision[13] (TTC) into our evaluation of the motorcyclist's PRT. This will lead us to a different PRT for the motorcyclist in the left lane versus the motorcyclist in the right lane. For a motorcyclist who needs to respond to the left-turning vehicle, the eccentricity for the motorcyclist in the left lane will be in the range of 3 to 6°. For the motorcyclist approaching in the right lane, the eccentricity will be in the range of 5° to 9°. For a motorcyclist in the left lane who needs to respond with braking, the TTC when the left turn begins would be in the range of 1.5 to 3 sec. For motorcyclists in the right lane who need to respond, the TTC would likely be in the range of 2.5 to 4 sec.

Table 10.1 lists average PRTs obtained from IDRR for this scenario using these ranges of eccentricity and TTCs. The values listed in parenthesis are the 85th percentile PRTs. The PRTs in Table 10.1 are parsed by TTCs that are either less than or greater than 3 sec. This 3-sec TTC is the dividing line used in IDRR, so it is also used in Table 10.1. Two observations can be made based on the data in Table 10.1. First, the parsing of the data into TTCs less than or greater than 3 sec creates a discontinuity in the data that may be too simplistic for analysis of some crashes. For example, these data suggest that, for an intruding vehicle turning from the next lane, approaching drivers have had an average PRT of 0.9 sec if the TTC was 2.9 sec, but 1.6 sec if the TTC was 3.1 sec, a 0.7 sec difference in PRT for an insignificant change in TTC. For the current illustration, the PRTs listed in Table 10.1 will be used without modification. However, for evaluating some crashes, the relationship between TTC and PRT may need to be evaluated in a still more nuanced fashion.

Second, the PRTs for approaching drivers more than one lane away from the intruding driver are consistently longer than those for approaching drivers in the next lane. This seems to suggest that drivers or motorcyclists in the right lane would be using at least some of the additional time they have available to evaluate the situation—they are waiting, perhaps, until it becomes clearer what the left-turning driver is going to do.

TABLE 10.1 Perception-response times (PRTs) for left turn across path/opposite direction (LTAP-OD) scenarios from Interactive Driver Response Research (IDRR)

Eccentricity (deg)	Perception-response time (PRT) (sec)			
	Next lane, time-to-collision (TTC) < 3 sec	Next lane, time-to-collision (TTC) > 3 sec	More than one lane away, time-to-collision (TTC) < 3 sec	More than one lane away, time-to-collision (TTC) > 3 sec
3	0.9 (1.2)	1.6 (2.3)	1.4 (1.9)	1.9 (2.7)
4	0.9 (1.3)	1.6 (2.3)	1.4 (2.0)	1.9 (2.7)
5	0.9 (1.3)	1.7 (2.3)	1.4 (2.0)	2.0 (2.8)
6	0.9 (1.3)	1.7 (2.4)	1.4 (2.0)	2.0 (2.8)
7	0.9 (1.3)	1.7 (2.4)	1.4 (2.0)	2.0 (2.8)
8	1.0 (1.3)	1.7 (2.4)	1.4 (2.0)	2.0 (2.8)
9	1.0 (1.4)	1.7 (2.4)	1.5 (2.1)	2.0 (2.9)

© SAE International

---

[13]. The TTC is calculated from the instant the left turn begins (first lateral movement) until the time a collision would occur assuming the approaching driver did not brake.

FIGURE 10.9 Avoidance zones for this scenario (red = unavoidable; yellow = avoidable with a deceleration between 0.35 and 0.60 g; gray = avoidable with a deceleration between 0.00 and 0.35 g; brown = braking not necessary but might be employed and could result in a crash).

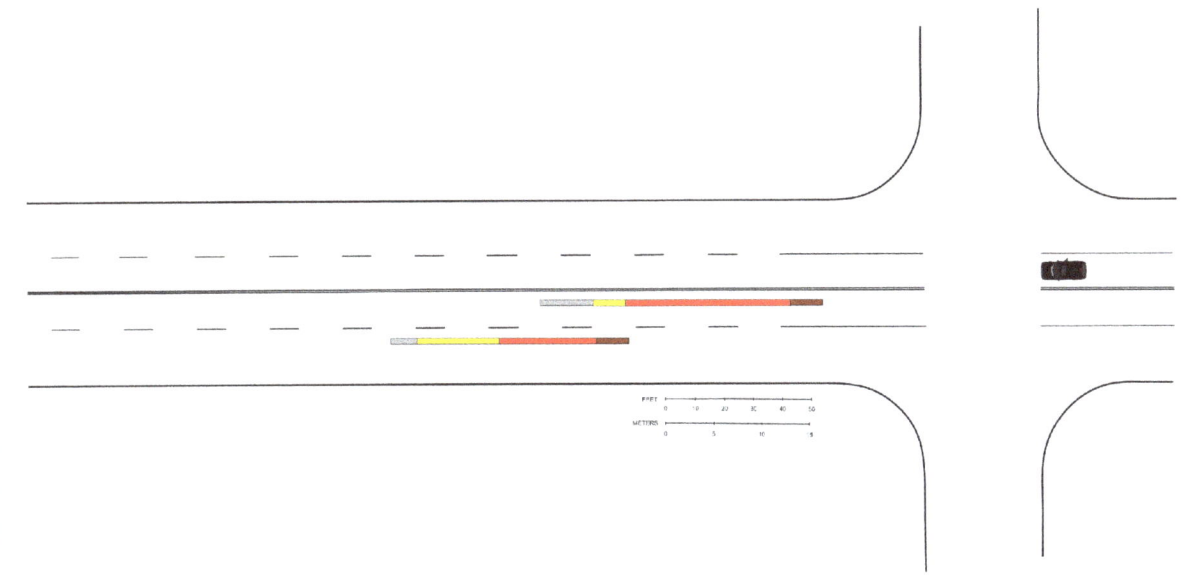

Incorporating the 85th percentile PRTs from Table 10.1 into the example of the prior section results in the modified zones depicted in Figure 10.9. With different PRTs for the two motorcyclists, the benefit to a motorcyclist of being in the right lane is not as great as in the prior illustration, but it is still present. The red zone is still smaller in the right lane. The combined size of the yellow and gray zones (the area where braking by the motorcyclist makes a difference) is still smaller for the rider in the left lane compared to the rider in the right lane. Thus, the rider who chooses the right lane still benefits from increasing the lateral distance between themselves and the left-turning vehicle. This would be reversed for a scenario where the Range Rover is attempting to cross the intersection from the motorcyclist's right.

Evidence of Ambiguity in Path Intrusion Scenarios

The influence of ambiguity in path intrusion scenarios shows up in driver-response studies. For example, Lechner and Malaterre [47] conducted a study in which the Daimler-Benz driving simulator was used to subject 49 drivers to an unexpected intersection incursion by another vehicle with timing that warranted an emergency response. Only 20% of the drivers avoided the intruding vehicle. The participants were between 22 and 59 years old, and the group consisted of 20 females and 29 males. Test subjects were not aware of the actual purpose of the experiment. They were instructed to drive the simulator at a speed of around 90 to 100 km/h (56 to 62 mph) and to familiarize themselves with the vehicle. They were told that after they completed the familiarization phase, they would be given further instructions. The intersection incursion scenario occurred 10 min into this familiarization phase, and no additional instructions were ever given.

The authors described the location of the incursion as "a four-legged intersection in the open country, protected by stop signs." The intruding vehicle pulled up to the right side of the intersection and stopped at a stop sign. The vehicle then accelerated into the intersection for 1.9 sec, and then braked to a stop having traveled 6 m. The intruding vehicle began accelerating into the intersection when the approaching vehicle was 2.0, 2.4, or 2.8 sec from impact (times calculated based on the speed the approaching vehicle was traveling when the incursion began). The authors reported that 88% of subjects resorted to braking at some point during their response. Some also swerved and some did not. Also, "only 10 or the 49 subjects were able to avoid colliding with the obstacle: 6 by braking, 3 by lateral swerving in front of the obstacle and 1 by combining braking and subsequent swerving behind the obstacle."

The average time from the beginning of the incursion to the approaching driver's first response (release of the throttle for many of the subjects) was approximately 0.8 sec, with a standard deviation of approximately two-tenths of a second. For the subjects who braked, the average time for them to reach the brake pedal after the start of the incursion was approximately 1.0 sec, with a standard deviation of 0.22 sec.[14] These reaction times imply that the approaching drivers were initiating their responses during the time when the intruding vehicle was still accelerating into the intersection. This means that these drivers were taking avoidance action during a time when they could not know if the intruding vehicle would continue through the intersection or stop.

This physical fact invalidates the authors observations that, first, the subject drivers would have been more successful in their avoidance attempts if they had preferred swerving over braking, and second, that for a driver to use a combination of steering and braking in this situation represents them using "two totally conflicting strategies." These driver responses flow from the fact that the ultimate actions of the intruding driver were unknown at the time the response had to be made. If the intruding drivers were to continue through the intersection—a possibility the subject drivers had to allow for—then braking would become the "correct" and only viable response. The same could be said for a scenario in which the intruding vehicle pulls further into the intersection and stops. The approaching driver cannot know that swerving would be a successful strategy, whereas braking at least has the advantage of buying more time for evaluation, even if it is ultimately unsuccessful. Thus, these researchers' judgments about the driver responses are the very definition of hindsight bias.

In line with this reasoning, Lechner and Malaterre carried out after-the-fact evaluation of how the scenario would have changed for the approaching drivers had the intruding vehicle continued to cross the intersection rather than stopping. The authors reported that they undertook this evaluation because, under the scenario drivers in their study were presented with, "it is clear that the choice of a given manoeuvre by the driver is not influenced by the sudden stopping of the obstacle." Based on their after-the-fact evaluation, these authors concluded that "16 of the subjects who braked then collided with the obstacle when it was stationary … would … have avoided the collision … This shows above all that the <u>result of an emergency situation is completely uncertain</u>, and that the behaviour of the obstacle to be avoided … is a determining factor" (emphasis in original).

[14.] In interpreting these reaction times, it is important to remember that they do not incorporate the time that elapses from the driver's action (steering or braking) to the build-up of the vehicle response; thus, these times do not capture the full perception-response process as defined in Muttart's research and in IDRR. In describing these times, I refer to them as reaction times, not PRTs.

Characteristics of Passenger Car Left Turns and Intersection Crossings

When analyzing a motorcycle collision involving a path intrusion or a left turn across path (LTAP) scenario, the reconstructionist will often need to analyze the characteristics of the intrusion or the left turn (i.e., the path, speed, lateral acceleration, and duration). In some instances, the path intrusion or left turn preceding the collision will be captured by an EDR or on video—perhaps from a surveillance camera on a nearby business or from a camera attached to the motorcyclist's helmet. In these cases, analysis of the EDR data or video can yield the actual motion of the passenger vehicle leading up to the collision. In other instances, the reconstructionist will need to generate a reasonable estimate for each of the characteristics of the intrusion or left turn based on the intersection geometry and on empirical data related to drivers crossing intersections or making left turns.

Fugger et al. [48] presented a study analyzing initial acceleration rates of vehicles at two signal-controlled intersections in Los Angeles, California. A total of 100 vehicles were analyzed. The vehicles were traveling straight through the intersections (not turning). Fugger et al. reported that "there appeared to be two distinct phases occurring as the vehicle accelerated through the intersection. The first phase appeared to be a slower acceleration rate leading up to the second phase, when the vehicle was accelerated at a significantly higher rate." The initial phase, which lasted 0.94 sec ± 0.48 sec, had an acceleration of 0.06 g ± 0.03 g. The second phase had an acceleration of 0.22 g ± 0.05 g. Kodsi and Muttart [49] presented a naturalistic study of the acceleration behavior of 244 drivers at two-way stop-controlled intersections in Toronto, Canada. The drivers captured in this study were traveling straight through the intersection. These authors also reported a two-phase acceleration profile, with the initial phase having an acceleration of 0.07 g ± 0.04 g. This phase lasted 0.9 sec ± 0.4 sec. The second phase had an average acceleration of 0.25 g ± 0.06 g.

Happer et al. [50] analyzed video of left-turning vehicles at a signalized intersection. This study reported that the average turning speed of vehicles that did not stop prior to turning was between 6.0 and 6.6 m/s (13.4 to 14.8 mph), and the average turning speed of vehicles that stopped was between 4.6 and 6.3 m/s (10.3 to 14.1 mph). These authors also reported that the average acceleration of vehicles that proceeded through the intersection after stopping was between 0.85 and 1.20 m/s^2 (0.087 to 0.122 g).

Carter et al. [42] reported of study of passenger car left turn characteristics at three intersections in Denver, Colorado. All the documented left turns in this study were completed on a permissive green turn phase, which allowed the drivers to turn after they yielded to oncoming traffic. The left turns were video recorded using a small unmanned aerial system (sUAS). Eighty-six left turns were analyzed. The authors reported that, for the entire dataset, the peak lateral accelerations for the turns had a mean of 0.17 g and a standard deviation of 0.08 g. The tangential accelerations for the entire dataset had a mean of 0.12 g and a standard deviation of 0.08 g. The authors selected the three intersections such that for one, the left-turning drivers had to cross one oncoming lane; for another, the left-turning drivers had to cross two oncoming lanes; and for the third, the drivers had to cross three oncoming lanes. They reported differences between the three intersections. For the intersection where the left-turning drivers had to cross one oncoming lane, the average peak lateral acceleration was 0.18 g with a standard deviation of 0.09 g. For the intersection where the left-turning drivers had to cross two oncoming lanes, the average peak lateral acceleration was 0.29 g with a standard deviation of 0.08 g. For the intersection where the left-turning drivers had to cross three oncoming lanes, the average peak lateral acceleration was 0.15 g with a standard deviation of 0.06 g.

Flynn et al. [51] reported typical longitudinal and lateral acceleration profiles for left turns captured in the SHRP2 naturalistic driving data. These researchers focused on lead vehicles making left turns from a stop at 15 signal-controlled intersections in North Carolina (8) and Washington (7). They analyzed 420 left turns. Across these left turns, the mean of the peak lateral accelerations was 0.315 g. The standard deviation was 0.071 g. Thus, in 68% of instances, the peak lateral acceleration was between 0.244 and 0.386 g. These authors observed: "The amplitude of the peak lateral acceleration was notably not affected by the turn radius, suggesting that drivers may adjust their longitudinal cornering speed to achieve similar levels of lateral acceleration across all turning radii. This choice may stem from optimizing occupant comfort and may be useful for estimating the speed drivers choose to negotiate turns of different radii."

Danaher et al. [52] reported lateral accelerations for heavy trucks making left turns at three intersections in Denver, Colorado. The motion of the trucks was obtained via covert surveillance with a drone. They reported that the average lateral accelerations experienced by the trucks were not dependent on the turn radius. For trucks that were stopped before initiating their left turns, the average lateral acceleration was 0.064 g. The range on this was generally 0.04 to 0.08 g. For trucks that did not stop before initiating their left turns, the average lateral acceleration was 0.109 g. The range on this was approximately 0.06 to 0.16 g. These authors did not consider the loading condition of the trucks.

Are Training and Licensing Effective?

In relationship to rider skill, licensing, and training, Hurt and DuPont [1] noted that "riding a motorcycle demands the development of a particular set of skills ... If the new motorcycle rider begins the learning process in traffic, there is a real threat to survival. The attention required to the details of coordinating clutch, throttle, shifter, steering, front brake, rear brake, etc., quickly saturates the inexperienced rider and leaves little attention to the surrounding traffic ... Lack of experience is a difficult thing for the new motorcycle rider to overcome or survive. The motorcycle rider with low experience is clearly overrepresented in the accident population ... The accident-involved motorcycle rider generally demonstrates a high level of primary control failure in the pre-crash circumstances. Usually the hazard is detected, but the rider is not capable of making the vehicle respond as wanted." Sometimes, a question arises about the degree to which training or licensing improve this situation.

Jonah et al. [53] reported a study that sought to determine whether graduates of the 20-h motorcycle training program (MTP) in Ontario, Canada, were less likely to have had an accident or commit a traffic violation on their motorcycles when compared to informally trained (IT) riders. The MTP included training in the following: brake applications, cold starting, moving off and stopping, gear shifting, hand signals, slow speed control, traffic behavior on a riding range, emergency braking, collision avoidance, emergency decision-making, advanced skills, obstacles, rules of the road, mechanical and electrical knowledge, and defensive riding. A sample of MTP graduates and IT riders were interviewed about their riding experiences in the previous 4 years. The interviews were carried out in 1978 and related to the motorcyclists' experiences between the beginning of 1974 and the end of 1977. These authors found that the MTP graduates were less likely than IT riders to have had accidents and violations during this 4-year period. However, they noted that "the graduates and IT riders differed in sex, age, time licensed, distance travelled, education and riding after drinking, all characteristics significantly related to accident and violation likelihood. Multivariate analyses, *controlling for the differences in these characteristics,* revealed that the MTP graduates and IT riders did not differ in accident likelihood but the MTP graduates were significantly less likely to have committed a traffic violation than the IT riders. Although the lower incidence of traffic violations among graduates could be attributed to the training program, it is possible that the graduates sought formal training because they were safety conscious, and this attitude also influenced their riding behavior" (emphasis added). They further noted that "MTP riders were more likely to be female, older, better educated, have higher family incomes and be married than IT riders. Although the MTP graduates were more likely to own their own motorcycle, they were licensed for a shorter period and had travelled less distance on a motorcycle in the last 4 yr … Finally, IT motorcyclists were more likely to have ridden a motorcycle after drinking alcohol than MTP graduates which is, perhaps, indicative of greater risk-taking among the IT motorcyclists. Riding after drinking was therefore used as a proxy variable for safety consciousness such that riders reporting that they rode a motorcycle after drinking were assumed to have a poorer attitude toward safety than those reporting never having ridden after drinking."

Mortimer [54] reported similar findings in 1984. He examined samples of riders who had and had not taken the 20-h Motorcycle Safety Foundation's motorcycle rider course within the prior 3 years. Mortimer reported that "when controlling for age and years licensed, those who took the course did not have a lower accident rate than the control group." There were also no differences in the violation rate between the two populations. Mortimer did report that "those who had taken the motorcycle rider course made significantly more use of boots, gloves and shirts or jackets with long sleeves when riding." Satten [55] had also reported that motorcyclists who had taken a motorcycle rider course made more use of protective clothing but did not have lower accident rates. Osga [56] similarly did not find that motorcyclists who took a motorcycle rider course had lower accident rates. Mortimer concluded that "there is apparently no evidence that the motorcycle rider course is effective in reducing accidents, but [some studies showed] graduates of the course had fewer traffic violations and made more use of safety clothing, which could be important in reducing the severity of injuries in the event of a crash." Mortimer also observed that "there may also be another, intangible benefit of the course … 44% of those who had passed the course did not ride a motorcycle. Perhaps, having learned to ride they decided against the use of this mode of transportation. If they had learned to ride on streets and highways, some of them would have become involved in accidents."

McDavid et al. [57] reported a study of the effectiveness of the British Columbia Safety Council's 37-h motorcycle safety training program. This study compared the driving records of two groups of riders from 1979 to 1984—one group that obtained their motorcycle license by taking the training course and the other that obtained their license without taking the Safety Council's training course or any other formal training. McDavid et al. observed that "a common methodological problem in previous studies is the lack of similarity between persons who seek motorcycle training and those who do not. Age and sex differences, as well as other uncontrolled differences between trained and untrained groups, could account for differences in key dependent variables (principally accident rates) … multivariate analysis does not necessarily compensate for selection biases and should not be used as a substitute for more rigorous research designs … In looking for the effects of training on motorcycle accidents, matching on age and sex is clearly desirable. In addition, matching on other variables that predict accident behavior after training, such as accident behavior before training, is important." McDavid et al.'s point can perhaps be summarized by noting that an important question not addressed by some of other studies is whether the likelihood a young, male motorcyclist (or any other category one might choose to look at) being involved in an accident is reduced by *that* young, male motorcyclist receiving formal training.

McDavid et al. addressed these issues by pairing each formally trained rider with an untrained or IT rider of similar age, sex, and driving record. They noted that "by controlling for differences in pre-training driving records, it is more likely that the two groups of riders are matched in driving attitudes, permitting a fairer comparison of trained and untrained riders after they are licensed." With this shift in methodology, McDavid et al. reported that there were "observable differences in the frequency and severity of accidents between the two groups. Trained riders tend to have fewer accidents of all kinds (all motor vehicle accidents combined), fewer motorcycle accidents, and less severe motorcycle accidents. Although these differences are not large in a statistical sense, they suggest that when care is taken to carefully match trained and untrained riders, training is associated with a reduction in accidents. Given that motorcycle accidents tend to be much more severe than automobile accidents, the evidence from the study supports the use of training as a means of reducing the human and material costs of motorcycle accidents."

Billheimer [58] examined the effectiveness of the California Motorcyclist Safety Program (CMSP). At the time he published his study, this program had been in use for more than 10 years and had trained more than 100,000 motorcyclists. Billheimer found that "fatal motorcycle accidents have dropped 69 percent since the introduction of the CMSP, falling from 840 fatal accidents per year in 1986 to 263 in 1995. If accident trends in California had paralleled those in the rest of the United States over this period, the state would have experienced an additional 124 fatalities per year." Billheimer noted that another factor that influenced this statistic was the introduction of a mandatory helmet law in 1992. He did not quantify the relative influence of training versus helmet use. For novice riders with less than 500 miles of experience prior to their training, Billheimer found that "trained riders experience fewer than half the accident rates of their untrained counterparts for at least 6 months after training. Beyond 6 months, riding experience begins to have a leveling effect on the differences between the two groups." For more experienced riders with more than 500 miles of experience prior to their training, "no significant differences in accident rates were detected between the two groups, either before or after riders took the basic training course. There was no evidence that riders electing to enter a safety course voluntarily rode any more safely than their untrained counterparts before taking training."

Daniello et al. [59] published a literature review of studies that examined the effectiveness of MTPs and the influence of licensing. In commenting on these studies, these authors make several observations relevant to interpreting the studies. For example, they observed that "accident rates are a common, but not necessarily ideal, measure of training effectiveness. Accidents are infrequent and may have many causes besides training or rider skill." They also noted that several studies have shown that "riders who choose training tend to be more conscious of safety than those who do not seek formal training." However, they also observe that "it is also possible that those who seek training are inherently not as good at motorcycling as those who do not seek training. Seeking training may then be a result of a lesser skill level, favoring the notion that those who are trained are more likely to be involved in an accident."

These authors also reported that "accident rates and the licensing system in place in a locality are correlated … states requiring a training course for licensing tended to have lower fatality rates based on the estimated vehicle miles traveled … the number of fatal accidents per mile traveled was significantly lower in states where a system with a restricted permit was implemented as opposed to states with an unrestricted permit. Also, states that (a) require a skills test to attain a permit, (b) mandate a longer duration of time between receiving a permit and obtaining a license, or (c) place three or more restrictions on permit holders have a lower motorcycle fatality rate than other states when the number of accidents per mile traveled is compared ...."

The Highway Loss Data Institute (HLDI) examined collision insurance claim frequency in states requiring motorcycle rider training [60]. They found that the requirement for training was not a statistically significant predictor of the claim frequency. They noted, however, that "contrary to the intent of training laws, it suggests a 10 percent increase in collision claim frequencies for riders younger than 21 in states where they are subject to an education requirement. The lack of statistical significance means it cannot be said with confidence that the collision claim frequencies of riders subject to a state education requirement actually are more likely to crash than riders of a similar age. However, if the increase is in fact real, one potential explanation might be that in some states, a participant is fully licensed upon completion of a course. This could, in practice, shorten the holding period for the permit and hasten riding."[15]

References

[1] Hurt, H.H. and DuPont, C.J., "Human Factors in Motorcycle Accidents," SAE Technical Paper 770103, 1977, doi:10.4271/770103.

[2] Association of European Motorcycle Manufacturers (ACEM), "MAIDS: Motorcycle Accidents In Depth Study," Final Report 2.0, April 2009.

[3] Helman, S., Weare, A., Palmer, M., and Fernandez-Medina, K., "Literature Review of Interventions to Improve the Conspicuity of Motorcyclists and Help Avoid 'Looked But Failed to See' Accidents," Published Project Report PPR638, Transportation Research Laboratory, 2012.

[4] Olson, P.L., "Motorcycle Conspicuity Revisited," *Human Factors*, 31(2): 141–146, 1989.

[5] Herslund, M.B. and Jørgensen, N.O., "Looked-But-Failed-to-See Errors in Traffic," *Accident Analysis and Prevention*, 35: 885–891, 2003, doi:10.1016/S0001-4575(02)00095-7.

[6] Hole, G., *Psychology of Driving*, Routledge, London, 2018, doi:10.4324/9781315516530.

[7] Sager, B., Yanko, M.R., Spalek, T.M., Froc, D.J. et al., "Motorcyclist's Lane Position as a Factor in Right-of-Way Violation Collisions: A Driving Simulator Study," *Accident Analysis and Prevention*, 72: 325–329, 2014.

[8] Ouellet, J.V., "Lane Positioning for Collision Avoidance: A Hypothesis," *Proceedings: The Human Element: 1990 International Motorcycle Safety Conference*, vol. 2, Orlando, FL, 9.58–9.80, 1990.

[9] Hole, G.J., Tyrrell, L., and Langham, M., "Some Factors Affecting Motorcyclists' Conspicuity," *Ergonomics*, 39(7): 946–965, 1996.

[10] Gershon, P., Ben-Asher, N., and Shinar, D., "Attention and search conspicuity of motorcycles as a function of their visual context," *Accident Analysis & Prevention*, 44(1): 97–103, 2012, doi:10.1016/j.aap.2010.12.015.

[11] Lenné, M.G. and Mitsopoulos-Rubens, E., "Drivers' decisions to turn across the path of a motorcycle with low beam headlights," *Proceedings of the Human Factors and Ergonomics Society Annual Meeting*, 55(1): 1850–1854, 2011, doi:10.1177/1071181311551385.

[15.] HLDI also notes: "It is important to emphasize that this analysis does not answer the question of whether riders who voluntarily take rider education courses have higher or lower crash risk. To conduct that analysis, HLDI would need to know which rated driver (riders) had training and which did not. This is not a data element currently in the HLDI database."

[12] Crundall, D., Crundall, E., Clarke, D., and Shahar, A., "Why do car drivers fail to give way to motorcycles at t-junctions?" *Accident Analysis & Prevention*, 44(1): 88–96, 2012, doi:10.1016/j.aap.2010.10.017.

[13] Magazzù, D., Comelli, M., and Marinoni, A., "Are car drivers holding a motorcycle license less responsible for motorcycle–car crash occurrence? A non-parametric approach," *Accident Analysis & Prevention*, 38(2): 365–370, 2006, doi:10.1016/j.aap.2005.10.007.

[14] Rogé, J., Douissembekov, E., and Vienne, F., "Low Conspicuity of Motorcycles for Car Drivers: Dominant Role of Bottom-Up Control of Visual Attention or Deficit of Tow-Down Control?" *Human Factors*, 54(1): 14–25, 2012, doi:10.1177/0018720811427033.

[15] Brenac, T., Clabaux, N., Perrin, C., and Van Elslande, P., "Motorcyclist Conspicuity Related Accidents in Urban Areas: A Speed Problem?" *Advances in Transportation Studies*, 8: 23–29, 2006.

[16] Labbett, S. and Langham, M., "What Do Drivers Do at Junctions?" *Road Safety Congress*, 2006.

[17] Olson, P.L. and Farber, E., *Forensic Aspects of Driver Perception and Response*, 2nd ed., Lawyers and Judges Publishing Company, Tucson, AZ, 2003.

[18] Hancock, P.A., Caird, J.K., and Shekhar, S., "Factors Influencing Drivers' Left Turn Decisions," *Proceedings of the Human Factors Society 35th Annual Meeting*, San Francisco, CA, September 2–6, 1991.

[19] Hancock, P.A. and Caird, J.K., *Factors Influencing Older Drivers' Left Turn Decisions*, Center for Transportation Studies, University of Minnesota, 1993.

[20] Pai, C.W., 2011. "Motorcycle Right-of-Way Accidents–A Literature Review," *Accident Analysis & Prevention*, 43(3): 971–982.

[21] Horswill, M.S., Helman, S., Ardiles, P., and Wann, J.P., "Motorcycle Accident Risk Could be Inflated by a Time to Arrive Illusion," *Optometry and Vision Science*, 82(8): 740–746, 2005.

[22] Horswill, M.S. and Helman, S., "A Behavioral Comparison Between Motorcyclists and a Matched Group of Non-motorcycling Car Drivers: Factors Influencing Accident Risk," *Accident Analysis and Prevention*, 35: 589–597, 2003, doi:10.1016/S0001-4575(02)00039-8.

[23] Zador, P.L., "Motorcycle Headlight-Use Laws and Fatal Motorcycle Crashes in the US, 1975-83," *American Journal of Public Health*, 75: 543–546, 1985.

[24] Hole, G.J. and Tyrrell, L., "The Influence of Perceptual 'Set' on the Detection of Motorcyclists Using Daytime Headlights," *Ergonomics*, 38(7): 1326–1341, 1995, doi:10.1080/00140139508925191.

[25] Yuan, W., "The Effectiveness of Ride-Bright Legislation for Motorcycles in Singapore," *Accident Analysis and Prevention*, 32: 559–563, 2000.

[26] Jenness, J., Huey, R., McCloskey, S., Singer, J. et al., "Perception of Approaching Motorcycles by Distracted Drivers may Depend on Auxiliary Lighting Treatments: A Field Experiment," *Proceedings of the Sixth International Driving Symposium on Human Factors in Driver Assessment, Training and Vehicle Design*, Lake Tahoe, CA, 2–8, 2011.

[27] Hurt, H.H. and DuPont, C.J., "Human Factors in Motorcycle Accidents," SAE Technical Paper 770103, 1977, doi:10.4271/770103.

[28] Olson, P.L., Halstead-Nussloch, R., and Sivak, M., "The Effect of Improvements in Motorcycle/Motorcyclist Conspicuity on Driver Behavior," *Human Factors*, 23(2): 237–248, 1981.

[29] Jenness, J., Jenkins, F., and Zador, P., "Motorcycle Conspicuity and the Effect of Fleet DRL: Analysis of Two-Vehicle Fatal Crashes in Canada and the United States 2001-2007," DOT HS 811 505, September 2011.

[30] Cavallo, V. and Pinto, M., "Are Car Daytime Running Lights Detrimental to Motorcycle Conspicuity?" *Accident Analysis and Prevention*, 49: 78–85, 2012, doi:10.1016/j.aap.2011.09.013.

[31] Rößger, L., Hagen, K., Krzywinskib, J., and Schlag, B., "Recognisability of different configurations of front lights on motorcycles," *Accident Analysis and Prevention*, 44: 82–87, 2012.

[32] Olson, P.L., "Driver Perception Response Time," SAE Technical Paper 890731, 1989, doi:10.4271/890731.

[33] Ayres, T. and Kubose, T., "Speed and Accuracy in Driver Emergency Avoidance," *56th Annual Meeting of the Human Factors and Ergonomics Society*, Boston, MA, 2012, ISBN 978-0-945289-41-8.

[34] Muttart, J.W., "Estimating Driver Response Times," Chapter 14 of *Handbook of Human Factors in Litigation*, CRC Press, Boca Raton, FL, 2005.

[35] Muttart, J.W., "Evaluation of the Influence of Several Variables Upon Driver Perception Response Times," *Proceedings of the 5th International Conference of the Institute of Traffic Accident Investigators*, Institute of Traffic Accident Investigators, York, England, 2001.

[36] Muttart, J., "Development and Evaluation of Driver Response Time Predictors Based upon Meta-Analysis," SAE Technical Paper 2003-01-0885, 2003, doi:10.4271/2003-01-0885.

[37] Muttart, J.W., "Quantifying Driver Response Times Based Upon Research and Real Life Data," *Proceedings of the 3rd International Driving Symposium on Human Factors in Driver Assessment, Training and Vehicle Design*, June 2005.

[38] Hurt, H., Thom, D., and Hancock, P., "The Effect of Hand Position on Motorcycle Brake Response Time," *28th Proceedings of the Human Factors Society*, 1984.

[39] Ecker, H., Wassermann, J., Ruspekhofer, R., Hauer, G. et al., "Brake Reaction Times of Motorcycle Riders," *International Motorcycle Safety Conference*, March 1–4, 2001, Orlando, FL.

[40] Prem, H., "The Emergency Straight-Path Braking Behaviour of Skilled versus Less-skilled Motorcycle Riders," SAE Technical Paper 871228, 1987, doi:10.4271/871228.

[41] Searle, J., "What if the Speed Had Been Less? Causation in Time Limited and Distance Limited Hazards," SAE Technical Paper 2020-01-0881, doi:10.4271/2020-01-0881.

[42] Carter, N., Beier, S., and Cordero, R., "Lateral and Tangential Accelerations of Left Turning Vehicles from Naturalistic Observations," SAE Technical Paper 2019-01-0421, 2019, doi:10.4271/2019-01-0421.

[43] Ising, K.W., Droll, J.A., Kroeker, S.G., D'Addario, P.M. et al., "Driver-Related Delay in Emergency Responses to a Laterally Incurring Hazard," *Proceedings of the Human Factors and Ergonomics Society 56th Annual Meeting*, 2012.

[44] Lee, D.N., "A Theory of Visual Control of Braking Based on Information about Time-to-Collision," *Perception*, 5: 437–459, 1976.

[45] Prynne, K. and Martin, P., "Braking Behaviour in Emergencies," SAE Technical Paper 950969, 1995, doi:10.4271/950969.

[46] Mazzae, E.N., Barickman, F.S., Forkenbrock, G., and Baldwin, G.H.S., "NHTSA Light Vehicle Antilock Brake System Research Program Task 5.2/5.3: Test Track Examination of Drivers' Collision Avoidance Behavior Using Conventional and Antilock Brakes," DOT HS 809 561, National Highway Traffic Safety Administration, 2003.

[47] Lechner, D. and Malaterre, G., "Emergency Maneuver Experimentation Using Driving Simulator," SAE Technical Paper 910016, 1991.

[48] Fugger, T., Wobrock, J., Randles, B., Stein, A. et al., "Driver Characteristics at Signal-Controlled Intersections," SAE Technical Paper 2001-01-0045, 2001, doi:10.4271/2001-01-0045.

[49] Kodsi, S. and Muttart, J., "Modeling Passenger Vehicle Acceleration Profiles from Naturalistic Observations and Driver Testing at Two-way-stop Controlled Intersections," *SAE International Journal of Passenger Cars—Mechanical Systems*, 3(1): 45–56, 2010, doi:10.4271/2010-01-0062.

[50] Happer, A., Peck, M., and Hughes, M., "Analysis of Left turning Vehicles at a 4-way Medium-Sized Signalized Intersection," *SAE International Journal of Passenger Cars—Mechanical Systems*, 2(1): 359–370, 2009, doi:10.4271/2009-01-0107.

[51] Flynn, T.I., McAllister, A.J., Wilkinson, C., and Siegmund, G.P., "Typical Acceleration Profiles for Left-Turn Maneuvers Based on SHRP2 Naturalistic Driving Data," SAE Technical Paper 2021-01-0889, 2021, doi:10.4271/2021-01-0889.

[52] Danaher, D., Donaldson, A., and McDonough, S., "Acceleration of Left Turning Heavy Trucks," *SAE International Journal of Advances and Current Practices in Mobility*, 2(4): 2019–2036, 2020, doi:10.4271/2020-01-0882.

[53] Jonah, B., Dawson, N., and Bragg, B., "Are Formally Trained Motorcyclists Safer?" *Accident Analysis and Prevention*, 14(4): 247–255, 1982.

[54] Mortimer, R, "Evaluation of the Motorcycle Rider Course," *Accident Analysis and Prevention*, 16(1): 63–71, 1984.

[55] Satten, R.S., "Analysis and Evaluation of the Motorcycle Rider Courses in 13 Northern Illinois Counties," *Proceedings of the International Motorcycle Safety Conference*, Washington, D.C., May 18–23, 1980.

[56] Osga, G.A., "An Investigation of the Riding Experiences of MSF Rider Course Participants in South Dakota," Report HFL-80-2, University of South Dakota, August 1980.

[57] McDavid, J.C., Lohrmann, B.A., and Lohrmann, G., "Does Motorcycle Training Reduce Accidents? Evidence from a Longitudinal Quasi-Experimental Study," *Journal of Safety Research*, 20: 61–72, 1989.

[58] Billheimer, J.W., "Evaluation of California Motorcyclist Safety Program," Transportation Research Record 1640, Paper No. 98-0652, 1998, doi:10.3141/1640-13.

[59] Daniello, A., Gabler, H.C., and Mehta, Y.A., "Effectiveness of Motorcycle Training and Licensing," *Transportation Research Record: Journal of the Transportation Research Board*, 2140: 206–213, 2009, doi:10.3141/2140-23.

[60] *Highway Loss Data Institute Bulletin*, "Motorcycle Collision Coverage Claims in States with Required Motorcycle Riding Training," Vol. 26, No. 12, December 2009.

11

Visualization of Motorcycle Crashes

View from the Motorcyclist's Perspective

There are three mounting options for cameras and other equipment that will record the view from a motorcycle or its motion: (1) the equipment can be mounted to the motorcycle; (2) the equipment can be mounted to the operator; or (3) the equipment can be mounted to another vehicle.

Mounting Equipment on the Motorcycle

Mounting cameras and other equipment on the motorcycle alleviates the need for the operator to handle or manage the equipment while riding. Camera equipment mounted on the motorcycle can have a fixed viewpoint, and with multiple cameras, a view ahead can be synchronized with a view of the speedometer. As an example, Figure 11.1 shows a camera mounted on the fuel tank. This rigid mount is designed to fit onto sport-type motorcycles, including Suzuki, Yamaha, Ducati, and Triumph, that have similar gas tank metal flanges with mounting fixtures around the gas cap. There is a rigid metal mount designed to screw into the gas tank screw holes for secure tank mounted connection.

Mounting equipment on a motorcycle can pose problems, though. There are many different types and styles of motorcycles, and the camera mounting hardware may need to be uniquely fitted to the specific make and model being tested. Generic mounts that are suctioned to the gas tank can be flexible enough to accommodate many motorcycles, but they may vibrate more, blurring the image, or even come detached. Suctioned mounts need to be further attached with adhesive or tape to make sure they are secure. Air, wind,

FIGURE 11.1 Sample mounting equipment for a motorcycle.

moisture, and debris may further affect the performance of a suction mount. The achievable height with some mounts may also be limited because the weight of a camera may put too much force on the mounting screws if the mount is too high. Typically, tank mounts are not suitable for larger cameras. They are better suited for lightweight, compact cameras, such as a GoPro. However, these compact action cameras may have limitations in focus, focal length, and field of view that higher end cameras do not have. Controls on higher end cameras allow the user to maximize the quality of the footage, specifying the settings that are most suitable for each specific photograph or video run. Compact action cameras can be especially limiting at night, when their limited control settings may be inadequate for calibrating the footage for the nighttime lighting.

Mounting Equipment on the Operator

Cameras can also be mounted to the motorcyclist via chest plates or helmet camera mounts. Examples are shown in Figure 11.2. One of the benefits of the chest plate is that it is more stationary than a helmet-mounted camera. The camera mounted on the helmet will turn and pivot as the operator turns his head, while the chest plate camera will continue facing generally forward. If a view straight ahead for the entire length of a motorcycle run is needed, then the chest plate camera can take care of this without interfering with the operation of the motorcycle. Because the video footage is on the body, there is a good opportunity to record smooth clear footage even on rougher roadways. The body acts like a shock absorber for the rough road, preventing vibrations from continuing into the mounting gear. As a result, footage from operator mounts is generally smooth. Like tank-mounted equipment, the operator-mounted equipment is easy to set up, lightweight, and simple to use.

Mounting equipment on the operator poses some similar issues to mounting equipment on the motorcycle. Because the mounting gear is lightweight, the use of high-end cameras is typically not feasible. The mounts will typically be too small to accommodate the weight and size of high-end cameras. Also, heavier equipment can cause weight balance issues while riding. In addition, mounting a camera on the operator's chest or helmet puts the camera at a different height off the ground than their eyes. Depending on the issue being

FIGURE 11.2 Sample operator-mounted camera equipment.

evaluated, this may or may not matter. In general, matching the exact eye height in a particular crash is irrelevant because the object or scene being viewed is far enough ahead that a change in eye height several inches up or down would not have any noticeable effect on the view. Nonetheless, there may be instances where this discrepancy between eye height and camera height matters.

Mounting Equipment on Another Vehicle

Sometimes high-quality footage can be obtained from the perspective of a motorcyclist without even utilizing a motorcycle. Typical issues with motorcycle or operator-mounted equipment can be remedied with a rig that is secured to a passenger vehicle. The photographs of Figure 11.3 show a sample set of equipment and subsequent mounting on a vehicle. The setup shown is for a nighttime visualization, and so, a special rig for the motorcycle headlamp is depicted. When using a setup like this, it will likely be important to use a vehicle where the headlamps can be turned off, so the only light source being recorded by the camera is from the headlamp rig.

The camera mounting rig depicted in Figure 11.3 includes three suction grips that are placed on the hood of the car to form a tripod for the camera. These suction grips have arm extensions that connect in the center of the tripod where a camera plate connects them together. This rig can be adjusted to various heights and distances behind the headlamp rig. When setting up a rig like this, be aware that many motorcycles position a rider with an eye height that is higher off the ground than that of a driver in a standard passenger car.

Another example of a camera mounting rig is shown in Figure 11.4. In this setup, an additional suction handle and extension arm were used to rigidly hold the back of the camera plate, as the plate sometimes tends to vibrate without the additional stiffening arm. If the view to be obtained is from a particular lane position, the camera mount could be set up on either side of the hood to accommodate this.

FIGURE 11.3 Rig for a passenger vehicle for replicating a motorcyclist's view.

FIGURE 11.4 Camera mounting equipment.

With the camera in place and adjusted, a headlamp rig can also be added to the vehicle. For nighttime motorcycle cases, it may be necessary to determine visibility using a substantially similar headlamp as the headlamp on the motorcycle in the accident. To accommodate this, a headlamp specific to the accident motorcycle can be obtained and used in the rig. To determine the correct mounting height and aiming, a comparison can be made to an inspection of an exemplar motorcycle with the same stock headlamp. Figure 11.5 illustrates this type of comparison.

FIGURE 11.5 Headlamp mounting equipment and comparison to exemplar motorcycle.

FIGURE 11.6 Motorcycle camera mount and frame from resulting footage.

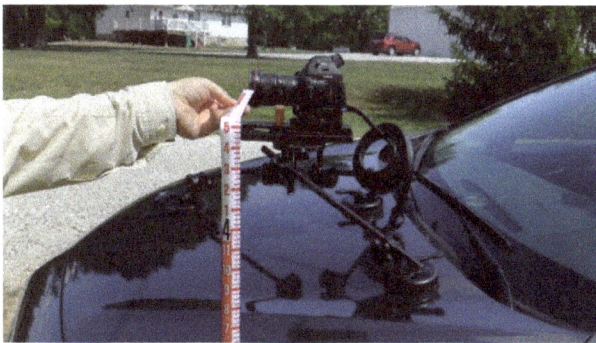

In this setup, the electrical power to the headlamp is supplied by the battery of the passenger vehicle. In the case of a motorcycle, a voltage level between 13.8 and 14.2 V would be typical. The actual voltage that is available from the battery of the car can be verified using a voltage meter. The controls to turn the headlamp on and off and to start and stop the video recording are inside the vehicle and attached with extension controls and cables connecting the unit to the controls. This setup can also include a video monitor in the vehicle showing the image being recorded. This enables checking that good quality video was captured. Figure 11.6 shows another sample camera mount for a daytime situation, along with a frame of the resulting footage.

View of the Motorcycle from Other Vehicles

The view other drivers have of a motorcycle may also be relevant to a reconstruction. This may include a vehicle traveling behind or approaching from the opposite direction of the motorcycle, though the most common two-vehicle collision with a motorcycle involves a vehicle turning left across the motorcyclist's path. Documenting the perspectives of other drivers often requires different techniques and equipment than what is required for collecting the motorcyclist's perspective. Two basic setups are common for capturing footage of a forward-looking scenario. The first is a rigid mount inside the vehicle. The second is a shoulder-mounted camera harness that is light weight and easy to set up and occupies a minimal amount of space. A photograph showing a rigid mount inside of a vehicle is included in Figure 11.7.

Sometimes, a driver who pulls out in front of a motorcyclist later says that they looked left, then right, then back left again before pulling out. It can be difficult to represent this scanning pattern with standard camera mounting. Swivel heads can be added to camera mounts, but this can increase the height of the camera, and maneuvering the swivel head from inside the vehicle can be difficult. Rotating the camera head by hand in a smooth, natural way can also be difficult. These issues can be addressed using a handheld camera grip with a motorized swivel. This equipment can be used with a range of camera sizes. Motorized control of the swivel results in motion that is smooth and consistent. If the video footage is good quality and smooth, the timing of the swivel motion can be adjusted in postprocessing video editing software. Figure 11.8 is an example of a handheld, motorized camera mount that can be positioned at the driver's height and used to mimic left and right glances.

FIGURE 11.7 Camera rig setups for other involved vehicles.

FIGURE 11.8 Motorized handheld camera mount.

© SAE International

Camera Setup and Representing a Driver's View

When capturing images that represent a vehicle operator's point of view, there are several variables to consider, including the height of the camera, what field of view to use, what settings best capture the view available to the operator, and how to represent the captured video to others. These considerations are addressed in this section.

Camera Height

Driver eye heights can vary. Seats can be adjusted, and drivers can be different heights. Even drivers of the same height can have differences in the length of their legs or torso that are significant enough to change the eye height by several inches. Seated posture can also make a difference. When representing driver views, the question would be: To what degree would these changes in height matter? The answer is that it depends on the location of the objects being viewed. A change in eye height of 6 in. makes very little difference for objects in the distance. For objects near the vehicle though, the eye height can influence how far in front of the vehicle a driver can see. For an overall view down the road, though, the eye height could be varied by a foot or more, and the camera would still capture a very similar view.

FIGURE 11.9 Cropped image to show view to the left.

FIGURE 11.10 Diagram showing viewing height differences for near and far objects.

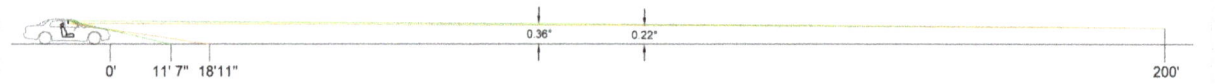

Figure 11.10 illustrates this graphically. This figure shows a scaled diagram with a cross section of a vehicle on a flat roadway with a 3-ft tall object placed 200 ft from the front of the vehicle. Two viewing positions are depicted. The one depicted with the green lines is 6 in. higher than the other. From each viewpoint, a line was drawn to the top of the 3-ft object 200 ft away. Lines of sight were also drawn from the top and bottom of the 6 in. range to the ground level immediately in front of the vehicle. This line was drawn to a point as close to the front of the vehicle as possible without having an intersection with the vehicle geometry, giving us a near range for the driver's line of sight.

For the view of the far object, there is negligible difference between the two viewpoints. One way to demonstrate this is to measure the difference in viewing angle that each viewpoint produces of the far target object. The difference in angle was only 0.14°. The angle of incidence from the top position driver's line of sight was 0.36°, and the angle from the bottom viewing position is 0.22°. However, the lines of sight for the nearer distances produce a larger difference. From the higher viewing position, the driver would be able to see the bottom of an object 11 ft, 7 in. in front of the vehicle. From the lower viewing position, the driver would be able to see the bottom of an object that was 18 ft, 11 in. from the front of the vehicle. For most accident situations, the view out ahead of the vehicle is the relevant one and a difference in eye height of 6 in. will be insignificant.

Figure 11.11 is another illustration of the influence of different viewing heights. These photographs have 6 in. differences in camera height. The photograph on the left was taken with the lowest viewpoint—6 in. lower than the middle photograph—and the photograph on the right had the highest viewpoint—6 in. higher than the middle photograph. The photographs were taken looking toward the center of a wall that had a height of approximately 4 ft and was approximately 200 ft in the distance. The viewing height changes what is depicted in the immediate foreground and the view of up-close objects like the steering

FIGURE 11.11 Diagram showing viewing height differences for near and far objects.

FIGURE 11.12 Diagram showing viewing height differences for near and far objects.

wheel and sun visor change. The overall view of the objects and scenery down the road is not significantly different, though.

To further demonstrate how the difference in 1 ft of height is insignificant in terms of the changing the view of the distant object, these three photographs were cropped to not include the interior roof and steering wheel. The results of this cropping are shown in Figure 11.12. Visually, there is no difference between these images, and when comparing each cropped image within Adobe Photoshop, all three images aligned almost perfectly, thus confirming that the view was virtually the same regardless of the differences in camera height within the vehicle.

©2022, SAE International

Field of View

The field of view is defined by the focal length of a camera lens, which is usually measured in mm. The focal length determines the angular extents of the view that will be captured. Higher focal lengths will make the angular extents of the image smaller (zoom), while lower focal lengths will provide an image with wider angular extents. The focal length is a physical property of a lens. It describes the numerical relationship of the optical distance between the center of the lens and where light converges to form an image on the sensor or film. When adjusting the focal length of the camera, it is important to recognize the physical fact that *changing the focal length does not change the perspective*. This means that zooming is the same as cropping.

A camera lens functions the same in analog and digital cameras. The film is what captures the image in an analog camera; the sensor captures the image in a digital camera. Figure 11.13 illustrates the basic geometry of a camera and illustrates how the field of view changes for several different focal lengths. In this figure, l_{focal} is the focal length, w_{sensor} is the width of the camera sensor, and θ_{fov} is the angular field of view. As the lens moves further away from the sensor plane, the angle θ_{fov} becomes smaller, creating a narrower field of view. As the lens moves closer to the sensor plane, this angle becomes larger, creating a larger field of view. Because the sensor size is not changing, however, there will be more detail included in the image when zooming into an area. Adjusting the zoom, which changes the focal length and field of view, will not affect the perspective of the image. To change perspectives, the photographer would have to change their position.

These concepts are illustrated with actual photographs in Figure 11.14. This image is an overlay of three different photographs taken from the same position but at different focal lengths. The images captured at longer focal lengths have been scaled to fit the overlaying image, demonstrating how zooming in or out on a scene from the same point of view is the same as cropping the amount of data that is captured. When zoomed in on an object,

FIGURE 11.13 Basic geometry of a camera and illustration of the influence of focal length on the field of view.

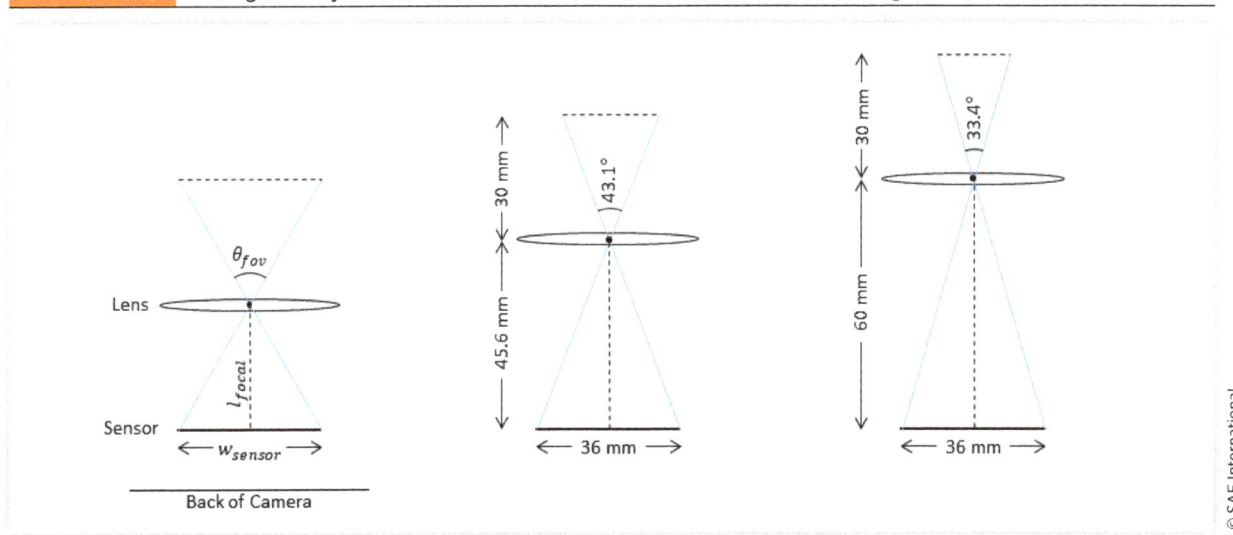

© SAE International

FIGURE 11.14 Illustration of the influence of focal length on the field of view (cropping).

there will be a higher resolution of image data captured of the object, when zoomed out there will be a lower resolution of data captured of the same object, but the perspective will not change.

Scaling the Images for Playback

In some instances, video will need to be played back or photographs shown on a screen that is sized and positioned to give the audience a correct depiction of the size of objects as a driver would have (or could have) seen them. Figure 11.15 shows the geometry relevant to determining the correct display size and positioning. In this figure, d_{view} is the distance between the viewer and the screen, w_{screen} is the width of the screen, and θ_{fov} is the field of view for this viewing geometry. This assumes that the image will fill the entire width of the screen. Thus, the field of view could be the field of view of the original image that is going to be displayed. Alternatively, if the width of the screen and distance to the viewer cannot be controlled, the image could be cropped down to the field of view that will generate objects of the correct size, assuming of course that the image will still have adequate resolution to display the objects in the image in a crisp, clear manner.

The following equation would yield the distance the screen should be placed from the viewer for a specified screen width and field of view:

$$d_{view} = \frac{w_{screen}}{2 \cdot \tan\left(\dfrac{\theta_{fov}}{2}\right)} \qquad (11.1)$$

FIGURE 11.15 Graphical depiction for determining playback scale.

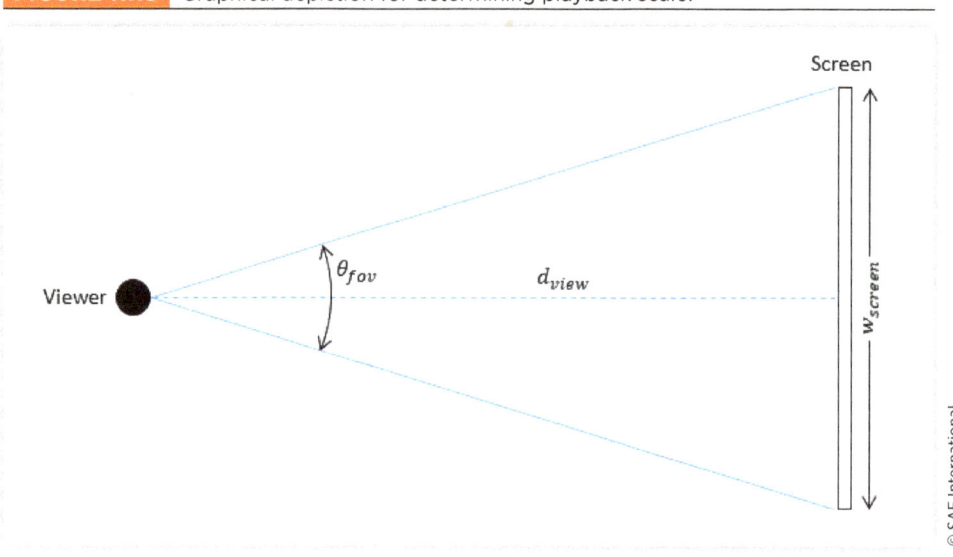

Computer Visualization of a Motorcycle Crash

When visually depicting a motorcycle crash or the view available to a driver or motorcycle operator, there are several types of visualizations and methods available. In general, these can be categorized into computer-generated (CG) visualizations, photo- or video-realistic visualizations, or a hybrid between the two. Technically, these are all generated in the computer to some degree, and hence could be classified as CG visualizations. However, there is a significant difference in the way the two types of visualizations are produced, how they look, how geometry that is visible in the visualization is created and rendered, and the flexibility to change views or interact within the environment.

In CG visualizations, all the components are contained in a three-dimensional (3-D) modeling and animating environment. The vehicles, the scenery, the motion, and the lighting are CG and controlled or modified through the software's tools and algorithms. Figure 11.16 shows still images from a CG animation of a crash involving a motorcycle. In these images, the geometry of the vehicles, the trees, the roadway, and other geometry were textured and modeled within the computer software's environment. The underlying animation could be played back at any rate, including slow motion. The lighting of the environment and the textures and colors of the objects are also generated within the computer environment and cameras can be placed anywhere in the environment without changing the underlying animation.

Another type of visualization utilizes video or photographs in addition to CG objects. Methodologies have been established for recording and representing photographic of video views of a scene, environment, or available view of the drivers and operators [1–6]. If the camera settings and setup are done correctly, then the resulting images can be fair and

FIGURE 11.16 Still images from a computer-generated (CG) animation of a motorcycle crash.

accurate representations of the view they intended to capture. Unlike a CG animation, these renderings get the size, shape, color, and lighting from the actual environment, recording it in a digital medium. Figure 11.17 shows frames from an example of this type of visualization. In this example, video was obtained to record the view available to a motorcyclist. Live vehicles were used and recorded at the actual accident scene. However, because recreating the accident is not feasible or safe, portions of the visualization were generated in the computer. The CG portions of this visualization rely on the photographic and video-realistic source recordings to determine how the textures, colors, and lighting should appear.

FIGURE 11.17 Still images from a video-based motorcycle crash visualization.

This type of video visualization can also be created for nighttime scenarios. Reference [7] presented a methodology for generating photo-realistic computer simulation environments of nighttime driving scenarios by combining nighttime photography and videography with video tracking and projection mapping technologies. Nighttime driving environments contain complex lighting conditions such as forward and signal lighting systems of vehicles, street lighting, and retro-reflective markers and signage. The high dynamic range (HDR) of nighttime lighting conditions makes modeling of these systems difficult to render realistically through CG techniques alone. Photography and video, especially when using HDR imaging, can produce realistic representations of the lighting environments. But because the video is only 2-D and lacks the flexibility of a 3-D, CG environment, the scenarios that can be represented are limited to the specific scenario recorded with video.

However, by combining the realistic imagery from video and photographs with the flexibility of a CG environment, it is possible to vary any number of factors such as the speed of vehicles and the driver lane position, and to vary the types of vehicles and lighting conditions involved in the scenario. The combination of projection mapping, video tracking, and nighttime video and photography methodologies allows this flexibility. In addition to presenting the methodology and the resulting computer simulation environment, the final simulation is compared to actual video recordings of the same driving scenario to evaluate how similar they are in value, tone, color, and visibility. The realistic simulation environment helps the user visualize the driving environment in a manner that more closely resembles the actual environment one would experience in the real world. Another advantage of the simulated environment is that, because it is completely CG, variables such as the roadway conditions, vehicle speeds and positions, and lighting conditions can all be changed, and a variety of factors that potentially contribute to accident causation can be visually represented for use in analysis, studies, or demonstrations. Some environments are more difficult to model than others, particularly low light level and nighttime environments where reflected light and artificial lighting sources create complex lighting situations. However, advances in digital photography and videography have made imaging these environments easier and more realistic.

While it is technically feasible to collect video-realistic recordings of real-world driving situations and even play back the recordings in high definition and in a calibrated manner where it represents what a driver would see, there are clear limitations. First, the video captured is linear in the sense that it can only be played forward or backward, but always in a prescribed sequence of images. Second, the conditions in which the video was captured represent the only set of conditions that can be played back. Without editing, compositing, or computer visualization, modifying the conditions of the driving situation that were recorded such as a driver's lane position, speed, or other traffic is limited. A third problem is that situations where accident conditions are of interest, such as driving through low-lit areas, or testing a driver's perception-response to unexpected situations, may be dangerous to conduct in a live setting. The methodology discussed here avoids these limitations without sacrificing the quality and visual realism that high definition and HDR video recording possess. This methodology sets forth steps that allow video-realistic footage of driving situations to be obtained in a manner that maximizes both safety and controllability of the driving variables that are of interest in the testing. Primarily this is accomplished by separating the vehicles from the driving environment when collecting video footage and then combining the separate data using computer modeling and visualization techniques. Because the data collected are maintained throughout the process as video-realistic imagery, the ending quality is also video realistic. Further, because the environment is eventually a CG environment, controlling and varying the driving parameters are safe and feasible.

Following is a general list of the steps involved in this methodology:

a) Collect video footage of the driving environment
b) Collect geometry data of the driving environment
c) Collect video footage of the vehicles under a variety of lighting conditions in a controlled area
d) Collect geometrical data of the vehicles involved
e) Use projection mapping techniques to create a video-realistic computer environment

f) Use computer visualization techniques for creating vehicles with varying parameters
g) Combine the environment and vehicle modeling systems into one system
h) Vary the parameters of the vehicle and scene to generate any number of video-realistic simulation scenarios

Case Study: CG Visualization of a Truck Versus Motorcycle Crash

This section illustrates a process for producing a physics-based animation of a collision between a heavy truck and a motorcycle. The subject collision occurred on an interstate highway with three lanes in each direction. The eastbound and westbound lanes were divided by a concrete center median barrier. According to the police, the crash occurred when the truck driver swerved from the right lane to avoid a slow-moving car that had experienced a blowout on its left rear tire. The truck driver swerved his tractor-trailer to the left, across the center and left lanes, onto the left shoulder, and then back to right, eventually bringing his vehicle to a stop straddling the left and center lanes. While swerving through the left lane, the tractor collided with the motorcyclist. The speed limit in the area was 70 mph. According to witnesses and involved parties, the tractor-trailer and the motorcycle were initially traveling approximately 70 mph, and the slow-moving vehicle was traveling approximately 15 mph.

The tractor involved in this crash was towing an empty, 53-ft Wabash National van trailer, which also had antilock brakes. According to testimony by the truck driver, the semitrailer was empty at the time of the crash. The collision damaged components on the tractor, including the driver's side fuel tank, the driver's side steps, and the hardware and tank bands that attached the steps to the fuel tank. There was black, brown, and red material transfer to the fuel tank that came either from the motorcyclist's clothing or from his motorcycle.

The electronic control module (ECM) on the tractor's engine was equipped with an event data recorder (EDR) capable of recording data from hard braking events. The recorded data from these systems include the vehicle-indicated speed, engine speed, engine load, throttle, brake status, clutch status, and cruise control status reported at 1-sec intervals for 59 sec preceding and 15 sec following the triggering event. An event is triggered when the system detects a change in wheel speed of 7 mph in a 1-sec interval. Data were imaged from the tractor's ECM 19 days after the subject crash. The obtained data included two hard brake events and a last stop record. The event designated by the ECM as "Hard Brake #1" was determined to be from the subject crash.

Figure 11.18 is a graph showing the vehicle-indicated speeds reported in Hard Brake #1 at 1-sec intervals for 59 sec prior to and 15 sec after the triggering deceleration event. The individual speed readings are designated on the graph with open black circles. This graph has time, in seconds, plotted on the horizontal axis and the triggering event occurs at a time of 0 sec. Times prior to this event are shown as negative and times after this event are shown as positive. The tractor's vehicle-indicated speeds, in miles per hour, are plotted on the vertical axis. These data show that the tractor-trailer was driving a speed of approximately 70 mph for the minute prior to this crash. The driver then applied his brakes with sufficient severity to trigger the recording of these data, ultimately coming to a stop around 12 sec later.

FIGURE 11.18 Event data recorder (EDR) speed data from the Freightliner.

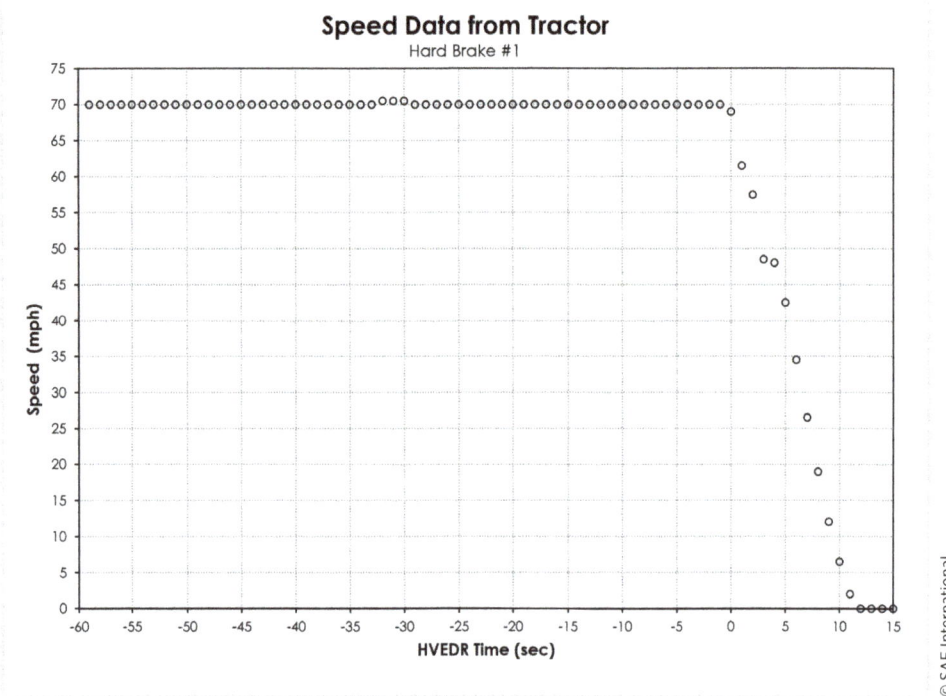

Figure 11.19 is similar with the exception that it focuses on the time from 5 sec before to 15 sec after the triggering brake event. Again, time is plotted on the horizontal axis and vehicle-indicated speeds are plotted on the vertical axis. On this graph, filled red circles designate times when the ECM indicated the driver was applying the brakes and filled blue circles designate times when the ECM indicated the driver was applying the throttle. These data indicated that the driver had the throttled applied up until 1 sec prior to the braking. The system indicated that sometime between time c1 sec and time 0 sec, the driver released the throttle. The system then indicated that the brakes were not applied at time 0, but they were applied by time 1 sec. Thus, the system indicated that the driver applied the brakes sometime in this 1-sec interval between 0 sec and 1 sec.

In the absence of heavy braking or swerving, the EDR on a heavy truck will capture the actual speed of the vehicle. This speed is typically accurate to within 1 mph. However, in the presence of heavy braking or swerving, the vehicle wheels can be traveling more slowly than the vehicle itself and this can cause the vehicle-indicated speeds to be lower than the actual vehicle speed. The speed data during the hard brake event were analyzed in accordance with techniques that take this into account [8–15]. The next graph (Figure 11.20) adds a black line indicating the probable speed of the tractor based on this analysis. This analysis showed that the driver of the tractor-trailer initially applied his brakes at a level sufficient to decelerate his vehicle at approximately 0.26 g. At an EDR time of 5 sec, the driver increased the level of his brake application, producing a deceleration of approximately 0.32 g.

CHAPTER 11 Visualization of Motorcycle Crashes **361**

FIGURE 11.19 Event data recorder (EDR) speed data from the tractor with throttle and brake inputs.

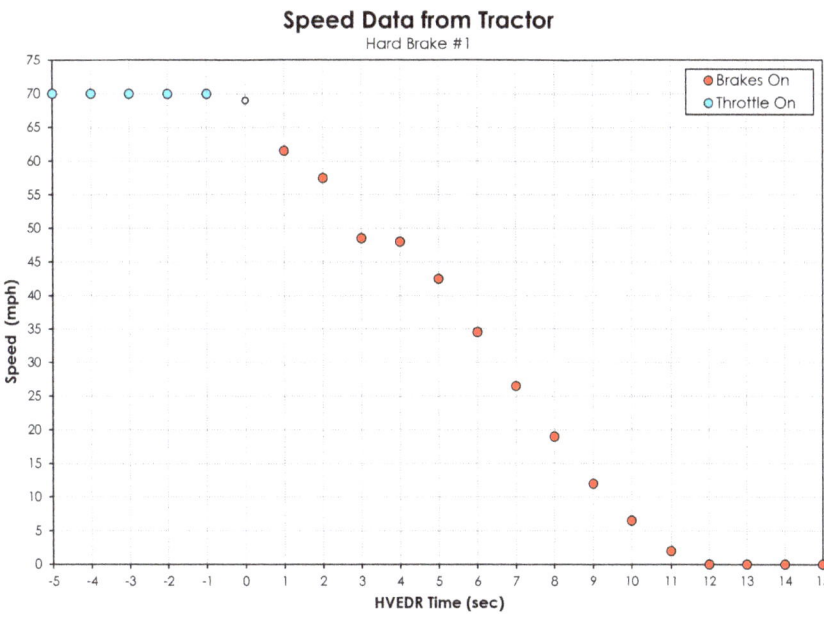

FIGURE 11.20 Event data recorder (EDR) speed data plotted with reconstructed speeds.

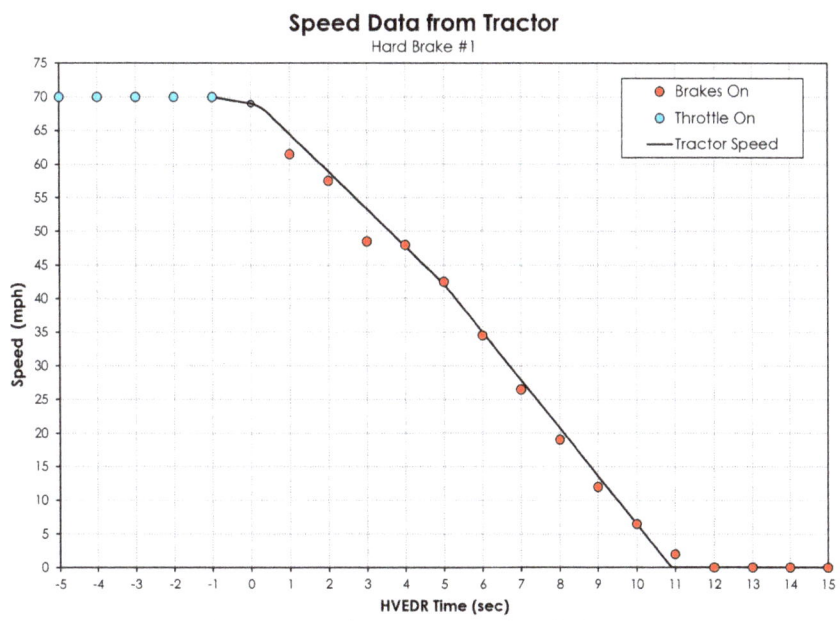

FIGURE 11.21 Faro scan data for the motorcycle.

The motorcycle involved in the subject crash weighed approximately 800 lb. The motorcycle was inspected, photographed, and mapped with a Faro laser scanner. The colorized scan data are depicted in Figure 11.21.

Damaged components on the motorcycle included the windshield, gauge cluster shade, both side mirrors, handlebars with associated hand control levers, front wheel cowling, and both turn signal indicators. Both radiators, which reside inward of the front tire, were also damaged. The front tire exhibited scuff marks aligned radially over the circumference of the treaded surface. The left side of the motorcycle showed damage that included the upper and lower faring, shifter pedal, passenger foot board, side luggage case, rear speaker cover, and elbow rest. The left side of the vehicle also had scratches aligned horizontally, spanning from approximately the front headlight to the left rear luggage case. The tip-over-bar on this side has also been displaced in an upward direction with scratches deposited on its underside. The right side of the motorcycle exhibited damage that included the side luggage case, speaker cover, upper and lower cowling, gas tank, brake pedal, tip-over-bar, and foot peg which was no longer attached to the vehicle. Both the tip-over-bar and shifter pedal had been displaced upward and exhibited scratching. Multiple scratch patterns were deposited on this side of the vehicle.

The accident site, which is shown in Figure 11.22, was inspected, documented, photographed, digitally mapped, and videoed. As the photograph shows, the roadway around this crash had three asphalt-paved travel lanes separated by dashed white lines. There was a paved shoulder on the left side of the travel lanes, separated from the travel lanes by a solid yellow edge line and bordered on the left by a concrete center median barrier. There was a wider paved shoulder on the right side of the travel lanes, separated from the travel lanes by a solid white edge line. The roadway in this area had a slight downgrade of around 1%. This accident occurred in a straightaway. A Sokkia total station was used to map the site.

Investigating officers took 40 photographs at the scene of this crash, which depicted physical evidence. The physical evidence included tire marks deposited by the tractor-trailer in the left lane and shoulder, paint transfer and scraping on the concrete center median that was deposited by the motorcyclist and his motorcycle, and scraping and debris on

FIGURE 11.22 Photograph of accident area.

the roadway from the motorcycle. The rest positions of the motorcycle, the rider, and the tractor-trailer were also documented. Camera-matching photogrammetric analysis was used to locate the evidence depicted in the police photographs. Figure 11.23 shows one of the accident scene photographs that I analyzed. Figure 11.24 shows this same photograph aligned with the mapping of the accident site. Figure 11.25 shows the evidence in the photograph traced, along with the vehicle rest positions. Several of the police photographs were analyzed using the same technique. Figure 11.26 shows portions of the physical evidence diagram that resulted from this photogrammetric analysis.

FIGURE 11.23 Accident scene photograph.

©2022, SAE International

FIGURE 11.24 Mapping data overlaid on accident scene photograph.

FIGURE 11.25 Physical evidence traced from camera-matching.

FIGURE 11.26 Portions of the physical evidence diagram from the photogrammetric analysis.

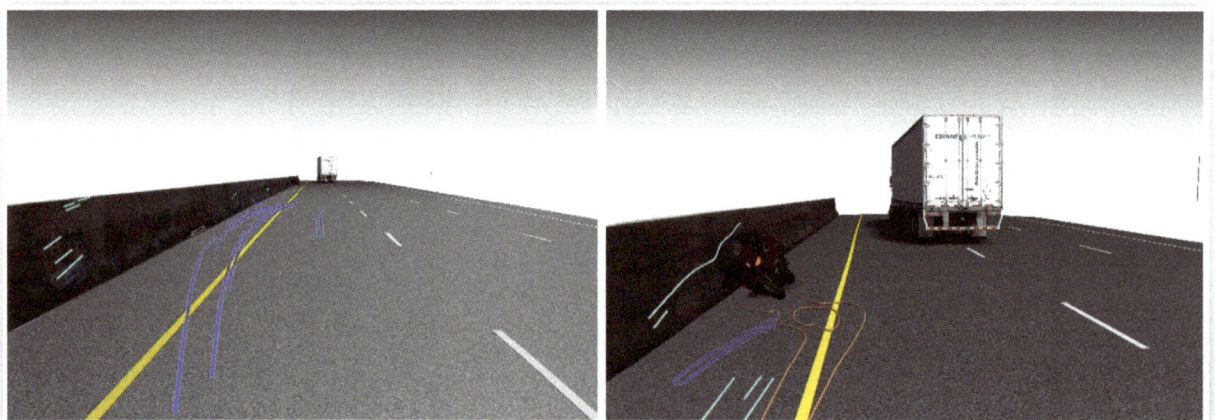

The motion of the vehicles during this crash was next analyzed based on the ECM data, the physical evidence, and the descriptions by the witnesses and involved drivers. To conduct this analysis, the accident analysis software called PC-Crash was used [16]. This software employs physics-based equations to calculate the motion of vehicles that result from steering, braking, and acceleration inputs of the drivers and from impacts. In addition to driver inputs, PC-Crash allows the analyst to specify the vehicle and scene geometries and the roadway surface conditions. This analysis resulted in a reconstruction that included the timing of the truck driver's steering and brake inputs, the timing of the collision, the speeds of the vehicles at the time the collision, and the motion of the vehicles.

Figure 11.27 is a graph that shows the speed of the tractor in the simulation compared to the speed data from the EDR of the tractor. This graph has time plotted on the horizontal axis and the speed of the tractor on the vertical axis. The black curve depicts the actual speed of the tractor, as determined from the reported speeds of the EDR, and the blue curve depicts the speed of the tractor in the simulation. This graph shows that the simulation exhibited excellent agreement with the speed data from the Freightliner EDR. The alignment of the simulation data with the EDR data also enabled identification of when within the EDR data certain events during the accident sequence took place. Based on this alignment, the tractor impacted the motorcyclist at a time of approximately 1.5 sec in the EDR data. The truck driver applied the brakes of his vehicle approximately 1.3 sec after he swerved.

FIGURE 11.27 Comparison between the event data recorder (EDR) data, the reconstructed speeds, and the PC-Crash simulation for the tractor-trailer.

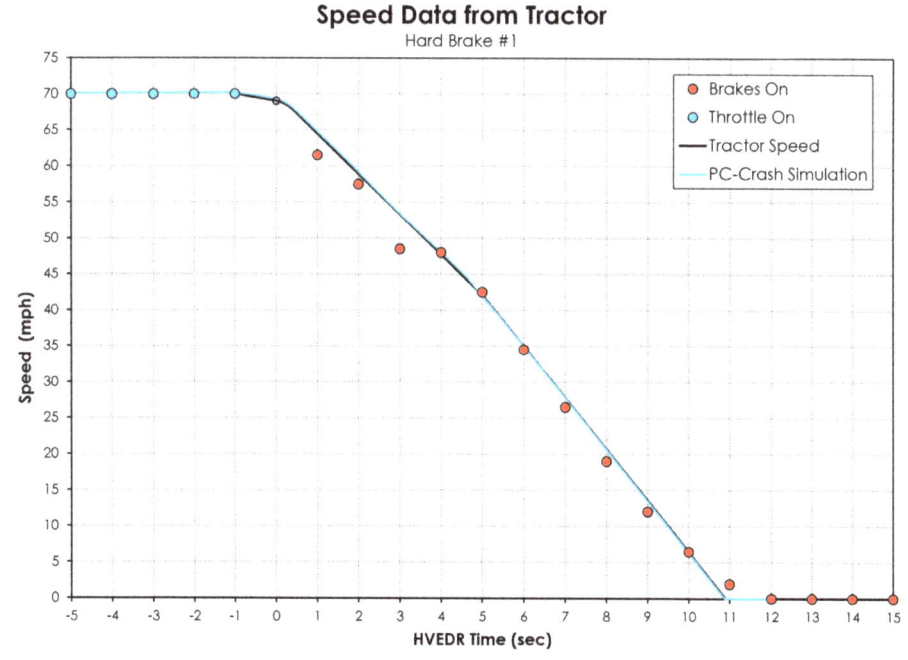

The postimpact motion of the motorcycle was not simulated with PC-Crash. However, the physical evidence and principles of physics were used to analyze and model this motion of the motorcycle and of the rider. As the tractor-trailer swerved into and through the left lane, the tractor struck the motorcycle. This impact occurred between the front of the motorcycle and the left side of the tractor. The collision forced the motorcyclist and his motorcycle into the center median barrier. The front left of the motorcycle shows direct impact damage to the upper portions of the vehicle consistent with impacting the center barrier. The evidence on the center median barrier from this impact (scratches and paint transfer) indicated that the motorcycle impacted the barrier in a near upright orientation. At this impact location, yellow paint markings were visible on the upper edge of the barrier from the motorcyclist's helmet. The height of this material transfer was indicative of the rider, also being upright on the motorcycle during this impact. This first impact with the barrier also bent the tip-over-bar upward against the motorcycle and caused scratching to its underside.

The left side of the motorcycle impacted the barrier again approximately 75 ft east of the first impact. This impact deposited evidence on the barrier indicative of the left handlebar grip and clutch lever contacting the barrier surface. These marks are visible as a black scuff mark and white scrape mark jointly traveling along the wall. Both surfaces on the motorcycle show contact damage consistent with this interaction with the barrier. A fluid trail began on the ground beyond the location of this impact with the barrier. The fluid trail ended near the rest position of the motorcycle against the barrier. This fluid trail demonstrates that the motorcycle was rotating as it traveled to rest. The right side of the motorcycle exhibited several scratch patterns in multiple orientations consistent with such rotation. A breach in the gas tank as well as a disconnected lower radiator hose was also found on the right side of the motorcycle. The damage evident on the right side of the motorcycle is consistent with the motorcycle impacting the ground on its right side and rotating while sliding to rest.

In the area where the fluid trail began, there were also yellow and orange streaks of material deposited on the barrier. The streak marks match the color of the helmet and orange shirt worn by the motorcyclist. The left side of the helmet exhibited scratch marks consistent with contacting the barrier at this point. In addition, the upper left portion of the rider's chest showed abrasions. The distance from the rider's helmet to the upper left chest abrasions was consistent with the gap between the yellow and orange streaks on the barrier and suggested the rider's head and chest were in contact with the barrier at that location. Because of the difference in rider and motorcycle motion after the second barrier impact, the rider and motorcycle likely began to separate at this location, with the motorcycle leaning to the right and the rider leaning to the left.

The motorcycle was photographed at the scene with its right side leaning against the barrier facing west. Within the photographs, fluid is seen covering the center of the front tire around the circumference of its treaded area. This area of the tire would not encounter the fluid trail the motorcycle was depositing while coming to rest because of the motorcycle sliding on its side. In addition, the fluid trail terminates a few feet south-west of the motorcycle leaning against the barrier. The fluid trail ending abruptly in addition to the wetted center of the tire indicated that the motorcycle was likely moved from its initial position of rest to the location at which it was ultimately photographed.

Using the motion from the PC-Crash and application of principles of physics to model the postimpact motion of the motorcycle and rider, animations were produced. Because these animations utilized the physics-based motion of the vehicles directly from PC-Crash, they depict motion that is physically realistic and determined directly from the physical evidence. Frames from this animation are included in Figure 11.28.

FIGURE 11.28 Frames from the physics-based animation.

The process described here illustrates several principles that can be used to produce physics-based animations that are likely to be admissible in a trial. These will typically be admissible as demonstrative exhibits illustrating an expert's opinions. *First, create properly scaled computer models of the scene and vehicles.* In a litigation context, animations like this one are often intended to be a demonstrative exhibit to help an expert explain their opinions. Modern jurors are sophisticated and accustomed to high-quality television, video, and animation production. They are accustomed to a high level of physical realism, and so, physical realism will help to build credibility for an animation. Physical realism begins with having objects in the animation that are correctly sized relative to one another. Jurors will perceive errors in scale, and this will give them cause to doubt the credibility of the animation.

Beyond the benefit to the jury, the process of putting together an animation in an accurately scaled computer environment can help the expert to develop their opinions by helping them to understand how objects fit together and interact [17, 18]. An example would be analysis of how two vehicles collided by creating accurately scaled models of the vehicles and their damage in a computer animation software package. Another example would be using correctly scaled objects to evaluate geometric visibility for a driver in an accident.

Second, connect the animation to the physical evidence. Car crashes typically leave physical evidence—tire marks, gouges, debris, and vehicle damage. The vehicle rest positions are also typically known. The vehicle motion in an animation will be credible to the extent that it is consistent with the physical evidence and explains how that evidence was created. The animation should reveal to the jury how that evidence was created and help them understand the case better. If an animation helps jurors understand the physical evidence better, then it will provide them with value. The value an animation provides to a jury is one of the criteria for its admissibility [19, 20].

Third, reconcile eye-witness accounts of the crash to the animation. Accident reconstructionists generally agree that eye-witnesses may not be good at estimating times, speeds, and distances. However, this should not discount the general account of a witness, or description of the event, as the overall story being and experience being described by a witness could be very credible. An accident reconstructionist should not dismiss the story that a witness tells without some systematic examination of that story. In certain situations, physical evidence and physical principles would compel an accident reconstructionist to dismiss what a witness says. In other situations when the only facts related to how a crash occurred come from the eye-witness, what is represented in the animation should reflect that testimony within the confines of physical evidence and physical principles.

Fourth, build the animation based on principles of physics. In an animation of a vehicular crash, the motion of the vehicles is going to be perceived as more realistic if that motion is grounded in physics. Physics-based motion gives the animation credibility because the motion of the vehicles looks and feels right. As Grimes [21] has noted: "Unfortunately, the word 'animation' is often associated with cartoons, where objects are not bound to the laws of physics. In contrast, an accurate depiction of a collision requires the animation to be consistent with the physical laws-of-motion. Computer animation requires sufficient data to produce all the images of the vehicle traveling through the collision scene. Therefore, credible animations must be based upon a detailed reconstruction of the collision sequence." In another article, Grimes [22] proposed the term "scientific animation" to describe an animation in which the objects are properly scaled and the depicted motion obeys the laws of physics. Day [23] used the term "scientific visualization" to refer to animations in which the underlying vehicle motion is generated by a physics-based simulation software package. Grimes et al. [24] defined scientific visualization as "a computer animation in which the motion of the primary objects is based on scientific analysis or scientifically accurate equations." Martin et al. [25] presented applications of this principle of building animations with principles of physics. Massa [26] demonstrated techniques for evaluating and critiquing the physical realism of an animation.

Fifth, incorporate the electronic evidence. Many vehicles now record electronic crash-related data and analysis of these data is usually a part of an accident reconstruction. The engine control module on the semitractor in this case study recorded crash-related data (speed, for instance) and these data were analyzed when producing the motion of the tractor-trailer for the animation. This relates back to the principle of tying the animation to the physical and testimonial evidence. The electronic data on modern vehicles are another source of evidence about the crash and an animation that is consistent with the known evidence will be more credible than one that is not.

Sixth, incorporate relevant secondary details. Car crashes occur at a particular time and place. Including accurate secondary details about the scene, such as road signs, vegetation, correct lighting, and logos on vehicles can add credibility to an animation. In the animation discussed here, several secondary details have been removed so that the case is not identifiable. However, if this animation was going to be shown to a jury, these details would be included. It is important to distinguish between primary and secondary details—or essential and nonessential details [20]. Animations can be reliable and admissible without the secondary (nonessential) details. However, including these secondary details will help orient the jury and will give the jury a sense that you understand the context in which a crash occurred [21, 22]. Grimes et al. [24] distinguish between primary and secondary details as follows: "The basic difference is that primary objects are important to the purpose of the presentation and secondary objects are only for helping orient the audience." Sound is another example of a secondary detail that can help to orient jurors and increase their understanding of what is happening in the animation. References [27] and [28] covered methods for scientifically incorporating sound into an animation.

Seventh, present the animation production process in a transparent way. While physical realism can add credibility to an animation, a jury should not be left with the impression that they are watching the real event. Animations are often a demonstrative exhibit illustrating an expert's opinions. A transparent presentation of the process through which the animation was created will help the jury understand the accident better but will not leave them with a misunderstanding of what they are watching. Present the physical evidence, present the analysis of the physical evidence, present how principles of physics were brought to bear on the physical evidence, and then present how physical evidence and physics flow directly into the animation. Transparent presentation of the process also helps lay the foundation for an animation, making it more likely to be admitted. Along these lines, Grimes [24] argues that "any presentation that is presumed to be based on scientific principles should be thoroughly documented such that a similarly qualified person can reproduce the findings." McLay et al. [29] discussed the process of laying the foundation for an animation. Fay [30] covered several examples of cases in which animations were either admitted or excluded.

References

[1] Holohan, R.D., Billing, A.M., and Murray, S.D., "Nighttime Photography – Show It Like It Is," SAE Technical Paper 890730, 1989, doi:10.4271/890730.

[2] Klein, E. and Stephens, G., "Visibility Study – Methodologies and Reconstruction," SAE Technical Paper 921575, 1992, doi:10.4271/921575.

[3] Ayres, T.J., "Psychophysical Validation of Photographic Representations," *Safety Engineering and Risk Analysis* (SERA-Vol. 6), American Society of Mechanical Engineers (ASME) International Mechanical Engineering Congress and Exposition, Atlanta, GA, November 17–22, 1996.

[4] Krauss, D.A. "Validation of Digital Image Representations of Low-Illumination Scenes," SAE Technical Paper 2006-01-1288, 2006, doi:10.4271/2006-01-1288.

[5] Allin, B.D. "Digital Camera Calibration for Luminance Estimation in Nighttime Visibility Studies," SAE Technical Paper 2007-01-0718, 2007, doi:10.4271/2007-01-0718.

[6] Ayres, T.J. and Kayfetz, P., *Calibration and Validation of Videographic Visibility Presentations*, American Academy of Forensic Sciences, Seattle, WA, 2010.

[7] Neale, W.T.C., Marr, J., and Hessel, D., "Nighttime Videographic Projection Mapping to Generate Photo-Realistic Simulation Environments," SAE Technical Paper 2016-01-1415, 2016, doi:10.4271/2016-01-1415.

[8] Bedsworth, K., Butler, R., Rogers, G., Breen, K. et al., "Commercial Vehicle Skid Distance Testing and Analysis," SAE Technical Paper 2013-01-0771, 2013, doi:10.4271/2013-01-0771.

[9] Bayan, F., Cornetto, A., Dunn, A., Tanner, C. et al., "Comparison of Heavy Truck Engine Control Unit Hard Stop Data with Higher-Resolution On-Vehicle Data," *SAE International Journal of Commercial Vehicles*, 2(1): 29–38, 2009, doi:10.4271/2009-01-0879.

[10] Plant, D., Cheek, T., Austin, T., Steiner, J. et al., "Timing and Synchronization of the Event Data Recorded by the Electronic Control Modules of Commercial Motor Vehicles – DDEC V," *SAE International Journal of Commercial Vehicles*, 6(1): 209–228, 2013, doi:10.4271/2013-01-1267.

[11] Reust, T., "The Accuracy of Speed Captured by Commercial Vehicle Event Data Recorders," SAE Technical Paper 2004-01-1199, 2004, doi:10.4271/2004-01-1199.

[12] Reust, T. and Morgan, J., "Commercial Vehicle Event Data Recorders and the Effect of ABS Brakes During Maximum Brake Application," SAE Technical Paper 2006-01-1129, 2006, doi:10.4271/2006-01-1129.

[13] Reust, T., Morgan, J., and Smith, P., "Method to Determine Vehicle Speed During ABS Brake Events Using Heavy Vehicle Event Data Recorder Speed," *SAE International Journal of Passenger Cars—Mechanical Systems*, 3(1): 644–652, 2010, doi:10.4271/2010-01-0999.

[14] Ruhl, R., Senalik, C., and Southcombe, E., "Numerical Methods for Evaluating ECM Data in Accident Reconstruction and Vehicle Dynamics," SAE Technical Paper 2003-01-3393, 2003, doi:10.4271/2003-01-3393.

[15] van Nooten, S. and Hrycay, J., "The Application and Reliability of Commercial Vehicle Event Data Recorders for Accident Investigation and Analysis," SAE Technical Paper 2005-01-1177, 2005, doi:10.4271/2005-01-1177.

[16] Rose, N. and Carter, N., "An Analytical Review and Extension of Two Decades of Research Related to PC-Crash Simulation Software," SAE Technical Paper 2018-01-0523, 2018, doi:10.4271/2018-01-0523.

[17] Hull, W. and Newton, B., "The Animation Computer as a 3-D Reconstruction Tool," SAE Technical Paper 920754, 1992, doi:10.4271/920754.

[18] Fay, R. and Gardner, J., "Analytical Applications of 3-D imaging in Vehicle Accident Studies," SAE Technical Paper 960648, 1996, doi:10.4271/960648.

[19] Jones, I., Muir, D., and Groo, S., "Computer Animation – Admissibility in the Courtroom," SAE Technical Paper 910366, 1991, doi:10.4271/910366.

[20] Hull, W., Newton, B., Macaw, C., and Miller, R., "Functional Classifications and Critique Methods for Litigation Support Forensic1 Accident Reconstruction Animations," SAE Technical Paper 960651, 1996, doi:10.4271/960651.

[21] Grimes, W., "Computer Animation Techniques for Use in Collision Reconstruction," SAE Technical Paper 920755, 1992, doi:10.4271/920755.

[22] Grimes, W., "Classifying the Elements in a Scientific Animation," SAE Technical Paper 940919, 1994, doi:10.4271/940919.

[23] Day, T., "The Scientific Visualization of Motor Vehicle Accidents," SAE Technical Paper 940922, 1994, doi:10.4271/940922.

[24] Grimes, W., Dickerson, C., and Smith, C., "Documenting Scientific Visualizations and Computer Animations Used in Collision Reconstruction Presentations," SAE Technical Paper 980018, 1998, doi:10.4271/980018.

[25] Martin, K., Banister, J., and Piziali, R., "Engineering Visualization of Vehicle Accidents: Data Sources and Methods of Production," SAE Technical Paper 910369, 1991, doi:10.4271/910369.

[26] Massa, D., "Using Computer Reverse Projection Photogrammetry to Analyze an Animation," SAE Technical Paper 1999-01-0093, 1999, doi:10.4271/1999-01-0093.

[27] Neale, W.T.C., "Evaluation of Discrete Vehicle Accident Sounds for use in Accident Reconstruction," Acoustical Society of America, *Proceedings of Meetings on Acoustics*, Vol. 5, 2008.

[28] Neale, W.T.C., "Methodology for Physics-Based Sound Composition in Forensic Visualization," Acoustical Society of America, *Proceedings of Meetings on Acoustics*, Vol. 1, 2007.

[29] McLay, R., Kiely, S., and Sheehan, M., "Case Studies in Animation Foundation," SAE Technical Paper 940920, 1994, doi:10.4271/940920.

[30] Fay, R., "Computer Images and Animations in Court," SAE Technical Paper 970965, 1997, doi:10.4271/970965.

Index

A
Acceleration capabilities, 62–63
Accident area, photograph of, 363
Accident reconstruction, 1, 3, 4
    causation, 8–10
    conservation of momentum, 18–21
    context of, 2
    energy conservation, 11
    investigation/analysis, 2–3
        avoidance scenarios, 10–11
        phases, 3
        uncertainty, 4–6
    Newton's second law, 12–17
    principle of impulse, 18–21
    theoretical/empirical modeling, 4
    two-dimensional particle impact model, 18
    witness statements/testimony, 6–8
    work/energy, principle of, 12–17
Accident scene
    photograph, 363, 364
    scan data, 134
Adobe Photoshop, 353
Advanced crash safety module (ACSM), 273, 277
    BMW, precrash data, 274
    CDR report, 278
    lateral accelerations, 278
AdvanceTrac system, 285
Adventure motorcycles, 30
Airbag control module (ACM), 254
    PC-Crash screen capture, 281
    uncorrected/corrected, 280
Alderson Research Labs CG-95 dummy, 151
Animation production process, 369
Anthropomorphic test devices (ATDs), 150, 238
Anti-lock braking systems (ABS), 25, 53, 59, 329
2016 ARC-CSI conference, 149
Association of European Motorcycle Manufacturers (ACEM), 9, 306
Automatic braking control (ABC), 67
Avoidance zones, 330, 333

B
Barrel distortion, 121
Barrier protection system, 245
Bartlett's motorcycle acceleration equations, 62
Bartlett, W., 52
BMW, 54, 271
    2003 BMW F650GS Dakar R13, 56
    1978 BMW R90, 50
    BMW R nineT, 62
    1995 BMW R1100RS, 102
    BMW X3 xDrive35i, 273
    CDR report, 276
    precrash data, 274
Bosch CDR system, 273
Bosch Crash Data Retrieval (CDR) system, 253
Braking systems
    decelerations, 48–57, 59, 60, 330
    dynamics, 43–47 (*See also* Motorcyclist braking)
    emergency, 331
    engine (*See* Engine braking)
    front-and rear-wheel braking, 50
    239 front-brake only tests, 54
    front-brake pressure, 43
    integrated/linked/antilock, 57–60
    maximal braking, 60
    motorcycles' speed and RPM, 62
    normalized gravitational constant, 61
    275 rear-brake, 54
    rear brakes/antilock brakes, 57
    rear-wheel load, 45, 46
    rigid body model, 44
    simple motorcycle braking model, 44
    skidding, 60
    speed/avoidance calculations, 60–61
    studies, 57
    suspension effects, 47
    throttle, 111
2003 Buell XB9R, 54, 102

C
California Association of Accident Reconstruction Specialists (CA^{2}RS), 182
California Motorcyclist Safety Program (CMSP), 338
Camera, basic geometry of, 354
Camera matching analysis, 137
    photogrammetric analysis, 129, 286, 363, 364
    video analysis, 132
Camera-matching technique, 117–119, 348
Camera mounting rig, 347
Camera setup
    driver's view, 351–356
    field of view, 354–355
    height, 351–353
    images scaling, for playback, 355–356
    rig, 350
Canadian National Collision Data Base (NCDB), 320
CAN traffic, 254
Capsize
    impact-induced, 167
    of motorcycle, 167
Car-DRL effect, 320
Car-to-car collisions, 189
Case study
    evidence documentation methods, 123–129
    gear, speed determination, 64–66
    high-side falls, 162–166
    speed determination, based on gear position, 63
    speedometer reading, 296–300
Center of gravity (CG), 25, 152, 169
    BMW scan, 274
    data, 33–36
    environment, 358
    longitudinal distance, 196
    motorcycle characteristics, 25
    struck vehicle, 202
Checklist
    physical evidence, on motorcycle, 110–112
    site inspection, 106
1982 Chevrolet G20, 151
Coefficient of friction, 4
Coefficient of restitution, 15, 19
Collins model, 174
Collisions
    car-to-car, 189
    violations of right, 309

Combined braking systems (CBS), 59
Computer-generated (CG) visualizations, 356
    case study, 359–369
    motorcycle and rider, 152
    of motorcycle crash, 356–369
    truck vs. motorcycle crash, 359–369
Computer Reconstruction of Automobile Speeds on the Highway (CRASH), 15, 189
Conservation of momentum, 19
Continental automotive, 297
Cornering motorcycle, forces applied, 73
Coulomb friction, 19, 20
Crashes
    reconstruction, 4, 9, 15
    simulation, 280
    types of, 1
Crash site
    document site characteristics, 106
    high-side crash videos, 164
    map, 106
    motorcycle front wheel and fork damage, 108
    photographs, 106
    physical evidence, 106
    physical evidence diagram, 165
CRASH2 User's Manual, 150
Cropped image, 352
Cruiser motorcycle, typical, 27
Crundall, D., 313, 314
Crystal Ball software, 280

D

Daytime running lights (DRL), 319
Debris, 167
Deformation
    BMW, 272
    wheel and fork deformation, 179
Dents, 113
Distortion correction, 121
DiTallo, M., 148, 149
DJI Phantom 2 sUAV, 133
1971 Dodge B200, 151
1982 Dodge B250, 151
Dodge crush, 224
1977 Dodge Sportman, 151
Drag tests, low-speed, 145
DriveCam units, 124–127
Drop test setup, 299
Dual-purpose motorcycle, typical, 31
Ducati Data Analyzer (DDA), 290

E

EDSMAC simulation, 204
EDSMAC4 simulation, 207, 241
Edwards Corner, 158–160
Electronic control module (ECM), 359
Electronic evidence, 368
Electronic stability control system (ESC), 286

Emergency braking, 331
Emergency straight-path braking tests, 49
Engine braking
    closed-throttle, 61
    gear, 61
Equipment mounting
    another vehicle, 347–349
    on motorcycle, 345–346
    on operator, 346–347
European Technical Specification CEN/TS 1317-8, 243
Event data recorders (EDRs), 209, 253, 359, 360
    data, 365
    heavy braking, 360
    PC-Crash, 276
    potential magnitude of error, 256–260
        lateral, 260–261
        longitudinal, 255–256
    speed data, 361
    sport utility vehicle (SUV), 283
    systematic error sources, 255–256
Evidence documentation methods, 112–114
    analytical reverse projection, 120
    case study, 123–129
    image-based scanning, 132–138
    lens distortion, removing, 121–123
    mapping
        with LIDAR, 114–115
        small unmanned aerial vehicles, 138
    motorcycle accident video, photogrammetric analysis, 123–132
    photogrammetry, 115–120
        camera reverse projection, 119–120
        "dummy" camera, 119
Exchangeable image file format (EXIF), 121

F

FARO Focus^{3D} X 330, 115
FARO Focus^{3D} X 130 laser scanner, 133
Faro Focus laser scanner, 130
Faro laser scanner, 114, 135, 137, 362
Fatal Analysis Reporting System (FARS), 318
Fatzinger, E., 56, 101
Fleet DRL Hypothesis, 320
Force-crush curve, 14, 16
1976 Ford Econoline 250, 151
Ford PCM, 286
Frictional-type energy losses, 11
Front-wheel lockup, 55, 101
Front-wheel over-braking
    falls of, 166
    lock-up, 166
Frozen speedometer, 295
FXDX speedometer dial, 293

G

Global Information Technology (GIT) system, 253

Google Earth, 138
Google Street View, 138
GoPro cameras, 82, 133, 296, 346
GoPro Cinema Studio, 135
GPS data acquisition systems, 148
GPS sensor, 291
Guardrail system, MADYMO _ finite element modeling of, 242

H

1968 Harley-Davidson FLH, 50
2009 Harley-Davidson FLHTPI, 92
2013 Harley Davidson FLTRU, 294
Harley-Davidson FXD, 89
2003 Harley-Davidson FXD motorcycle, 88
1999 Harley Davidson FXDX, 293
2019 Harley-Davidson Heritage Softail Classic, speedometer testing, 296–300
Harley-Davidson (H-D) motorcycles, 184, 208
1990 Harley Davidson Road King FLHTPI, 59
Harley-Davidson XL 1200 Sportster Custom, 54
2005 Harley-Davidson XL 1200 Sportster Custom, 102
2013 H-D Dyna Street Bob FXDBA motorcycle, 212
2013 H-D FXSB 103B Softail Breakout motorcycle, 208
2003 H-D Sportster 883, 232
Headlamp mounting equipment, 349
High dynamic range (HDR), 357
High-side falls, 161
    case study, 162–166
    physical evidence diagram, 165
Highway Loss Data Institute (HLDI), 26, 339
Hill crestion, 308
Honda CB500, 52–53
1967 Honda CB305, 143
1974 Honda CB360G, 91
Honda CBR1000RR, 162
Honda CB400T, 49
2015 Honda GL 1800 Gold Wing, 294
Honda 350 impacting, 178
Hugemann, W., 51
Human factors, in crashes
    categories of, 305
    daytime running lights, 318–321
    fail to "see" motorcyclists, 309–316
    left turn across path (LTAP) scenario, 325, 328–331
    licensing, 336–339
    Motorcycle Accidents In Depth Study (MAIDS), 306
    motorcyclists' responses, 325–334
    nuanced example, 332–333
    path intrusion scenarios, ambiguity evidence, 333–334

©2022, SAE International

Index **373**

perception-response process, 321–325
size-arrival effect, 316–318
training, 336–339
visibility, physical factors, 306–308
HVE, 12
Hyundai Sonata, 226, 229, 233
2006 Hyundai Sonata, 224

I
Institute for Police Technology and Management (IPTM), 54, 146
  motorcycles, 148
  Suzuki Katana, 148
Interactive Driver Response Research (IDRR), 322, 323, 332
Investigation phase, 2

K
Kawasaki motorcycles, 80, 81, 289
2010 Kawasaki Ninja 250, 92
Kawasaki Ninja ZX-6R/ZX-10R motorcycles, 161, 290
Kawasaki 1000 police motorcycles, 51, 180
1983 Kawasaki 1000 police motorcycles, 51
1985 Kawasaki 1000 police motorcycles, 51
2007 Kawasaki VN900-D, 76
2002 Kawasaki ZRX1200R, 166
1992 Kawasaki ZX-7 Ninja, 146
2012 Kawasaki ZX-14R, 294
KTM Adventure 790, 31
2020 KTM Adventure 790 motorcycle, 25

L
Lane change, 90–96
Lane positions zone
  brown zone, 329
  gray zone, 329, 333
  red zone, 329
  yellow zone, 329
Lange, F., 51
Laser scan data, 138
Left turn across path (LTAP) scenario, 325, 332, 335
  motorcycle, 325
  numerical example, 328
  opposite direction (LTAP-OD) scenarios, 332
Lens distortion, 116
Licensing, 336–339
Lifan motorcycle, 64
  gearing parameters, 65
  gear train calculations, 65
  physical testing, 64
  testing, 66
Linear force-crush characteristics, 16
Load indexes, for motorcycle tires, 39
Locked-wheel coefficients, 40
Locking
  high-side fall, 162
  rear tire, 158
Longitudinal pavement edge, 167–171

Low-side falls, 157
  case study, 158–161
  right-side leading sideslip, 161

M
Mass multipliers, 15, 190
Mechanical Impact Dynamics, 18
Metzeler ME Z2 sport touring radial tire, 38
Meyers turnaway testing, 94
Mock accident scene, 134
Mock scene, 136
Moments of inertia (MOI), 31, 36, 37
Monte Carlo analysis, 5
Motorcycle accelerations, 173
Motorcycle Accidents in Depth Study (MAIDS), 9, 306
  attention failure, 10
  comprehension failure, 10
  decision failure, 10
  perception failure, 10
Motorcycle autonomous emergency braking (MAEB), 67
Motorcycle braking systems, 60
Motorcycle camera mount, 349
Motorcycle characteristics, 25, 46
  adventure motorcycles, 30
  center of gravity (CG), 25
  off-road motorcycles, 28
  scooter, 28
  sport motorcycles, 28
  three-wheeled motorcycles, 28
Motorcycle collisions
  angular displacement *vs.* postimpact yaw velocity, 203
  ARC-CSI 2016, 185
  crash test points, 187, 188
  Empirical coefficients, 214
  postimpact yaw velocity, 200–202
  precollision positions, 132
  real-world collision, 113
  roadside barriers, 241–248
  staged collisions, 181, 183
  struck vehicle, 186
    damage, 112
    yaw moment of inertia, 215
  WREX 2016, 184
Motorcycle controls, 37
Motorcycle crash, computer-generated (CG) animation of, 357
Motorcycle Crash Reconstruction, 146
Motorcycle dimensions, 31–37
Motorcycle-guardrail crash, 241
Motorcycle Industry Council Tire Guide, 39
Motorcycle, inertial properties, 31–37
Motorcycle Operator Skill Test (MOST), 49
Motorcycle Safety Foundation (MSF), 76
Motorcycle sliding, 109
  on roadway, 105
  scrapes and tire marks, 104

Motorcycle tires, 38
  load indexes, 39
  marks, 112
  softer and stickier, 40
  speed ratings, 39
Motorcycle training program (MTP), 336
Motorcycle transmissions, 37
Motorcycle traversing
  analysis of, 71–90
  assumptions, 75–76
  friction-limited speed, 86–87
  geometric limit on speed, 87–90
  lean angle calculation, 74–75
  lean angle equations, 76–86
  roadway superelevation, 72–74
  willingness to lean, 90
Motorcycle types, 26–31
  cruisers, 26, 27
  dual-purpose
    dual-sport, 28–31
    typical, 31
  off-road, 28–31
  scooter, typical, 28, 29
  sport, typical, 28
  standard, typical, 26
  three-wheeled, 28
    type of, 30
    typical, 29
  touring, 27
Motorcycle velocity change, 214
Motorcycle, view from other vehicles, 350–351
Motorcyclists' visibility, 314
Motorized handheld camera mount, 351
Motor speedometers, 293
Motor tachometers, 293
Moving vehicles, impacts, 240–241

N
Newton's second law, 12, 72
Newton's third law, 14
1981 NHTSA motorcycle, 103
2013 Ninja EX300, 166
Nissan crush, 211
2006 Nissan Maxima, 262
Nissan tire forces, 270
North American driving configuration, 310
Northbound traffic signal, 128

O
Off-road motorcycles, 28
Operator-mounted camera equipment, 347
Oval-shaped abrasions, 107

P
Passenger car
  characteristics of, 335
    left turn across path (LTAP) scenario, 335
    longitudinal and lateral acceleration profiles, 336
  occlusion of, 307

Passenger vehicle drivers, 168
Pavement friction tester, 40
PC-Crash, 12, 213, 217, 218, 261, 275, 276
  file, 276
  moment of inertia, 219
  roadway coefficient of friction, 219
  simulations, 200, 210–212, 216, 267
  simulation software, 287
Peck, L., 56
Pedestrian crashes, 3
Perception-response times (PRTs), 321, 323, 332
Phantom 2 UAV, 135
Photogrammetric analysis, physical evidence diagram, 364
Photogrammetric technique, 116, 132
Photographs
  camera matching, 287
  colorized scan data, 130
  crash site, 106
  east-west roadway, 123
  forward-looking camera, 126
  grid photographing, 122
  mock scene, 136
  motorcycle and SUV, 282, 283
  motorcycle, reconstructed positions, 131
  north-east corner, 125
  overlaid onto video image, 131
  physical evidence, 288
  physics-based animation, 367
  police, 287
  scan data, 287
  surveillance footage, 130
  up-close objects, 352
Photomodeler Scanner, 133
Photoscan, 133
Physical evidence, on motorcycle, 107–112
  damage/abrasions, 109
  front wheel and fork damage and deformation, 108
  inspection checklist, 110–112
  oval-shaped abrasions on rear tire, 107
PIX4D, photogrammetry software, 133, 135, 138
Planar impact mechanics (PIM), 191
  center of gravity, 192
  moving motorcycle, collision of, 195
  variable, definitions of, 192
1980 Plymouth D100, 151
Potholes, 171–173
  left longitudinal edge, 169
  motorcycle interactions with, 167
Powertrain control module (PCM), 284, 285
Principle of work and energy, 12

R
Racelogic VBOX, 82
Racelogic VBOX III 100Hz, 93
Range Rover, 328–330
Rear brake application, 166

Rear-wheel skid, 166
Remote sensing, 113
Restraint control module (RCM), 284, 285
Restraint Deployment Signal (RDS), 285
Riders
  center of gravity, 47
  hand-and foot-position, 324
  2003 Harley-Davidson FXD, 89
  informally trained (IT), 336
  novice, 51
  projected, motion of, 173–174
  real crash scenarios, 51
  rider-to-car impact, 189
  weight, 238–240
Roadside barriers
  capsized, 246–248
  concrete barrier, 245–246
  MADYMO simulations, 242
  motorcycle collisions, 241–248
  steel guardrail, 243–245
  W-beam barriers, 243
Roadway deterioration, 167
Roadway superelevation, 74, 86
Rotational friction factors, 204–206
Rule 702 of the Federal Rules of Evidence, 2

S
Safety systems ON, 26
Scene evidence, from motorcycle crashes, 99
  evidence diagram, 99
  gouges, 104–106
  scrapes, 104–106
  scuffs, 104–106
  site inspection checklist, 106
  skid mark, 101–103
  tire marks, 104–106
Scooter, 28
  barriers, 189
  crush damage, 189
  motorbikes, 28
  typical, 29
Searle's model, 174
Shifting pattern, for motorcycles, 37
SHRP2 naturalistic driving data, 336
Siemens VDO, 297
Signal-controlled intersections, 336
SIMON, 241
Simulation Model of Automobile Collisions (SMAC) program, 15, 16
Skid marks, 101–103
Sliding/tumbling motorcycle
  average sliding deceleration, 149, 150
  decelerations, 143–150
  initial speed determination, 152
Sliding/tumbling rider, decelerations, 150–151
Small unmanned aerial system (sUAS), 335
Smith's tests, 241
Society of Automotive Engineers (SAE), 90

Speed
  bump, 173
  case study, 64–66
  driver judgement of, 316
  event data recorder (EDR), 360
  gear, 63–64
  geometric limit, 87–90
  high-end motorcycle speed, 281
  PC-Crash screen, 282
  westbound vehicles, 129
Speed determination
  based on gear position, 63
  Kloberdanz equation, 203–207
  Ogden equation, 203–207
  planar impact mechanics (PIM), 191–196
  rotational displacement, to angular velocity, 196–203
  sliding motorcycle/rider, 152
  speed loss, for sliding motorcycle, 152–153
  struck vehicle, postimpact translation/rotation, 191
Speedometer, 298
  2015 BMW R1200GS, 294
  case study, 296–300
  drop testing, 294
  frozen speedometer, 295
  2019 Harley-Davidson Heritage Softail Classic, 296–300
  2015 Honda GL 1800 Gold Wing, 294
  residual, 292–300
  straight-line maximum effort braking, 296
  tachometer reading, 292–293
Speed testing, in gear, 66
Sport motorcycles, 28
  super-sport motorcycles, 32
  typical, 28
  wheelbases, 28
Sport utility vehicle (SUV), 271, 284, 285
  event data recorder (EDR), 283
  powertrain control module (PCM), 284
  yaw rotational speed, 289
Standard motorcycle, typical, 26
Struck/striking vehicle
  crash data, 254–261
  damage, 112–114
  EDR data, from passenger vehicle, 271–289
  event data, 253, 289–291
  PC-Crash screen capture, 282
  Precrash EDR data, 253–254
  tire forces, accounting, 261–271
  video, as data source, 291–292
Surveillance footage, 130
Suspension effects, 45
Suspension loading, 87
2012 Suzuki DR650SE Enduro, 76
Suzuki GS 550, 49

2003 Suzuki GSF1200, 56
2007 Suzuki GSX-R750 motorcycle, 81
Suzuki motorcycle, 79
Suzuki's tire, 77
Swerve calculation, 90–96

T
Testing
    Austrian motorcyclists, 324
    Bartlett and Meyers testing, 93
    Carter, T., 147
    curve
        aerial photograph, 83
        lean angle, 85
    lean angle equation, 79, 85
    motorcycle and rider at apex of curve, 78
    motorcycles used, 50, 52, 55, 77
    motorcycle-to-vehicle tests, 178
    parking lot and range, 77
    Suzuki motorcycle, 82
    timeline, 180
Three-wheeled motorcycle, typical, 29, 30
Time-to-collision (TTC), 332
Time-to-intersection (TTI), 331
Tires
    friction coefficients, 40–41
    load indexes, 39
    markings, 38–39, 112
    speed ratings, 39
    typical markings, 38
TM-Easy tire model, 198, 209, 274
Touring motorcycles, 27
Tractor's vehicle-indicated speeds, 359
Tractor-trailer, PC-Crash simulation, 365

Traffic Collision Investigation, 119
Traffic Safety Center at the University of Southern California, 305
Transport Canada (TC), 59
2014 Triumph T100 Bonneville, 294
Turn-away calculation, 90–96

U
Unmanned aerial vehicles (UAVs), 99
    imagery, 138
    photography, 133
Upside down (USD) forks, 186
USC-DOT Motorcycle Accident Research Teams, 319
Utility pole obstruction, 308
U-turn, 164

V
Vavryn, K., 58
VBOX data, 78
VBOX sensors, 84
V-Crash, 261
VCRware, 191
Vehicle–pedestrian collisions, 4
Vehicle-to-vehicle frontal impact, 17
Video-based motorcycle crash visualization, 357
Video footage, 133, 164, 313, 346
Video, from camera, 291
Virtual CRASH, 12, 191
VisualSFM, 133

W
W-beam barriers, 243
Wheelbase reduction, analysis of, 177–191

Winkelbauer, M., 58
World Reconstruction Exposition (WREX), 107, 207
World Reconstruction Exposition in 2000 (WREX 2000), 5, 223
World Reconstruction Exposition in 2016 (WREX 2016), 13, 112, 184, 213
    actual/calculated speed, 237
    collisions methods, 236
    crash tests, 207–238
    H-D impact accelerations, 264, 268
    impact damage/rest positions, 217, 221, 225, 228, 232, 263, 267
    motorcycle-to-vehicle collisions, 184
    Nissan impact accelerations, 264, 268
    PC-Crash
        moment of inertia, 236
        simulation, 216, 224, 228, 231, 235, 270, 271
    sport utility vehicle (SUV), 271
    velocity time histories, 265, 269
    video frames, 262
        from motorcycle-to-car crash test, 13

X
XS-1100 motorcycle, 58

Y
Yamaha FJ1100, 49
2004 Yamaha FJR1300, 59
2005 Yamaha R6, 92
Yamaha 550 Vision, 49
1979 Yamaha XS-400, 58
2000 Yamaha YZF-R6, 47
2006 Yamaha YZF-R6, 55, 166

www.ingramcontent.com/pod-product-compliance
Lightning Source LLC
LaVergne TN
LVHW070124080526
838200LV00087B/338